Contributions to Statistics

The series **Contributions to Statistics** contains publications in theoretical and applied statistics, including for example applications in medical statistics, biometrics, econometrics and computational statistics. These publications are primarily monographs and multiple author works containing new research results, but conference and congress reports are also considered.
Apart from the contribution to scientific progress presented, it is a notable characteristic of the series that publishing time is very short, permitting authors and editors to present their results without delay.

More information about this series at http://www.springer.com/series/2912

Ignacio Rojas · Héctor Pomares
Olga Valenzuela
Editors

Advances in Time Series Analysis and Forecasting

Selected Contributions from ITISE 2016

Springer

Editors
Ignacio Rojas
CITIC-UGR
University of Granada
Granada
Spain

Olga Valenzuela
CITIC-UGR
University of Granada
Granada
Spain

Héctor Pomares
CITIC-UGR
University of Granada
Granada
Spain

ISSN 1431-1968
Contributions to Statistics
ISBN 978-3-319-55788-5   ISBN 978-3-319-55789-2 (eBook)
DOI 10.1007/978-3-319-55789-2

Library of Congress Control Number: 2017943098

Mathematics Subject Classification (2010): 62-XX, 68-XX, 60-XX, 58-XX, 37-XX

Printed on acid-free paper

This Springer imprint is published by Springer Nature
The registered company is Springer International Publishing AG
The registered company address is: Gewerbestrasse 11, 6330 Cham, Switzerland

Preface

This book is intended to provide researchers with the latest advances in the immensely broad field of time series analysis and forecasting (more than 200,000 papers published in this field since 2002 according to Thomson Reuters' Web of Science, see Fig. 1). Within this context, not only will we consider that the phenomenon or process where the values of the series come from is such that the knowledge of past values of the series contains all available information to predict future values, but we will also address the more general case in which other variables outside the series, also called external or exogenous variables can affect the process model. It should also be noted that these exogenous variables can be discrete variables (day of the week), continuous variables (outside temperature), and even other time series.

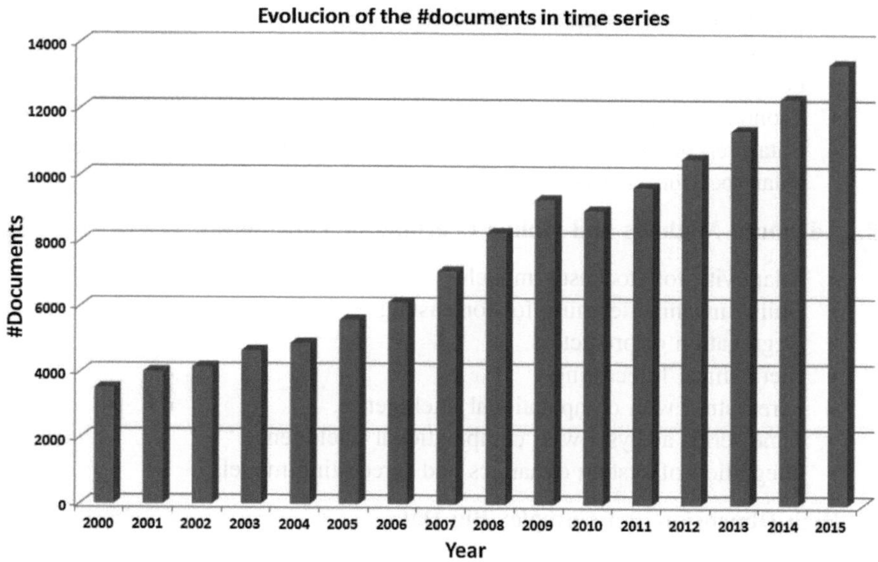

Fig. 1 Evolution of the number of documents in time series

The applications in this field are enormous, from weather forecasting or analysis of stock indices to modeling and prediction of any industrial, chemical, or natural process (see Fig. 2). Therefore, a scientific breakthrough in this field exceeds the proper limits of a certain area. This been said, the field of statistics can be considered the nexus of all of them and, for that reason, this book is published in the prestigious series "Contributions to Statistics" of the Springer publishing house.

The origin of this book stems from the International work-conference on Time Series, ITISE 2016, held in Granada (Spain) in June, 2016. Our aim with the organization of ITISE 2016 was to create a friendly discussion forum for scientists, engineers, educators, and students about the latest ideas and realizations in the foundations, theory, models, and applications for interdisciplinary and multidisciplinary research encompassing disciplines of statistics, mathematical models, econometrics, engineering, and computer science in the field of time series analysis and forecasting.

The list of topics in the successive Call for Papers has also evolved, resulting in the following list for the last edition:

1. **Time Series Analysis and Forecasting.**
   - Nonparametric and functional methods.
   - Vector processes.
   - Probabilistic approach to modeling macroeconomic uncertainties.
   - Uncertainties in forecasting processes.
   - Nonstationarity.
   - Forecasting with many models. Model integration.
   - Forecasting theory and adjustment.
   - Ensemble forecasting.
   - Forecasting performance evaluation.
   - Interval forecasting.
   - Econometric models.
   - Econometric forecasting.
   - Data preprocessing methods: data decomposition, seasonal adjustment, singular spectrum analysis, and detrending methods.
2. **Advanced Methods and Online Learning in Time Series.**
   - Adaptivity for stochastic models.
   - Online machine learning for forecasting.
   - Aggregation of predictors.
   - Hierarchical forecasting.
   - Forecasting with computational intelligence.
   - Time series analysis with computational intelligence.
   - Integration of system dynamics and forecasting models.
3. **High Dimension and Complex/Big Data**
   - Local versus global forecast.
   - Techniques for dimension reduction.

- Multi-scaling.
- Forecasting from Complex/Big data.

4. **Forecasting in Real Problems**

- Health forecasting.
- Telecommunication forecasting.
- Modeling and forecasting in power markets.
- Energy forecasting.
- Financial forecasting and risk analysis.
- Forecasting electricity load and prices.
- Forecasting and planning systems.
- Real-time macroeconomic monitoring and forecasting.
- Applications in other disciplines.

At the end of the submission process of ITISE 2016, and after a careful peer review and evaluation process (each submission was reviewed by at least 2, and on the average 2.7, program committee members or additional reviewers), 124 contributions were accepted for oral, poster, or virtual presentation, according to the recommendations of reviewers and the authors' preferences.

High-quality candidate papers (28 contributions, i.e., 22% of the contributions) were invited to submit an extended version of their conference paper to be considered for this special publication in the book series of Springer: Contributions to Statistics. For the selection procedure, the information/evaluation of the chairman

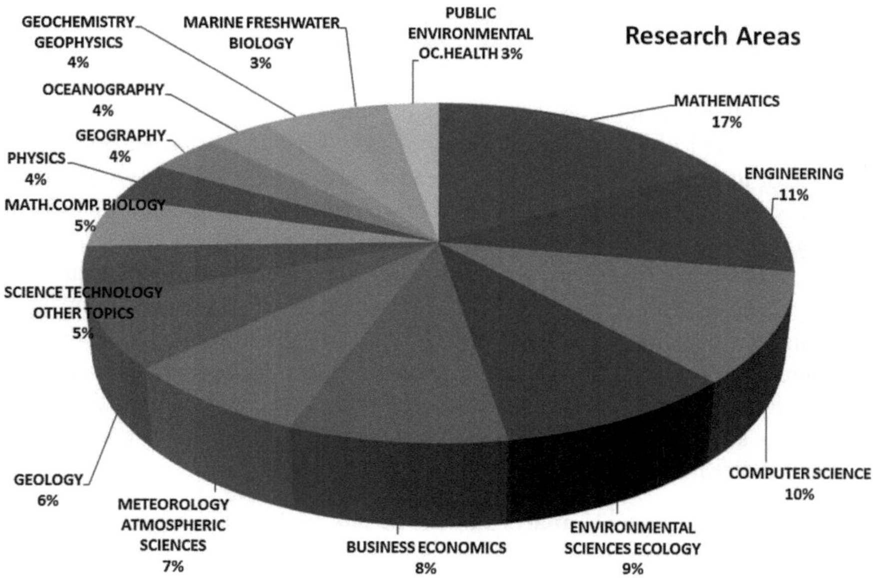

Fig. 2 Main research areas in time series

of every session, in conjunction with the review comments and the summary of reviews, were taken into account.

So, now we are pleased to have reached the end of the whole process and present the readers with these final contributions that we hope, will provide a clear overview of the thematic areas covered by the ITISE 2016 conference, ranging from theoretical aspects to real-world applications of Time Series Analysis and Forecasting.

It is important to note that for the sake of consistency and readability of the book, the presented papers have been classified into the following chapters:

- **Chapter 1: Analysis of irregularly sampled time series: techniques, algorithms, and case studies**. This chapter deals with selected contributions in the field of Analysis of irregularly sampled time series (topics proposed by Prof. Eulogio Pardo-Igúzquiza and Prof. Francisco Javier Rodríguez-Tovar), being the main objective the presentation of methodologies and cases studies dealing with the analysis of scales of variability of times series and joint variability between time series. As discussed by the organizers, "Unevenly spaced time series are very common in many scientific disciplines and industry applications. Missing data, random sampling, gapped data, and incomplete sequences, among other causes, give origin to irregular time series. The common approach to deal with these sequences has been interpolation in order to have an evenly sampled sequence and then to apply any of the many methods that have been developed for regularly sampled time series. However when the spacing between observations is highly irregular, interpolation introduces unwanted biases. Thus, it is desirable to have direct methods that can deal with irregularly sampled time series. This session welcomes contributions on this problematic: quantification of sampling irregularity in time series, advanced interpolation techniques and new techniques of analysis that can be applied directly to uneven time series. The main objective of this session is the presentation of methodologies and cases studies dealing with the analysis of time series with irregular sampling. The contributions can be on any area of time series analysis. Among others, areas of interest are: event analysis, trend and seasonality estimation of uneven time series, smoothing, correlation, cross-correlation and spectral analysis of irregular time series; non-parametric and parametric methods, non-linear analysis; bootstrap, neural networks and other soft-computing techniques for irregular time series; expectation-maximization and maximum entropy algorithms and any other technique that deals with uneven time series analysis. New theoretical developments, new algorithms implementing known methodologies, new strategies for dealing with uneven time series and case studies of time series analysis are appropriate for this special session." A total of four contributions were selected in this chapter.

- **Chapter 2: Multi-scale analysis of univariate and multivariate time series**. This chapter deals with selected contributions of an special session organized by Prof. Eulogio Pardo-Igúzquiza and Prof. Francisco Javier Rodríguez-Tovar during ITISE 2016, being the main objective the presentation of methodologies and cases studies dealing with the analysis of scales of variability of times series and joint variability between time series. As discussed by the organizers: "every physical, physiological and financial time series has a characteristic behavior with respect to the scale at which it is observed. That is so because the time series is the output of a physical, biological or market system with a given dynamics resulting in one of two extremes, scale-invariant properties on one hand and scale-dependent properties on the other. In any case, each time series has variabilities at different temporal scales. Also the joint variability between each pair of variables may also be a function of scale. There are different approaches for doing such scale analysis, from classical spectral analysis to wavelets and from fractals to non-linear methods. The choice of a given approach may be a function of the question that one wants to answer or may be a decision taken by the researcher according to his/her familiarity with the different techniques." A total of four contributions were selected in this chapter.

- **Chapter 3: Lineal and Nonlinear time series models (ARCH, GARCH, TARCH, EGARCH, FIGARCH, CGARCH, etc.)**. In this chapter, classical methods such as ARIMA or ARMA methods, in conjunction with nonlinear model (such as ARCH, GARCH, etc) are analyzed. There are, for example, contributions which take into account that in time series analysis, noise is a relevant element which determines the accuracy of the forecasting and prediction, and in order to face with this problem, present an automatic, auto-adaptive, partially self-adjusting data-driven procedure, created to improve the forecast performances of a linear prediction model of the type ARIMA by eliminating noisy components within the high-frequency spectral portion of the series analyzed. In this chapter, comparison of different model Autoregressive Moving Average models and different Generalised Autoregressive Moving Average models is also discussed to forecast financial time series. Furthermore, examples of using linearl and nonlinear model for specific problem (for example, outlier detection) is presented.

- **Chapter 4: Advanced time series forecasting methods**. This chapter analyzed specific aspect of time series analysis and its hybridization with other paradigms (such as, for example, computer science and artificial intelligence. A total of five contribution were selected, where the reader can learn, for example, about:
  - how recurrent and feedforward models can be used for turbidity forecasting, predicting peaks of turbidity with 12 h lag time, presenting of a new architecture which take into account explicitly the role of evapotranspiration in this problem,
  - how to analyze and predict the productivity of the public sectors in the US across the states, using several methodologies, with combine exploratory (for understanding clusters of similar dynamics in an precise way, Self-Organizing Maps, SOM, clustering methods were employed (both raw time series and

unobservable components, such as trend and slope), and empirical techniques (e.g., panel model) via the Cobb–Douglas production function.

- how to categorize/analyze multivariate time series (MTS) data. For the classification or clustering, a similarity measure to assess the similarity between two multivariate time series data is presented. There exist several similarity measures presented in the bibliography (dynamic time warping (DTW) and its variants, Cross Translational Error (CTE) based on multidimensional delay vector (MDV) representation of time series, Dynamic Translational Error (DTE), etc). Improved version of nowadays available similarity measures have been done using available benchmark data sets with simulation experiments.
- how to model nonlinear relationships in complex time-dependent data, using the Dantzig-Selector convex optimization problem to determine the number and candidate locations of the Radial Basis Function Neural Networks, and analyze the performance of the methodology using the well-known Mackey-Glass chaotic time series (exploring time-delay embedding models in both three and four dimensions).

- **Chapter 5: Applications in time series analysis and forecasting**.
Finally, we wanted to finish this book showing that it is really very important the application of the new methodologies development to real problems. No theory can be considered useful until it is put into practice and it is scientifically demonstrated the success of its predictions. That is what this chapter is about. It is shown how multiple and rather different mathematical, statistical, and computer science models can be used for so many analyses and forecasts of time series in fields such as wind speed and weather modeling, determining the pollen season (start and end), analysis of eye-tracking data for reading ability assessment, and forecasting models to predict the data of Malaysia KLCI price index, in which statistical models and artificial neural networks as machine learning techniques are simultaneously analyzed. The selection here was very strict, only four contributions, but we are confident to give a clear enough vision of what we have just said.

Last but not least, we would like to point out that this edition of ITISE was organized by the University of Granada together with the Spanish Chapter of the IEEE Computational Intelligence Society and the Spanish Network on Time Series (RESeT). The Guest Editors would also like to express their gratitude to all the people who supported them in the compilation of this book, and especially to the contributing authors for their submissions, the chairmen of the different sessions and to the anonymous reviewers for their comments and useful suggestions in order to improve the quality of the papers.

We wish to thank our main sponsors as well: the Department of Computer Architecture and Computer Technology, the Faculty of Science of the University of Granada, the Research Centre for Information and Communications Technologies (CITIC-UGR), and the Ministry of Science and Innovation for their support and grants. Finally, we wish also to thank Prof. Alfred Hofmann, Vice President Publishing—Computer Science, Springer-Verlag and Dr. Veronika Rosteck, Associate Editor, Springer for their interest in editing a book series of Springer based on the best papers of ITISE 2016.

We hope the readers can enjoy these papers the same way as we did.

Granada, Spain
November 2016

Ignacio Rojas
Héctor Pomares
Olga Valenzuela

Contents

Part I
Analysis of Irregularly Sampled Time Series: Techniques, Algorithms and Case Studies

Small Crack Fatigue Growth and Detection Modeling with Uncertainty and Acoustic Emission Application

Reuel Smith and Mohammad Modarres

Abstract In the study of fatigue crack growth and detection modeling, modern prognosis and health management (PHM) typically utilizes damage precursors and signal processing in order to determine structural health. However, modern PHM assessments are also subject to various uncertainties due to the probability of detection (POD) of damage precursors and sensory readings, and due to various measurement errors that have been overlooked. A powerful non-destructive testing (NDT) method to collect data and information for fatigue damage assessment, including crack length measurement is the use of the acoustic emission (AE) signals detected during crack initiation and growth. Specifically, correlating features of the AE signals such as their waveform ring-count and amplitude with crack growth rate forms the basis for fatigue damage assessment. An extension of the traditional applications of AE in fatigue analysis has been performed by using AE features to estimate the crack length recognizing the Gaussian correlation between the actual crack length and a set of predefined crack shaping factors (CSFs). Beside the traditional physics-based empirical models, the Gaussian process regression (GPR) approach is used to model the true crack path and crack length as a function of the proposed CSFs. Considering the POD of the micro-cracks and the AE signals along with associated measurement errors, the properties of the distribution representing the true crack is obtained. Experimental fatigue crack and the corresponding AE signals are then used to make a Bayesian estimation of the parameters of the combined GPR, POD, and measurement error models. The results and examples support the usefulness of the proposed approach.

Keywords Fatigue crack damage · Gaussian process regression · Probability of detection · Measurement error · Model error · True crack length

R. Smith (✉) · M. Modarres
Center for Risk and Reliability, University of Maryland, College Park, MD 20742, USA
e-mail: smithrc@umd.edu

M. Modarres
e-mail: modarres@umd.edu

I. Rojas et al. (eds.), *Advances in Time Series Analysis and Forecasting*, Contributions to Statistics, DOI 10.1007/978-3-319-55789-2_1

1 Introduction

Fatigue crack propagation and detection research has been a vital part of PHM and the engineering industry as a whole for many years. Modern applications of PHM for example include correlation of crack lengths a and certain time series markers [1], as well as correlation between small crack propagation and various AE signal indices [2]. Similarly, many crack POD models have been proposed as cumulative density functions (CDF) such as the cumulative lognormal and the logistic distributions [3]. Such variety in both crack propagation and detection modeling has resulted in a lot of options for PHM; however several of these options are based on empirical models which can possess certain uncertainties stemming from gathered data and observations.

Because empirical models are often assumed as a form of a behavior (in this case crack propagation), many PHM assessments include three principle types of uncertainties: 1. Data uncertainty, 2. Physical variability uncertainty, and 3. Modeling uncertainty/error [4]. Sankararaman et al. [4] proposed several methodologies for accounting these uncertainties in order to improve existing empirical models, including: measurement error correction on gathered data, representing the physical variation in material properties as distributions, and selection of an appropriate crack propagation model that best represents historical fatigue crack data. However, Moore and Doherty [5] cite that unless model input properties that have a direct bearing on the output are considered, predictions made by that model may still possess model error. That is the most suited crack propagation model will be based on time series data as well as relevant test and material properties that affect the crack length.

A model was developed by Mohanty [6, 7] correlates test and material properties to the detected crack propagation by way of a machine learning tool called multivariate GPR [8]. The advantage that the GPR model has over most crack propagation models is a stricter adherence to the characteristics of the source data depending on the kernel functions used to train the GPR model [8]. This approach forms the central methodology described in this paper resulting in a representation of crack propagation, detection, and crack path to failure that is more realistic than the existing empirical models. The outline of this paper is as follows: Sect. 2 briefly defines the crack propagation and detection models as well as the likelihood function that predicts the model parameters for those models. Section 3 covers the steps in preprocessing the data used. Section 4 goes through an example of the outlined procedure and the results therein. Finally Sect. 5 will draw conclusions from the analysis.

2 Structure of Models and Likelihood

The probabilistic crack propagation models in this study will be expressed in integrated form, where crack length is represented by the variable a, and a set of *crack shaping factors* (CSFs) represented by the vector $\vec{x}$. These CSFs define correlated properties that directly or indirectly affect the size, shape, and growth rate of a crack. The probability of crack detection model, by contrast, is a function of a. By way of the Bayesian parameter estimation approach, the vector of the crack length (propagation) model parameters $\vec{A}$, crack detection parameters $\vec{B}$, and probability of false crack detection P_{FD}, are estimated for each *crack propagation and detection* (CPD) model set. Three crack propagation assessment models and four detection models are discussed further in the remaining sections.

2.1 Crack Propagation Models

The crack propagation models specifically adhere to a time series. The first model is based on a log-linear or exponential relation [1, 20, 23],

$$\ln[a(N)] = b + mN \tag{1}$$

where N represents the number of load cycles at crack length $a(N)$, and $\vec{A}$ is the vector of parameters $[m, b]$ where the initial crack length a_0 is e^b. Several studies support the position that crack propagation curves can be expressed in exponential form [20, 23]. The second model is based on the AE Intensity $I(N)$ [2], defined as a weighted measure that is a function of two AE measures: cumulative count[1] and signal amplitude (both functions of fatigue cycles N). A more detailed definition of the AE Intensity is available in other literature [2, 9]. The relation between I and a may be expressed as a linear model or a power model as shown in Eqs. (2) and (3), respectively.

$$a(N) = \alpha I(N) + \beta \tag{2}$$

$$a(N) = \alpha I(N)^{\beta} \tag{3}$$

In this case $\vec{A}$ can be defined as the vector $[\alpha, \beta]$. The third model is based on a multivariate GPR [7] correlating the CSF input variables $\vec{x}$ to the crack length output variable a. The model is a complex function of time-based, material-based, and test-based CSFs written as follows,

$$a(\vec{x}) = g([\,CSF_1 \quad CSF_2 \quad \ldots \quad CSF_Q\,]) \tag{4}$$

where Q is the number of CSFs being correlated to a. The general GPR input/output relation is stated as,

[1]An AE count is the number of times the AE signal amplitude exceeds a given threshold amplitude level.

$$a \sim NOR\left[0, K\left([X], \vec{A}\right)\right] \tag{5}$$

where $K(\cdot)$ is the $M \times M$ covariance matrix or kernel matrix that correlates a and $[X]$, the complete set of input data represented as an $M \times Q$ matrix with M being the number of data points. Kernel matrices are made up of kernel functions $k\left(\vec{x}_i, \vec{x}_j, \vec{A}\right)$ which take two sets of CSF data $\vec{x}_i$ and $\vec{x}_j$ and the Gaussian crack length model parameters $\vec{A}$ to produce one element (i, j) of the kernel matrix. In the Gaussian modeling the objective is to develop a kernel function $k(\cdot)$ based on the assumptions of the input and output relation being modeled [7].

2.2 Crack Detection Models

The lognormal distribution is the first POD model [3],

$$POD\left(a|\vec{B}, a_{lth}\right) = \int_{a_{lth}}^{a} \frac{1}{(x - a_{lth})\sqrt{2\pi\zeta_1^2}} \exp\left\{-\frac{1}{2}\left[\frac{\ln(x - a_{lth}) - \zeta_0}{\zeta_1}\right]^2\right\} dx \tag{6}$$

where $\vec{B}$ is the vector of the parameters $[\zeta_0, \zeta_1]$ of the lognormal POD model, and the value a_{lth} is the smallest crack length that can be detected through a specific non-destructive test (NDT) method. The random variable a is adjusted as $(a - a_{lth})$ for all POD models because $0 \leq POD(a) \leq 1$ for crack lengths greater or equal to the specific NDT's lowest detectable crack length threshold, a_{lth}.

The log-logistic distribution is the second POD model [3],

$$POD\left(a|\vec{B}, a_{lth}\right) = \frac{\exp[\beta_0 + \beta_1 \ln(a - a_{lth})]}{1 + \exp[\beta_0 + \beta_1 \ln(a - a_{lth})]} \tag{7}$$

where $\vec{B}$ is the vector of parameters $[\beta_0, \beta_1]$ of the log-logistic model. This model is usually assumed for most POD models, because of its mathematical simplicity and ease of use with censored data [3, 26].

The third model is the logistic distribution model [27],

$$POD\left(a|\vec{B}, a_{lth}\right) = 1 - \frac{1 + \exp(-\eta_0\eta_1)}{1 + \exp[\eta_0(a - \eta_1 - a_{lth})]} \tag{8}$$

where $\vec{B}$ is the vector of parameters $[\eta_0, \eta_1]$ of the logistic model.

The final model is the Weibull distribution model [10],

$$POD\left(a|\vec{B},a_{lth}\right)=1-\exp\left[-\left(\frac{a-a_{lth}}{\alpha_0}\right)^{\alpha_1}\right] \tag{9}$$

where $\vec{B}$ is the vector of parameters $[\alpha_0, \alpha_1]$.

2.3 Likelihood Function for Bayesian Analysis

The Bayesian framework in this study takes one model from each group (propagation and detection) and obtains the parameter estimates for the combined CPD model by performing Bayesian parameter estimation on the following CPD likelihood function [9],

$$\begin{aligned} l\Big(D=0,\,1;\,a_{i=1},\,\ldots,\,a_{n_D},\,\vec{x}_{i=1},\,\ldots,\,\vec{x}_{n_D},\,\vec{x}_{j=1},\,\ldots,\,\vec{x}_{m_{ND}}|\vec{A},\vec{B},P_{FD}\Big) \\ =\prod_{i=1}^{n_D}\left[(1-P_{FD})POD\left(D=1|\vec{B},a_i\rangle a_{lth}\right)f\left(a_i|\vec{A},\vec{x}_i\right)\right] \\ \prod_{j=1}^{m_{ND}}\left[1-(1-P_{FD})\int_{a_{lth}}^{\infty}POD\left(D=1|\vec{B},a\rangle a_{lth}\right)f\left(a|\vec{A},\vec{x}_j\right)da\right] \end{aligned} \tag{10}$$

This represents the likelihood of a set of n_D NDT detection data points $(\vec{x}_{i=1},\ldots,\vec{x}_{n_D};a_{i=1},\ldots,a_{n_D})$ and m_{ND} non-detection (missed) data points $(\vec{x}_{j=1},\ldots,\vec{x}_{m_{ND}};a_{j=1}=0,\ldots,a_{m_{ND}}=0)$, where detection state D is 1 for positive detection and 0 for non-detection. Any of the detection equations (Eqs. 6 through 9) may be used in place of the POD terms in Eq. 10. Likewise the $f(\cdot)$ term is the crack propagation PDF, modeled as a lognormal distribution,

$$f\left(a|\vec{A},\vec{x}\right)=\frac{1}{a\sigma\sqrt{2\pi}}\exp\left[-\frac{1}{2}\left(\frac{\ln a-\ln g\left(\vec{A},\vec{x}\right)}{\sigma}\right)^2\right] \tag{11}$$

where any of the crack propagation models (Eqs. 1 through 5) may be used in place of $g\left(\vec{A},\vec{x}\right)$. Finally, the false crack detection probability P_{FD} represents the probability of an NDT method detecting a crack that is not present (i.e., false detection). Bayesian inference for the posterior is written according to the Bayes' Theorem [11] as follows,

$$\pi_1(\vec{A},\vec{B},P_{FD}|D=0,1;a_{i=1},\ldots,a_{n_D},\vec{x}_{i=1},\ldots,\vec{x}_n,\vec{x}_{j=1},\ldots,\vec{x}_{m_{ND}})$$
$$\propto l\left(D=0,1;a_{i=1},\ldots,a_{n_D},\vec{x}_{i=1},\ldots,\vec{x}_n,\vec{x}_{j=1},\ldots,\vec{x}_{m_{ND}}|\vec{A},\vec{B},P_{FD}\right)\pi_0\left(\vec{A},\vec{B},P_{FD}\right) \tag{12}$$

where $\pi_1(\vec{A},\vec{B},P_{FD}|D=0,1;a_{i=1},\ldots,a_n,\vec{x}_{i=1},\ldots,\vec{x}_n,\vec{x}_{j=1},\ldots,\vec{x}_{m_{ND}})$ is the posterior PDF for the CPD model parameters $\vec{A}$,$\vec{B}$, and P_{FD}, and $\pi_0\left(\vec{A},\vec{B},P_{FD}\right)$ is the joint prior PDF for the CPD model parameters.

3 Data Preprocessing

The crack length data for this study is obtained from a series of fatigue tests at the Center for Risk and Reliability, where the crack lengths are detected from high frequency photographs of the crack propagation area [2]. Detection and sizing of the cracks is conducted both based on visual identification, and based on the AE signals that are measured concurrently [2]. The photographs include in-test images of the area taken at 100× magnification and post-test images taken at a higher magnification of 200×. All detected cracks are measured using a Java-based software called ImageJ [12]. However, the raw data is processed prior to the Bayesian analysis using the likelihood stated in Sect. 2.3. Reduction of data uncertainty and model error is a key step in this research. In order to do that three different crack length are defined: 1. Crack length measured experimentally (contains experimental measurement errors), 2. Crack length computed using a crack propagation model (contains model error), and 3. the “true” crack length.

3.1 Crack Length Definitions and Correction

Experimental crack lengths a_e are detected lengths that are used to process the CPD models. The first of these is the measured crack length $a_{e,m}$ which is the in-test crack length measured at 100× magnification. This measurement, since it is done while the test is being conducted, involves detection and measurement error due to the shortness of time in observing a crack, errors in measurement tools, and the blurry images of the tiny cracks due to the vibration of the sample during fatigue test. Model crack lengths a_m are measures obtained from crack propagation models defined in Sect. 2.1. The AE crack length $a_{m,AE}$ for example is based on the $a_{e,m}$ measurement, but obtained from AE intensity I versus $a_{e,m}$ models as depicted by Eqs. 2 and 3. As a result of both blur and the small base magnification in crack pictures, both the $a_{e,m}$ lengths and the $a_{m,AE}$ lengths involve measurement error.

Measurement error correction, with respect to an estimated crack length a_{est} (either a_e or a_m), is determined multiplicatively as,

$$E_a = \frac{a_{est}}{a} \tag{13}$$

where a is the true crack length.[2] True crack length a in this study is the post-test path-wise crack length measurement taken at 200× magnification. "True" is defined as a length that is more accurate and more precise than the previously obtained real-time length measurement $a_{e,m}$. Because of the absence of motion-blur and higher amplification, measurements of the after-test images at 200× allows for higher precision and accuracy in measurements than the previous in-test images. A percent error analysis was conducted comparing crack length measurements between 200× and higher magnifications of 400× and 1000× and it was found to be roughly a 2% error difference [9]. As a result of the low percent error, 200× magnification assumed as the scale in which a would be considered as "true".

3.2 *Probability of Detection Definitions*

The measurement error corrected crack length a_e provides the basis for computing the prior crack propagation model parameters $\vec{A}$ for the models outlined in Sect. 2.1. It also provides the AE intensity I versus crack length a relation that is used to calculate the prior estimate for the POD parameters $\vec{B}$ for the models defined in Sect. 2.2. To estimate $\vec{B}$, only lengths a between a_{lth} and a_{hth} are considered in deference to the POD boundaries $0 \leq POD(a) \leq 1$. As with the lower threshold for detection a_{lth}, a_{hth} is defined as the largest crack length that can be missed using an NDT technique [3]. True crack length data a is converted into POD data via a signal response function for AE intensity [3],

$$POD(a) = 1 - F\left[\frac{\ln I_{th} - \ln I}{\sigma_a}\right] \tag{14}$$

where the AE intensity I may be assumed to follow a linear or power form as stated by Eqs. 2 and 3 respectively and σ_a represents the standard deviation associated with the error between log forms of model AE intensity $\widehat{I}$ and the true AE intensity I [3].

$$\ln \widehat{I} = \ln I + NOR(0, \sigma_a) \tag{15}$$

[2]The reciprocal of Eq. 13 $E_a = a/a_{\text{est}}$ is the measurement error with respect to true crack length a.

The I_{th} term in Eq. 16 is the AE intensity threshold, above which flaws are detected and below which flaws go undetected.

4 Application Example

The following example is an analysis performed on data gathered from a series of fatigue life tests in order to test the effectiveness of the previously outlined procedure. Note that this specimen's I versus a data showed a higher R^2 value (96.0%) when a linear model was assumed versus a power model (93.5%). Therefore I versus a will assume a linear relation (Eq. 2) for this example.

4.1 Fatigue Life Test Description

The fatigue test was conducted on eight flat dog-bone Al 7075-T6 specimens whose geometry is depicted by Fig. 1 (all dimensions and geometries in millimeters).

The first six specimens (designated as DB3, DB4, DB5, DB6, DB7, and DB15) fit the geometry presented in Fig. 1a and the last two specimens (designated as 1A2 and 1B3) fit the geometry presented in Fig. 1b. This "dog-bone" geometry was selected based on ASTM-E466-2007 [18]. A small notch of radius 0.5 mm is milled to instigate crack initiation. The fatigue tests were conducted on a uniaxial 22 kN Material Testing System (MTS) 810 load frame. Each sample was tested at varying test frequencies, load ranges, and load ratios listed in Table 1.

These four test conditions are considered as CSFs, since they are known to be directly correlated to the propagation of the crack [7, 9, 13, 14, 21]. Microscopy at

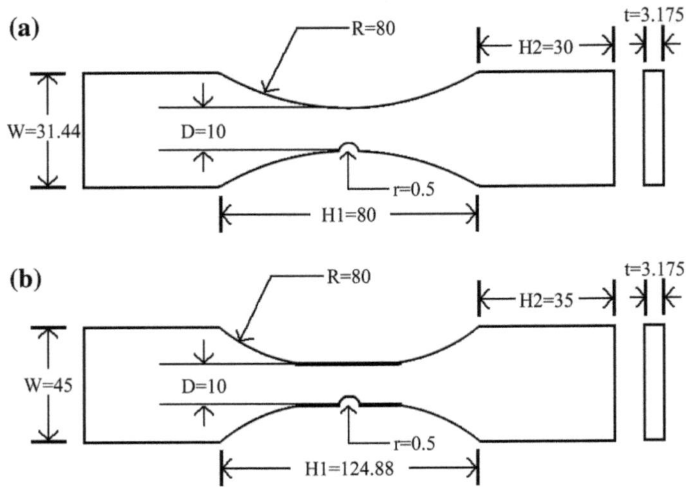

Fig. 1 Schematic of dog-bone specimens used for fatigue tests

Table 1 Testing conditions for the Al 7075-T651 dog-bone specimens

Specimen name	DB3	DB4	DB5	DB6	DB7	DB15	1B3	1A2
Loading frequency (Hz)	3	3	2	3	2	2	5	5
Load ratio	0.1	0.1	0.5	0.1	0.5	0.3	0.1	0.1
Min force (kN)	0.8	0.8	6.5	0.8	6.5	3	0.8	0.75
Max force (kN)	8	8	13	8	13	10	8	7.5

200× magnification is used to obtain the material-based CSFs considered for this study, the mean grain diameter and the mean inclusion diameter. Niendorf et al. among other researchers have cited a known correlation between grain size (or diameter) and crack propagation [24]. That is, overall crack propagation is inversely proportional to the grain diameter [22]. While MacKenzie cites that the concentration and size of material inclusions reduce the ductility of steels [15]. To account for physical variability [4, we model the material CSFs in form of Weibull distributions. For example, for specimen DB7 the mean grain diameter distribution is $WBL(69.69\,\mu m, 1.61)$ and the mean inclusion diameter distribution is $WBL(10.44\,\mu m, 1.68)$. Additional material-based CSF distributions are available in other literature [9]. Both lognormal and Weibull distributions were considered for modeling these CSFs, but the Weibull distribution was selected based on a goodness-of-fit analysis. Including the variable fatigue cycles N, the four test conditions, and two Weibull parameters for each of the two material property distributions, there are a total of nine CSFs ($Q=9$) to represent crack propagation [9].

4.2 Kernel Definition

For the GPR model (Eqs. 4 and 5) a kernel function is developed such that it best models the data. This kernel function assures that the crack length function mean and confidence bounds are always monotonically increasing, and that the mean is a good fit to the data. For this study, the best kernel function was found to be,

$$k\left(\vec{x}_i, \vec{x}_j, \vec{A}\right) = A_1 + \sum_{q=1}^{Q} A_{q+1}\vec{x}_{i,q}\vec{x}_{j,q} + A_{2+2Q}\exp\left[-\sum_{q=1}^{Q} A_{q+Q+1}\left(\vec{x}_{i,q} - \vec{x}_{j,q}\right)^2\right] + A_{4+2Q}\sin^{-1}\frac{A_{3+2Q}\sum_{q=1}^{Q}\vec{x}_{i,q}\vec{x}_{j,q}}{\sqrt{\left(A_{3+2Q}\sum_{q=1}^{Q}\vec{x}_{i,q}\vec{x}_{j,q}\right)^2}} + A_{5+2Q}\delta_{i,j} \quad (16)$$

where $\delta_{i,j}$ is a Dirac function that equals 1 when $i=j$ and 0 elsewhere [19] and $\vec{A}$ is a 23 parameter crack propagation vector. The complete design and validation of this kernel function is detailed in another publication by Smith et al. [9].

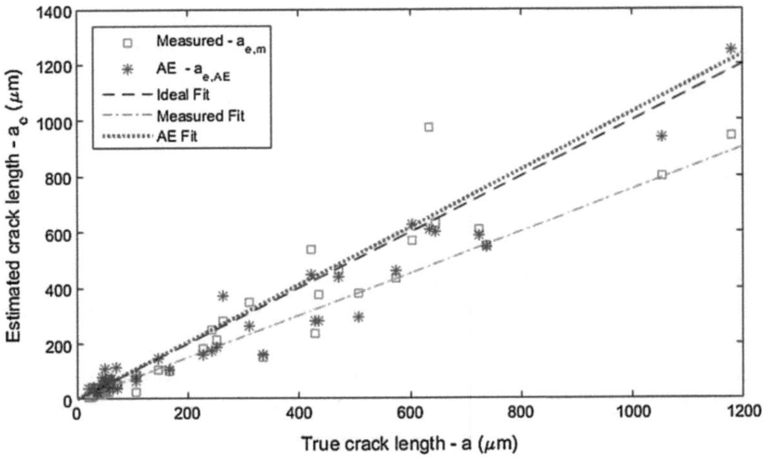

Fig. 2 Comparison of the true crack length to two estimated crack lengths: measured crack length and AE crack length

4.3 Measurement Error Analysis

As stated in Sect. 3.1, measurement error is performed to correct estimated crack length measurements. About 40 measurements are taken to compute the overall measurement error [9]. These are all presented in Fig. 2 to draw upon two implications of their behavior.

Figure 2 presents the comparison between the true crack length a and two estimated crack lengths of reference the measured crack length $a_{e,m}$ and the AE crack length $a_{e,AE}$. The first implication is that the true length a is often greater than both estimated lengths except for $a_{e,AE}$ for early fatigue cycles. The second implication is that crack measurements obtained by way of AE signals are much closer to the true crack measurements than those obtained by visual means alone [9]. The mean measurement error for the data in Fig. 2 is 1.03 for the AE length measurement error and 0.75 for the measured length measurement error [9]. This is a significant finding because it means that there is a 24.7% error between the measured and true lengths, but only a 2.9% model error between the AE and true lengths [9].

4.4 Bayesian Analysis

For the Bayesian parameter estimation procedure outlined in Sect. 2.3, data from each specimen was broken into two classes of data: non-detection data and detection data. Approximately nine tenths of the detection data are used as the training data set, and the detection data remaining (about 44 total) are used for validation of the model. That is the training data is used for the Bayesian estimation procedure while the validation data is used to check against the posterior models.

The Bayesian estimation of the model parameters is done by a MATLAB [16] routine developed that uses the standard Metropolis-Hastings (MH) Markov Chain Monte Carlo (MCMC) analysis of complex likelihood functions such as the one stated in Eq. 10. The routine also makes use of Rasmussen's GPML code [25] where estimation of GPR crack length model parameters is required. Let $\vec{\Theta}$ be the hyper-parameter set being updated which is made up of the CPD parameter set of interest: $\vec{A}$, $\vec{B}$, and P_{FD}.

$$\vec{\Theta} = \begin{bmatrix} \vec{A} & \vec{B} & P_{FD} \end{bmatrix}^T \tag{17}$$

When the crack propagation model under study is either the log-linear (Eqs. 1 and 11) or the AE model (Eqs. 2 or 3 and 11), $\vec{\Theta}$ is made up of six hyper-parameters, while for the GPR model (Eqs. 5, 11, and 16) $\vec{\Theta}$ is made up of 26 hyper-parameters due to the number of CSFs under study.

The MATLAB routine was executed 12 times per specimen to obtain the posterior PDF of the hyper-parameters for each CPD model pair. It was discovered that the hyper-parameters for the crack propagation models throughout the test results don't show much difference from one result to another. In the case of specimen DB7 for example, the standard deviation between the AE crack propagation model hyper-parameters α and β are 3.1×10^{-3} and 1.6×10^{-2}, respectively. The resulting mean crack propagation models for specimen DB7 is presented in Fig. 3.

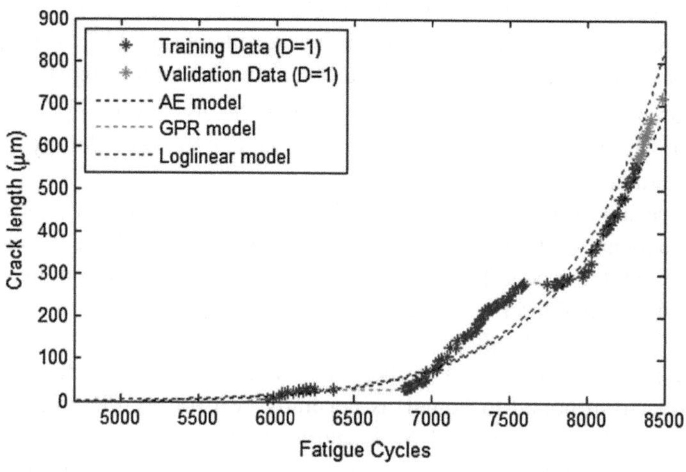

Fig. 3 The crack propagation curves for the log-linear, AE, and GPR models against the original DB7 specimen data

This study assumes that crack initiation takes place at an aberration near the small notch where milling takes place. Visually the GPR crack propagation model fits the best to the training and validation data where its posterior parameters $\vec{A}$ is,

$$\vec{A} = \begin{bmatrix} 0.08, 9.8 \times 10^2, 3.6 \times 10^5, 8.3 \times 10^5, 5.6, 25, 1.1 \times 10^3, 25, 1.7 \times 10^2, \ldots \\ 19, 0.08, 1, 1, 1, 1, 1, 1, 1, 1, 0.0046, 86, 2.9, 0.03 \end{bmatrix}^T \quad (18)$$

This is further verified by a model validation methodology [17] that applies the measurement errors $E_{a'}$ (reciprocal of Eq. 13) of the model a_m and experimental crack lengths a_e with respect to the true crack length a. The variability of the measurement errors is addressed by representing them as log-logistic distributions,

$$f(E_{a',e}) \sim LOGLOGIST(\mu_e, \sigma_e) \text{ and} f(E_{a',m}) \sim LOGLOGIST(\mu_m, \sigma_m) \quad (19)$$

where μ_e and σ_e are the log-logistic parameters for $E_{a',e}$ and μ_m and σ_m are the log-logistic parameters for $E_{a',m}$. Therefore, a combined effect measurement error $E_{a',t}$,

$$a = E_{a',e} a_e = E_{a',m} a_m \Rightarrow \frac{E_{a',m}}{E_{a',e}} = \frac{a_e}{a_m} = E_{a',t} \quad (20)$$

would also fit to a log-logistic distribution.

$$f(E_{a',t}) \sim LOGLOGIST\left(\mu_m - \mu_e, \sqrt{\sigma_m^2 + \sigma_e^2}\right) \quad (21)$$

The 40 measurements described in Sect. 4.3 and their conjoining model and expected lengths are used to obtain these experimental log-logistic parameters by way of Bayesian parameter estimation using the likelihood $\prod_{i=1}^{40} f(E_{a',e,i}|\mu_e, \sigma_e)$. Then the 44 validation points are used to obtain the model log-logistic parameters, the measurement error, and model error using the likelihood $\prod_{i=1}^{44} f(E_{a',t,i}, \mu_e, \sigma_e|\mu_m, \sigma_m)$. More detailed information on this validation methodology is available in other literature [9, 17].

The Bayesian estimation of the first likelihood function produces a mean μ_e of 0.074 and a mean σ_e of 0.071 which provides the means to validate the twelve CPD models through the second likelihood. Table 2 presents the 95% confidence bounds and median for validation model error of all CPD model pairs.

The validation points show that for this series of tests, the accuracy of all propagation models are acceptable (between 1 and 7%). However the precision for model error (from bound to bound) is generally best for a GPR propagation (55–57%), followed by log-linear (58–59%) and then AE (77–85%), the latter of which has been exhibited previously [2]. This is nearly a 30% improvement in model error precision between GPR and AE propagation.

Table 2 CPD model validation presenting the 95% confidence bounds for the model error only

Propagation model	GPR (%)			Log-linear (%)			AE (%)		
Confidence	2.5	50	97.5	2.5	50	97.5	2.5	50	97.5
Logistic	29	4	−27	32	7	−26	31	1	−45
Log-logistic	29	4	−27	30	7	−28	35	2	−47
Lognormal	28	4	−28	32	6	−27	37	3	−48
Weibull	29	4	−27	29	5	−28	36	2	−50

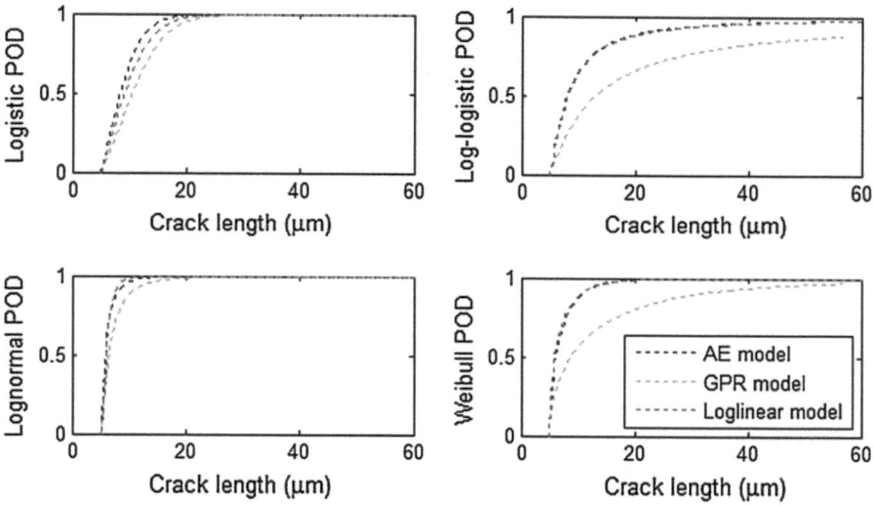

Fig. 4 The mean POD curves for all 12 CPD model evaluations on specimen DB7

Based on both the model error precision and the propagation curve example in Fig. 2, the GPR model is the most realistic representation of crack propagation of the three models. Assessment of these results confirms that the number of CSFs used for modeling is directly proportional to model error precision. For instance, the GPR model has nine CSFs while both the AE model and the log-linear model have from one to three CSFs. This means that additional CSFs, guided by material, test, and/or time-series based properties, improve the realism of the crack propagation model. However, it is still likely that there are missing or extraneous CSFs in the GPR model causing additional model error. Further analysis of this effect will be done for future studies.

It is also noted that the pair with the lowest model error spread is the GPR/log-logistic CPD set where the mean values of the log-logistic and false detection hyper-parameters for specimen DB7 are $[2.1, -0.07]$ and $[0.06]$, respectively. The POD plots are presented in Fig. 4 and visually the results are very different depending on the nature of the crack propagation model.

In general the detectable crack lengths a at $POD(a)$ are the largest when the detection model is related to a GPR propagation model. That is, the posterior POD curves are more conservative for the GPR propagation model.

5 Conclusions

The approach outlined in this paper has led to a number of important findings in support of PHM assessments of fatigue damage. The example presented shows the results of a fatigue life test and the preconditioning of the data for use in model prediction. In comparing model error between true and estimated crack lengths, it was shown that the AE detections have a much lower model error than visual detections. Furthermore, this research was able to improve the AE-based propagation model error bounds from previous averages of about 46% [2, 9] to a lower value of about 29%. By implementing a powerful Bayesian estimation technique on 12 CPD model pairs it was found that the number of CSFs used directly impact the fitness and realism of the crack propagation model. It is important to underline the importance of identifying most relevant CSFs in order to further improve model error.

References

1. Rusk, D.: Model Development Plan: Hinge Inspection Reliability (2011)
2. Keshtgar, A.: Acoustic Emission-Based Structural Health Management and Prognostics Subject to Small Fatigue Cracks. University of Maryland, College Park, MD (2013)
3. Georgiou, G.A.: Probability of Detection (PoD) Curves: Derivation, Applications and Limitations. Crown, London (2006)
4. Sankararaman, S., Ling, Y., Shantz, C., Mahadevan, S.: Uncertainty quantification in fatigue damage prognosis. In: Annual Conference of the Prognostics and Health Management Society. San Diego (2009)
5. Moore, C., Doherty, J.: Role of the calibration process in reducing model predictive error. Water Resour. Res. **41**, W05020 (2005)
6. Mohanty, S., Chattopadhyay, A., Peralta, P.: Bayesian statistic based multivariate gaussian process approach for offline/online fatigue crack growth prediction. Exp. Mech. **51**, 833–843 (2011)
7. Mohanty, S., Chattopadhyay, A., Peralta, P., Das, S., Willhauck, C.: Fatigue life prediction using multivariate gaussian process. AIAA (2007)
8. Rasmussen, C.E.: Evaluation of Gaussian Processes and other Methods for Non-Linear Regression. University of Toronto, Toronto (1996)
9. Smith, R., Modarres, M., Droguett, E.L.: A recursive Bayesian approach to small fatigue crack propagation and detection modeling (2017, in Review)
10. Bencala, K.E., Seinfeld, J.H.: On frequency distributions of air pollutant concentrations. Atmos. Environ. **10**, 941–950 (1976)
11. Bayes, T.: An essay towards solving a problem in the doctrine of chances. Philo. Trans. **53**, 370–418 (1763)
12. NIH: ImageJ Version 1.50c. http://imagej.nih.gov/ij/index.html (2015). Accessed 14 Oct 2015

13. Paris, P., Erdogan, F.: a critical analysis of crack propagation laws. J. Basic Eng. **85**(2), 528–534 (1963)
14. Walker, K.: The effect of stress ratio during crack propagation and fatigue for 2024-T3 and 7075-T6 aluminum. Eff. Environ. Complex Load Hist. Fatigue Life ASTM STP **462**, 1–14 (1970)
15. MacKenzie, S.: Overview of the mechanisms of failure in heat treated steel components. Fail. Anal. Heat Treat. Steel Compon. (ASM International), 43–86 (2008)
16. Mathworks: MATLAB 2014a. (2014)
17. Ontiveros, V., Cartillier, A., Modarres, M.: An integrated methodology for assessing fire simulation code uncertainty. Nucl. Sci. Eng. **166**(3), 179–201 (2010)
18. ASTM E466-07: Standard practice for conducting force controlled constant amplitude axial fatigue tests of metallic materials. ASTM International, West Conshohocken, PA (2007)
19. Chen, T., Morris, J., Martin, E.: Gaussian process regression for multivariate spectroscopic calibration. Chemometr. Intell. Lab. Syst. **87**, 59–71 (2007)
20. Davidson, D.L., Lankford, J.: Fatigue crack growth in metals and alloys: mechanisms and mircomechanics. Int. Mater. Rev. **37**(2), 45–76 (1992)
21. Forman, R.G., Kearney, V.E., Eagle, R.M.: Numerical analysis of crack propagation in cyclic loaded structures. J. Basic Eng. **89**, 459–464 (1967)
22. Hanlon, T., Kwon, Y.N., Suresh, S.: Grain size effects on the fatigue response of nanocrystalline metals. Scripta Mater. **49**, 675–680 (2003)
23. Molent, L., Barter, S., Jones, R.: Some practical implications of exponential crack growth. Solid Mech. Appl. **152**, 65–84 (2008)
24. Niendorf, T., Rubitschek, F., Maier, H.J., Canadinc, D., Karaman, I.: On the fatigue crack growth-microstructure relationship in ultrafine-grained interstitial-free steel. J. Mat. Sci. **45** (17), 4813–4821 (2010)
25. Rasmussen, C.E., Nickisch, H., Williams, C.: Documentation for GPML Matlab Code version 3.6. http://www.gaussianprocess.org/gpml/code/matlab/doc/ (2015). Accessed 21 Oct 2015
26. Singh, K.P., Warsono, Bartolucci, A.A.: Generalized log-logistic model for analysis of environmental pollutant data. In: MODSIM 97 IMACS Proceedings. Hobart (1997)
27. Yuan, X., Mao, D., Pandey, M.D.: A Bayesian approach to modeling and predicting pitting flaws in steam generator tubes. Reliab. Eng. Syst. Saf. **94**(11), 1838–1847 (2009)

Acanthuridae and *Scarinae*: Drivers of the Resilience of a Polynesian Coral Reef

Alizée Martin, Charlotte Moritz, Gilles Siu and René Galzin

Abstract Anthropogenic pressures are increasing and induce more frequent and stronger disturbances on ecosystems especially on coral reefs which is one of the most diverse on Earth. Long-term data series are increasingly needed to understand and evaluate the consequences of such pressures on ecosystems. This 30-years monitoring program allowed a description of the ability of the coral reef of Tiahura (French Polynesia) to recover after two main coral cover declines, due to *Acanthaster planci* outbreaks. The study is divided in two distinct periods framing the drop of coral cover and analyze the reaction of two herbivorous family: *Acanthuridae* and *Scarinae*. First we compared the successive roles they played in the herbivorous community, then we evaluated the changes in species composition that occurred for both *Acanthuridae* and *Scarinae* between these two periods. The long-term study of this coral reef ecosystem provided a valuable study case of the resilience over 30 years.

Keywords Resilience · Long-term analysis · Coral reef · Herbivorous fish · Shift

1 Introduction

The ability of an ecosystem to recover or shift to another state after acute disturbance is still difficult to predict [1–3]. Resilience refers to the capacity of an ecosystem to face disturbance and absorb changes without losing its key functions [4]. This implies a reorganization of ecosystem components before adapting to its surrounding changing environment, which can lead to a redefinition of the communities' structure toward a new stable state. This multi-equilibrium conception of resilience is also called ecological resilience [4–6] and has been studied in many ecosystems [7–10] such as savannahs [9, 11], grasslands [12, 13], forests [14, 15] or

A. Martin (✉) · C. Moritz · G. Siu · R. Galzin
EPHE, PSL Research University, UPVD, CNRS, USR 3278 CRIOBE,
Laboratoire d'Excellence "CORAIL", 98729 Moorea, French Polynesia
e-mail: alizee.martin@outlook.com

I. Rojas et al. (eds.), *Advances in Time Series Analysis and Forecasting*,
Contributions to Statistics, DOI 10.1007/978-3-319-55789-2_2

lakes [7, 16–18]. Coral reefs are also considered as complex systems which experience diverse stable states [19–22] that are mainly characterised by different substrate composition: coral or algal-dominated systems [23, 24].

Coral reefs are one of the most diverse ecosystems on Earth [25] and provide a valuable example for resilience studies in oceans [19, 20, 26–28]. Natural and anthropogenic disturbances have dramatically increased during the last three decades triggering fundamental changes and high rates of mortality in coral reefs ecosystems [29–31]. Seventy five percent of coral reef are currently reported as acutely threatened by anthropogenic pressures and this number should raise to 90% in 2050 [32, 33]. The principal reasons for the degradation of reefs ecosystems are climate change, habitat destruction and overfishing [34–36]. These threats have already been pointed out as responsible for phase shifts in coral reefs ecosystems [31, 34, 37–39], for example from a coral-dominated to a macroalgae-dominated system, as reported in the Caribbean [19, 24, 26]. Such shifts are generated by a perturbation (seastar outbreaks, mortality of sea urchins, cyclones) which induce a development of macroalgae that becomes the dominant substrate, avoiding coral development. In the Caribbean, ecosystems that have gone through this kind of shift exhibit no recovery [40] whereas a recent study establish that 46% of Indo-Pacific reefs do [41].

The coral reef ecosystem of Moorea, a Polynesian island located in the central south Pacific, has been the core of many studies for more than 30 years [42–49]. This island, quite populated compared to other Pacific islands, may be considered as a model island for the study of Indo-Pacific reef resilience. Moorea's coral reef went through many disturbance theses last decades such as coral bleaching, cyclones or Crown-of-thorn seastar *Acanthaster planci* outbreaks [44, 46–50]. *A. planci* is considered as the major enemy of reef-building corals [51] and their outbreaks are one of the most destructive disturbance faced by tropical reefs [52, 53]. Outbreaks occurred in Moorea in 1980, 1981, 1987 and between 2006 and 2010 [44, 46, 48–50]. On the north coast of Moorea, *A. planci* destroyed 35% of the 3000 m^2 of living substrate in 1983 [54] and a maximum density of about 151 ind.km^{-2} was recorded in 2010 [50]. In 2010, cyclone Oli increased the damages due to *A. planci* by breaking and displacing many coral skeletons [50]. Despite all these destructive events, Moorea coral reef successfully avoided shift and showed relatively high resilience. One reason for this may be the important herbivorous biomass that supports Indo-Pacific reefs, three times greater than in Caribbean and mainly due to *Scarinae* (parrot fish), a sub-family of *Labridae*, and *Acanthuridae* (surgeon fish) which have a biomass respectively twice and fourfold higher than in Caribbean [41]. Our study focused on these two families of herbivorous fish that are among the most abundant encountered in coral reefs ecosystems [41, 55].

Our study monitored the substrate cover and the abundance of herbivorous fish, specifically *Acanthuridae* and *Scarinae*, over more than 30 years. We focused this long term survey on two main *A. planci* outbreak periods, separated by more than 10 years, which provided a comparison of the changes that occurred in the ecosystem over time. Here we compared the abundance and the relative abundance (within the herbivorous fish) of *Acanthuridae* and *Scarinae* between these two periods to reveal the successive roles they played in the herbivore community. We further aimed to evaluate the changes of community composition that occurred for both *Acanthuridae* and *Scarinae*. Then the analysis of relationship between the two fish families and different substrate allowed us to propose a hypothesis regarding the resilience of Tiahura's coral reef.

2 Materials and Methods

2.1 Study System

Moorea is one of the 118 islands of French Polynesia, located in the Pacific Ocean. Its coral reef extends along 61 km around the island and 750 m from the coast to the fore reef, the oceanic side of the reef crest (Fig. 1). The study site of Tiahura is located on the north-western part of Moorea, and is one of the most heavily studied reefs in the world since 1971 [43, 45, 47, 56, 57]. Three major habitats can be defined in Tiahura: the fringing reef (<2 m depth), the barrier reef (<3 m depth) inside the lagoon and on the reef crest, and the fore reef on the outer slope (<40 m depth). The fore reef of Tiahura, located further from the inhabited coastal areas than the fringing reef, is less impacted by anthropogenic pressures [58] but is regularly exposed to acute natural disturbances such as cyclones or *A. planci* outbreaks [44, 46, 48, 50].

2.2 Data Sampling

Tiahura fore reef was sampled for the first time in 1983, and, from 1987, every year in the context of a long-term monitoring program conducted by the CRIOBE (Centre de Recherches Insulaires et Observatoire de l'Environnement). A permanent fish counting site of 50 × 2 m, located at 12 m depth and parallel to the barrier reef, was surveyed twice a year during wet and dry seasons (in March

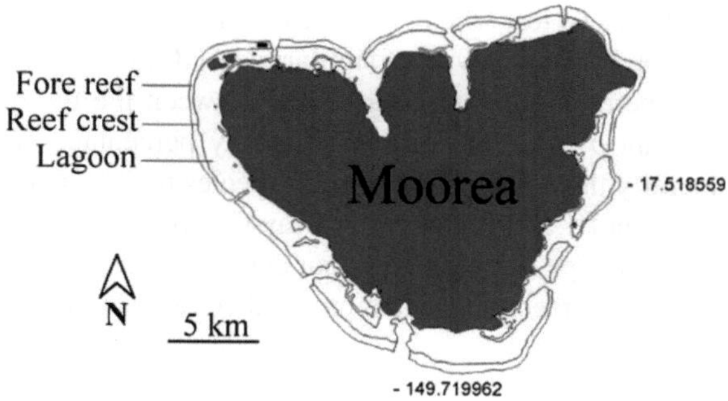

Fig. 1 Map of the study site representing Tiahura location

and September, respectively). Fish were counted and identified to the species level in this corridor by visual census, repeated 4 times in the day. In this study we focused on two families of herbivorous fish: *Acanthuridae* (surgeon fish) and *Scarinae* (parrot fish), sub-family of *Labridae*.

Benthic cover was determined along 3 transects parted along Tiahura's fore reef between 6 and 20 m depth, framing the fish counting site at 12 m depth. Transects were 50 m long and substrate cover was evaluated using the Point Intercept Transect (PIT) method by determining substrate every meter along the transect [59]. Categories of substrate were defined as follows: coral, macroalgae, pavement, CCA (crustose coralline algae), rubble, sand, turf and others. CCA, pavement, rubble and turf cover were grouped as "cropped cover" as they are all substrates possibly grazed by herbivorous fish. This counting was repeated every year since 1987. Percent coral cover collected on the outer slope of Tiahura reef in 1979 [43] and 1983 [54] was retrieved from the publications.

2.3 Data Analysis

We defined two time windows of 8 years framing the drop and recovery of coral cover due to *A. planci* outbreaks, with the lowest point in 1983 and 2010. Due to the limited extent of our data (no substrate sampling before 1987 and in 2012), the periods were chosen as follows: P1 from 1987 to 1994 and P2 from 2007 to 2015. We analysed fish community and substrate cover using R version 3.2.3 [60] including vegan [61] and stats packages.

A t-test was used to compare the real and relative abundances (i.e. percentage among herbivorous fish) of *Acanthuridae* and *Scarinae* between P1 and P2, choosing a minimum significant p-value of 0.05. To apply this parametric test based on comparison of the means, we first checked the normal distribution of data and the homogeneity of variances. Nonmetric multidimensional scaling (NMDS) was used to visualise the temporal variations in species composition of both families. This ordination based on a Bray-Curtis dissimilarity matrix was represented in two dimensions after checking for the stress value. Bray-Curtis distance is a dissimilarity index based on species abundance which varies between 0 and 1 [62] and is one of the most commonly applied measures to express composition similarity in ecology. To test whether species composition for each family was different between P1 and P2, an analysis of similarity (ANOSIM) based on Bray-Curtis distance using 999 permutations was performed between the two periods, using a minimum significant p-value of 0.05. A similarity percentage analysis (SIMPER) was used to identify the contribution of each species to the change that occurred in communities between P1 and P2 based on Bray-Curtis dissimilarity [63]. We finally analysed the relationships between the abundance of *Acanthuridae* and *Scarinae* and the percentage of macroalgae and cropped cover using linear models (minimum significant *p*-value of 0.05). An analysis of variance (aov function) was used to determine whether *Acanthuridae* and *Scarinae* abundance were different with macroalgae and cropped cover. A minimum significant *p*-value of 0.05 was chosen.

3 Results

3.1 Substrate Cover and Species Abundance

Between 1979 and 2015, Tiahura's coral reef faced two main *A. planci* outbreaks, starting in 1980 and 2006 (Fig. 2a). These declines in coral cover were followed by a recovery period: 5 years after the lowest point of P1, coral cover reached 30.0 ± 13.9 and 30.7 ± 19.4% after the lowest point of P2. Macroalgae percentage cover varied between 3.3 ± 5.8% (1994) and 20.6 ± 27.6% (1987) during P1 whereas it varied between 0.7 ± 1.2% (2007) and 8.7 ± 1.2% (2014) during P2 (Fig. 2b). Cropped cover increased during P1 and P2, raising from 37.3 ± 11.6% in 1987 to 67.3 ± 14.0% in 1994 and reaching 92.0 ± 24.0% in 2011. For both periods this increase was due to turf cover (Fig. 2c), but in 2013 an important increase in pavement cover occurred between 2011 (0 ± 0%) and 2015 (36 ± 7.2%). Increase in rubble cover was observed after two cyclones in 1993 (32 ± 12.5%) and 2010 (38 ± 32.2%). *Acanthuridae* remained more abundant than *Scarinae* (Fig. 2d), but their abundance dropped during P2 from 121.6 ± 12.7 (2009) to 72.5 ± 15.6 individuals (2015). On the contrary, *Scarinae* abundance increased during P1 from 15.4 ± 8.7 individuals in 1987 to 48.5 ± 23.1 individuals in 1994 and from 20.5 ± 12.1 individuals in 2007 to 45.8 ± 20.6 individuals in 2015 during P2. Raw abundance of *Acanthuridae* did not change between P1 and P2 ($p = 0.086$) (Fig. 3a), whereas *Scarinae* abundance was significantly higher during P2 (Fig. 3b). However, the mean relative abundance of *Acanthuridae* within herbivorous fish was significantly higher in P1 (32.0 ± 4.2%) than in P2 (25.3 ± 5.6%) (Fig. 3c). Conversely, the relative abundance of *Scarinae* was different ($p < 0.05$) and lower in P1 (8.4 ± 4.9%) than in P2 (12.0 ± 5.8%) (Fig. 3d).

3.2 Community Structure Variations

The NMDS revealed changes in community structure for both *Acanthuridae* and *Scarinae* between P1 and P2 (Fig. 4). The major change was captured by the first dimension of the NMDS (NMDS 1). For both families, a drastic change in community occurred in 2009, 2 years after the beginning of period P2, and when coral cover was almost the lowest (Fig. 2a). Two different communities could be distinguished with the ANOSIM between P1 and P2 for both *Acanthuridae* ($R = 0.6172$, $p < 0.05$) and *Scarinae* ($R = 0.7309$, $p < 0.05$). SIMPER analysis

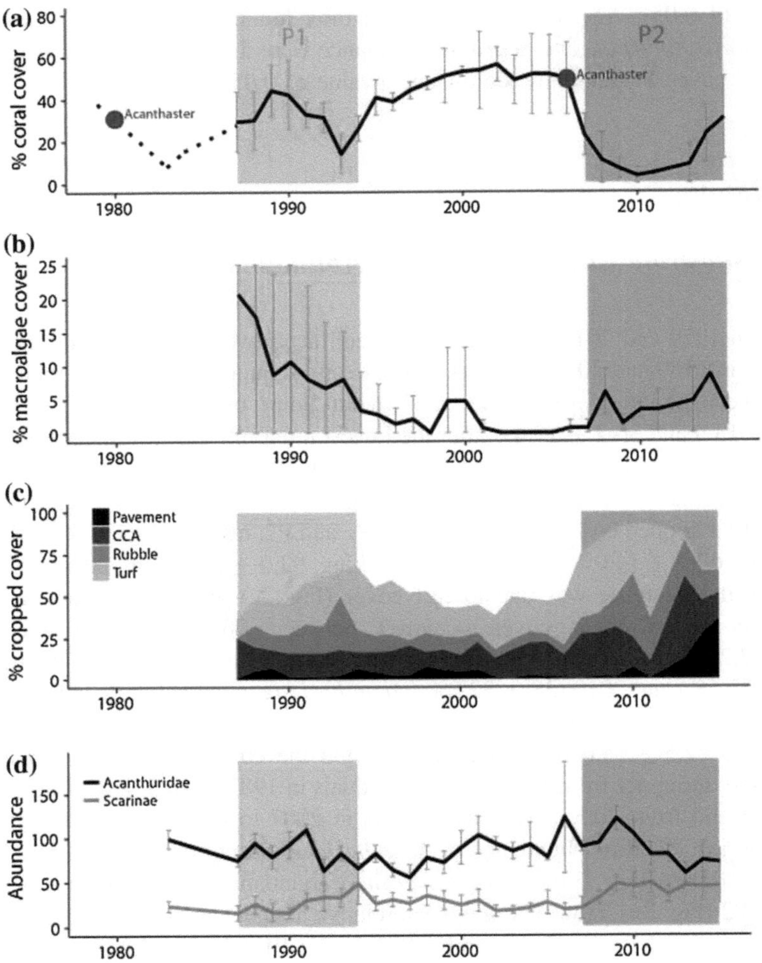

Fig. 2 Percentage of **a** coral cover, **b** macroalgae cover, **c** cropped cover, including different categories and **d** abundance of *Acanthuridae* and *Scarinae*. Periods P1 and P2 are represented in *orange* and *blue* respectively. Onset dates of outbreaks are represented by a *red point* (see color figure online)

singled out *Ctenochaetus striatus* (average = 0.25 ± 0.14), *Acanthurus nigricans* (average = 0.09 ± 0.06), and *Acanthurus nigrofuscus* (average = 0.07 ± 0.06) as the three major species responsible for the *Acanthuridae* community change between P1 and P2. For the *Scarinae*, *Chlorurus sordidus* (average = 0.20 ± 0.09), *Scarus psittacus* (average = 0.05 ± 0.04) and *Scarus oviceps* (average = 0.05 ± 0.03) were singled out. All these species showed a decrease in abundance during P2. The analysis of species composition of each family revealed that there were more species but with fewer individuals each during P2 for both *Acanthuridae* and *Scarinae*. *Acanthuridae* were mainly composed of *Ctenochaetus*

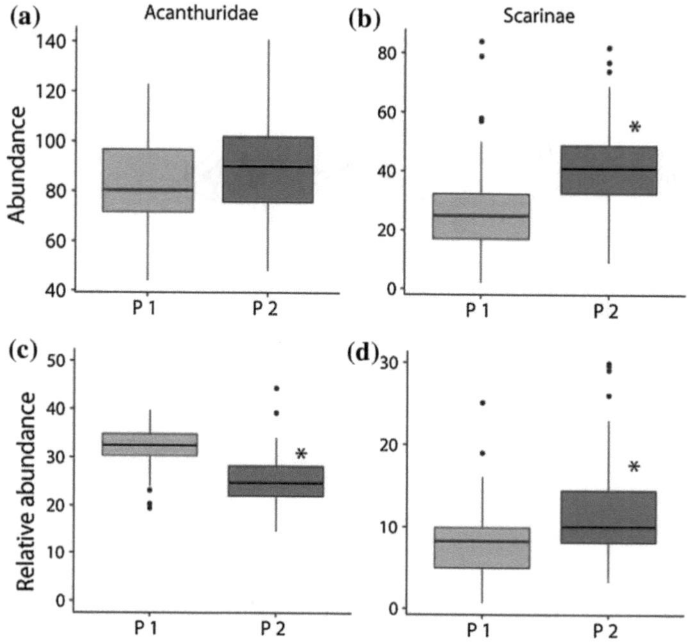

Fig. 3 Boxplot of the abundance of **a** *Acanthuridae* and **b** *Scarinae* and of their relative abundance **c, d** for periods P1 and P2. Relative abundance is abundance of a given family within the abundance of all herbivores. Error bars are 95% confidence limits. Significantly different boxplot are identified by a *black star*

striatus during P1, with a mean of 7 species, whereas there were around 13 different and rare species during P2 with no dominating species. The same patterns were observed for *Scarinae*, dominated by *Chlorurus sordidus* during P1 and comprising a mean of 6 species, and comprising 13 species with less individuals during P2.

3.3 Relationship Between Abundance of Acanthuridae and Scarinae and Macroalgae and Turf Cover

The ANOVA showed that *Scarinae* abundance was different with different values of macroalgae ($p < 0.05$) and cropped cover ($p < 0.05$) (Table 1) whereas *Acanthuridae* abundance was related to any of them. The relationships between *Acanthuridae* and *Scarinae* abundance and substrate cover showed opposite patterns. While *Acanthuridae* tended to become more abundant with increasing macroalgae cover during P1 (Fig. 5a), *Scarinae* abundance decreased with macroalgae cover ($p < 0.05$) (Fig. 5c). The opposite trend was observed during P2: *Acanthuridae* were negatively correlated with the percentage of macroalgae cover ($p < 0.05$),

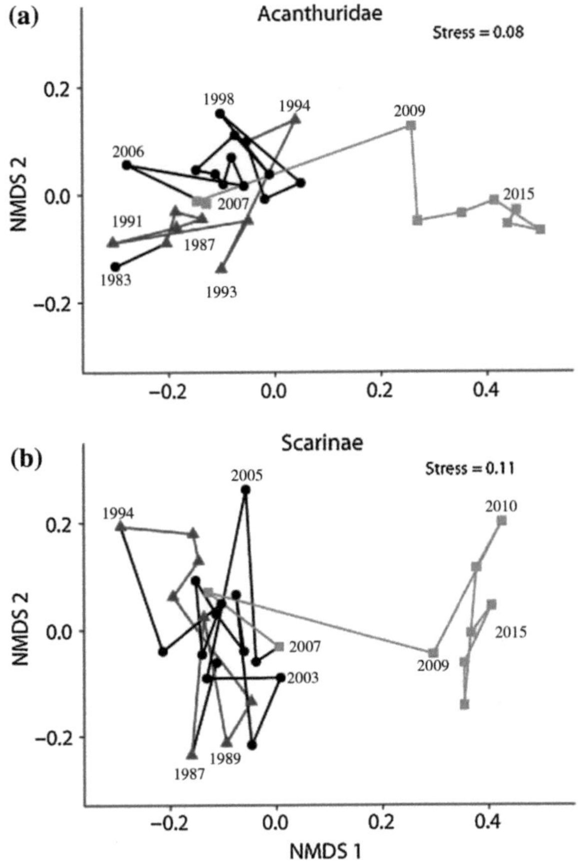

Fig. 4 Temporal non-metric multidimensional scaling of the **a** *Acanthuridae* and **b** *Scarinae* communities. Periods P1 and P2 are indicated by *orange* and *blue dots*, respectively. Stress corresponds to the disagreement between 2-D configuration and predicted values from the regression (see color figure online)

Table 1 Results of linear models (Fig. 5), Bold *p*-values are significant

		Algae		Turf	
		P1	P2	P1	P2
Acanthuridae	p	0.2911	**<0.05**	0.3698	**<0.05**
	R^2	0.004969	0.15	−0.005545	0.1531
Scarinae	p	**<0.05**	0.3017	**<0.05**	0.2604
	R^2	0.3152	0.003361	0.5904	0.01009

whereas *Scarinae* tended to be positively correlated with it. Opposite patterns were observed for the relationship between fish and cropped cover: *Scarinae* percentage increased with cropped cover during P1 ($p < 0.05$), tended to increase during P2 (Fig. 5d), and *Acanthuridae* percentage tended to decrease during P1 but increased during P2 ($p < 0.05$) (Fig. 5d).

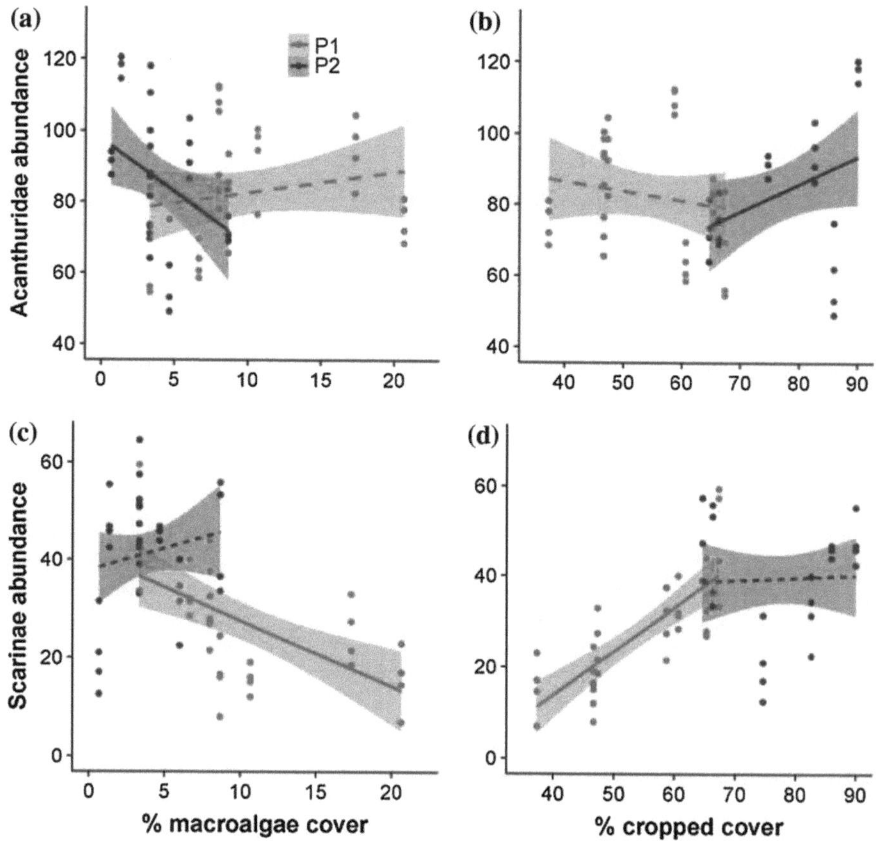

Fig. 5 Relationship between the abundance of *Acanthuridae* and *Scarinae* and the percentage of macroalgae and cropped cover. Linear models are plotted with 95% confidence intervals. *Dotted lines* represent non-significant relationships

4 Discussion

4.1 Herbivory and Resilience

Changes that Tiahura's coral reef went through may be perceived from the perspective of ecological resilience: *A. planci* outbreaks that occurred in 1980 and 2006 might have led to ecosystem re-organization. The notion of ecological resilience refers to a multi-equilibrium system due to changes of the environment [19, 20, 26] inducing a reorganization of the system that experiences different states. In Tiahura, this reorganization led to two different states: herbivorous fish, usually known to promote reef recovery [3, 64–67], were dominated by *Acanthuridae* during both period, but *Scarinae* importance increased during P2. In numerous studies changes between different trophic levels were expounded (e.g. [68–70]), whereas our study showed transition within the herbivorous fish community. Whether other changes between different trophic levels occurred in the coral reef of Tiahura remains to be assessed.

4.2 Comparison with Shifting Systems

Many studies were carried out on coral reefs phase shifts from coral to macroalgae-dominated systems in the Caribbean [19, 24, 26], more than twice as much as in the Indo-Pacific [41]. Even though some shifts occurred in Indo-Pacific coral ecosystems [20, 27], an analysis of long-term studies showed that declines without ensuing recovery occurred in 57% of studied cases in Caribbean and in only 27% in Indo-Pacific [40]. Our study supports these results since the Indo-Pacific coral reef of Tiahura showed a long-term resilience, without shifting to algae-dominated state for more than three decades. Some hypothesis were suggested to explain this major difference between Caribbean and Indo-Pacific reefs. One of them is the high herbivore diversity supported by Indo-Pacific reefs compared to Caribbean ones. For example *Acanthuridae* counts 88 species in the Indo-Pacific and only 4 in the Caribbean where Naso genus, consumer of macroalgae, does not occur [41].

4.3 Species Richness Increase

Increasing fish diversity may imply a complexification of coral reef ecosystem which is then more adapted to buffer disturbances [3, 24] and hence increase the resilience of the system [71]. Although an increase in species richness can be observed just after a disturbance, resulting from new species attracted by new available resources or free habitats [72], species richness in Tiahura expanded over 30 years and beyond many disturbances. The increase of species richness of both *Acanthuridae* and *Scarinae* was not likely to be an observer bias (increasing scientific expertise) since only three different observers monitored the fish communities, but could be a real diversification of these two herbivorous families. Although climate change has been predicted to conduce to species extirpation [73] as it has already been observed [74], local increase in fish species richness was observed in Tiahura and in other parts of the world [75, 76]. Usually, a loss of coral cover that exceeds 20% induces a decline in species richness of fish communities [77]. In Tiahura, strong coral cover loss did not affect species richness of the main herbivorous families, and this may be one of the key components of its high resilience.

4.4 Changes in Community Composition

Herbivorous fish can be classified into two functional groups: grazers and browsers [78] and grazers are subdivided into different types: scrapers, excavators and detritivores [79]. *Scarinae* are mainly scrapers or excavators that let large grazing scars in the substrate, removing both hard substratum and algal recruit developing on it [79]. Such a feeding mode could explain the low percentage of macroalgae during P2 as their increasing abundance would have removed many algal recruits and then prevented macroalgae development during this period. This is the main hypothesis that could explain the relationship between *Scarinae* and macroalgae since *Scarinae* are mainly grazers and only *Calotomus carolinus* is considered as browser and thus feed on macroalgae. Two species of *Acanthuridae* (*Naso lituratus* and *Naso unicornis*) may feed on macroalgae on the outer slope, but their abundance stayed really low even when macroalgae developed during P1. Hence the opposite patterns of *Acanthuridae* and *Scarinae* related to substrate cover were not due to their diet since they are both mainly grazers feeding on cropped cover. Nonetheless, the abundance of *Scarinae* presented a threshold when cropped cover reached 60%, which may confirm the results of [80] who have found that herbivorous fish and specifically *Scarinae* are able to maintain 40–60% of reef substratum in cropped state. To explain the shift between the role played by *Acanthuridae* and *Scarinae*, we can consider that if dominant species decline under changing conditions then minor species are able to substitute for them, which maintain ecosystem function over time [81]. In Tiahura, *Scarinae* were considered as a minor species in the nineties (P1) but then played a crucial role when *Acanthuridae* decreased around 2009 (P2). However, both states were dominated by cropped cover but the first one had higher percentage of macroalgae. This slight change in substrate cover may have attracted different families of herbivores and thus explain the different controls by *Acanthuridae* then *Scarinae*. The increasing importance of *Scarinae* may also have been due to fish recruitment that was not affected by *A. planci* outbreaks since juveniles settled in the lagoon, less impacted by the seastar. Growing *Scarinae* could then benefit from new available food, such as turf developing on the outer slope [82]. All these hypothesis assume that *A. planci* outbreaks triggered changes in the herbivorous fish community, and within both *Acanthuridae* and *Scarinae* communities. This post-disturbance reorganization allowed a quicker recovery of the coral ecosystem, as suggested by [83].

4.5 Implication of Different Herbivores

Acanthuridae, *Scarinae* and sea urchins are the major herbivores that control algal development in coral reefs [84, 85]. However, urchins were not taken into account in this study since they encountered an important wave of mortality in 1983–1984 reported by Caribbean studies [86, 87] which have concerned the Indo-Pacific since

2000 [88] and thus may distort the dataset. Urchins, *Acanthuridae* and *Scarinae* are competing for benthic algae [89], which means that the mortality wave of sea urchins could have let an available gap for the development of herbivorous fish in the eighties. On the other hand, such a mortality may have led to a reduction of the herbivory power which could have decreased the resilience capacity of the ecosystem. In the Caribbean, the presence of both sea urchins (*Diadema antillarum*) and parrot fish maintained the system in a coral-dominated state, without urchins, the system turned to macroalgae-dominated state [90]. In our case, it remains complex to assess whether the presence of both *Acanthuridae* and *Scarinae* was necessary to maintain resilience over time, and whether urchins were significantly involved in controlling algal development. Nevertheless we can assume that, at least, these two herbivorous families have been involved in the preservation of the coral reef of Tiahura from a shift toward a macroalgae dominated state, as it has been reported in the Red Sea [55].

5 Conclusion

The coral reef of Tiahura persisted over more than 30 years and underwent many disturbances without shifting toward a macro-algal ecosystem. *Acanthuridae* in the nineties and *Scarinae* around 2010 have played a crucial role in maintaining a low rate of algal cover, and this was made possible by the changes that occurred within communities. It follows that reef fish community of Tiahura was highly resistant to natural disturbances, as it has been observed around the whole island [48, 49]. Nevertheless we did not take into account the implication of human pressure in our study although Moorea hosts more than 16.000 habitants and face many anthropogenic disturbances (tourism, overfishing, habitat destruction) which undoubtedly affect the ecosystems [34–36]. The long-term analysis of the reef ecosystem of Tiahura provided us a valuable case of the expression of coral reef resilience, and the impact of the marine protected area established in 2004 at Tiahura should be further studied in order to evaluate whether the limitation of human activities could have improved this resilience.

References

1. Brown, J.H., Valone, T.J., Curtin, C.G.: Reorganization of an arid ecosystem in response to recent climate change. Proc. Natl. Acad. Sci. USA **94**, 9729–9733 (1997)
2. Standish, R.J., Hobbs, R.J., Mayfield, M.M., et al.: Resilience in ecology: abstraction, distraction, or where the action is? Biol. Conserv. **177**, 43–51 (2014)
3. Graham, N.A.J., Jennings, S., MacNeil, M.A., et al.: Predicting climate-driven regime shifts versus rebound potential in coral reefs. Nature **518**, 1–17 (2015)
4. Holling, C.S.: Resilience and stability of ecological systems. Annu. Rev. Ecol. Syst. **4**, 1–23 (1973)

5. Walker, B.H.: Is succession a viable concept in african savanna ecosystems? In: For Succession, Springer Adv Texts Life Sci, pp. 431–447 (1981)
6. Gunderson, L.H.: Ecological resilience—in theory and application. Annu. Rev. Ecol. Syst. **31**, 425–439 (2000)
7. Scheffer, M., Hosper, S., Meijer, M., et al.: Alternative equilibria in shalow lakes. Trends Ecol. Evol. **8**, 275–279 (1993)
8. van de Koppel, J., Rietkerk, M., Weissing, F.J.: Catastrophic vegetation shifts and soil degradation in terrestrial grazing systems. Trends Ecol. Evol. **12**, 352–356 (1997)
9. Carpenter, S., Walker, B., Anderies, J.M., Abel, N.: From metaphor to measurement: resilience of what to what? Ecosystems **4**, 765–781 (2001)
10. Nyström, M., Folke, C., Moberg, F.: Coral reef disturbance and resilience in a human-dominated environment. Trends Ecol. Evol. **15**, 413–417 (2000)
11. Anderies, J.M., Janssen, M.A., Walker, B.H.: Grazing management, resilience, and the dynamics of a fire-driven rangeland system. Ecosystems **5**, 23–44 (2002)
12. Wang, G., Eltahir, E.A.B.: Role of vegetation dynamics in enhancing the low-frequency variability of the Sahel rainfall. Water Resour. Res. **36**, 1013–1021 (2000)
13. Foley, J.A., Coe, M.T., Scheffer, M., Wang, G.: Regime shifts in the Sahara and Sahel: interactions between ecological and climatic systems in Africa. Ecosystems **6**, 524–539 (2003)
14. Peterson, G.: Forest dynamics in the Southeastern United States: managing multiple stable states. Gunderson Pritchard, pp. 227–246 (2002)
15. O'Dowd, D.J., Green, P.T., Lake, P.S.: Invasional "meltdown" on an oceanic island. Ecol. Lett. **6**, 812–817 (2003)
16. Scheffer, M., Szabó, S., Gragnani, A., et al.: Floating plant dominance as a stable state. Proc. Natl. Acad. Sci. **100**, 4040–4045 (2003)
17. Blindow, I., Andersson, G., Hargeby, A., Johanson, S.: Long-term pattern of alternative stable states in two shallow eutrophic lakes. Freshw. Biol. **30**, 159–167 (1993)
18. Jackson, L.J.: Macrophyte-dominated and turbid states of shallow lakes: evidence from Alberta lakes. Ecosystems **6**, 213–223 (2003)
19. Hughes, T.P.: Catastrophes, phase shifts, and large-scale degradation of a Caribbean coral reef. Science **80–265**, 1547–1551 (1994)
20. Done, T.J.: Phase shifts in coral reef communities and their ecological significance. Hydrobiologia **247**, 121–132 (1992)
21. Bruno, J.F., Sweatman, H., Precht, W.F., et al.: Assessing evidence of phase shifts from coral to macroalgal dominance on coral reefs. Ecology **90**, 1478–1484 (2009)
22. Nyström, M., Folke, C.: Spatial resilience of coral reefs. Ecosystems **4**, 406–417 (2001)
23. Gardner, T., Gill, J., Grant, A.: Hurricanes and Caribbean coral reefs: immediate impacts recovery trajectories and contribution to longterm decline. Ecology **86**, 174–184 (2005)
24. McClanahan, T.R., Muthiga, N.A.: An ecological shift in a remote coral atoll of Belize over 25 years. Environ. Conserv. **25**, 122–130 (1998)
25. Parravicini, V., Kulbicki, M., Bellwood, D.R., et al.: Global patterns and predictors of tropical reef fish species richness. Ecography (Cop) **36**, 1254–1262 (2013)
26. Knowlton, N.: Thresholds and multiple stable states in coral reef community dynamics. Am. Zool. **32**, 674–682 (1992)
27. McCook, L.J.: Macroalgae, nutrients and phase shifts on coral reefs: scientific issues and management consequences for the Great Barrier Reef. Coral Reefs **18**, 357–367 (1999)
28. Eakin, C.M.: Where have all the carbonates gone? A model comparison of calcium carbonate budgets before and after the 1982–1983 El Nin?o at Uva Island in the eastern Pacific. Coral Reefs **15**, 109–119 (1996)
29. Salvat, B.: Death for the coral reefs. Oryx **15**, 341–344 (1980)
30. Hughes, T.P., Bellwood, D.R., Folke, C., et al.: New paradigms for supporting the resilience of marine ecosystems. Trends Ecol. Evol. **20**, 381–386 (2005)
31. Hoegh-Guldberg, O.: Climate change, coral bleaching and the future of the worldís coral reefs. Mar. Freshw. Res. **50**, 839–866 (1999)

32. Wilkinson, C., Souter, D.: Status of caribbean coral reefs after bleaching and hurricanes in 2005 (2008)
33. Burke, L., Reytar, K., Spalding, M., Perry, A.: Reefs at risk. World Ressources Inst (2011)
34. Bellwood, D.R., Hughes, T.P., Folke, C., Nyström, M.: Confronting the coral reef crisis. Nature **429**, 827–833 (2004)
35. Pandolfi, J.M., Jackson, J.B.C., Baron, N., et al.: Are U.S. coral reefs on the slippery slope to slime? Science **80–30**, 1725–1726 (2005)
36. De'ath, G., Lough, J.M., Fabricius, K.E.: Declining coral calcification on the Great Barrier Reef. Science **80–323**, 116–119 (2009)
37. Gardner, T.A., Côté, I.M., Gill, J.A., et al.: Long-term region-wide declines in Caribbean corals. Science **301**, 958–960 (2003)
38. Hughes, T.P., Baird, A.H., Bellwood, D.R., et al.: Climate change, human impacts, and the resilience of coral reefs. Science **80–301**, 929–934 (2003)
39. Pandolfi, J.M., Bradbury, R.H., Sala, E., et al.: Global trajectories of the long-term decline of coral reef ecosystems. Science **80–301**, 955–958 (2003)
40. Connell, J.H.: Disturbance and recovery of coral assemblages. Coral Reefs **16**, S101–S113 (1997)
41. Roff, G., Mumby, P.J.: Global disparity in the resilience of coral reefs. Trends Ecol. Evol. **27**, 404–413 (2012)
42. Galzin, R.: Structure of fish communities of French Polynesian coral reefs. I Spatial scales. Mar. Ecol. Prog. Ser. **41**, 129–136 (1987)
43. Bouchon, C.: Quantitative study of scleractinian coral communities of Tiahura reef (Moorea Island, French Polynesia). In: Proceedings of the Fifth International Coral Reef Congress, Tahiti 6 (1985)
44. Adjeroud, M., Chancerelle, Y., Schrimm, M., et al.: Detecting the effects of natural disturbances on coral assemblages in French Polynesia: a decade survey at multiple scales. Aquat. Living Resour. **18**, 111–123 (2005)
45. Pratchett, M.S., Trapon, M., Berumen, M.L., Chong-Seng, K.: Recent disturbances augment community shifts in coral assemblages in Moorea, French Polynesia. Coral Reefs **30**, 183–193 (2011)
46. Trapon, M.L., Pratchett, M.S., Penin, L.: Comparative effects of different disturbances in coral reef habitats in Moorea, French Polynesia. J. Mar. Biol. **2011**, 1–11 (2011)
47. Berumen, M.L., Pratchett, M.S.: Recovery without resilience: persistent disturbance and long-term shifts in the structure of fish and coral communities at Tiahura Reef, Moorea. Coral Reefs **25**, 647–653 (2006)
48. Lamy, T., Galzin, R., Kulbicki, M., et al.: Three decades of recurrent declines and recoveries in corals belie ongoing change in fish assemblages. Coral Reefs (2015)
49. Galzin, R., Lecchini, D., Lison, T., et al.: Long term monitoring of coral and fish assemblages (1983–2014) in Tiahura reefs, Moorea, French Polynesia. Cybium, 1–11 (2016)
50. Kayal, M., Vercelloni, J., Lison de Loma, T., et al.: Predator crown-of-thorns starfish (Acanthaster planci) outbreak, mass mortality of corals, and cascading effects on reef fish and benthic communities. PLoS One (2012)
51. Rotjan, R.D., Lewis, S.M.: Impact of coral predators on tropical reefs. Mar. Ecol. Prog. Ser. **367**, 73–91 (2008)
52. Bruno, J.F., Selig, E.R.: Regional decline of coral cover in the Indo-Pacific: timing, extent, and subregional comparisons. PLoS One (2007)
53. Osborne, K., Dolman, A.M., Burgess, S.C., Johns, K.A.: Disturbance and the dynamics of coral cover on the Great Barrier Reef (1995–2009). PLoS One (2011)
54. Faure, G.: Degradation of coral reefs at Moorea island (French Polynesia) by Acanthaster planci. J. Coast. Res. **5**, 295–305 (1989)
55. Khalil, M.T., Cochran, J.E.M., Berumen, M.L.: The abundance of herbivorous fish on an inshore Red Sea reef following a mass coral bleaching event. Environ. Biol. Fishes **96**, 1065–1072 (2013)

56. Galzin, R., Legendre, P.: The fish communities of a coral reef transect. Pac. Sci. **41**, 158–165 (1987)
57. Adjeroud, M., Augustin, D., Galzin, R., Salvat, B.: Natural disturbances and interannual variability of coral reef communities on the outer slope of Tiahura (Moorea, French Polynesia): 1991 to 1997. Mar. Ecol. Prog. Ser. **237**, 121–131 (2002)
58. Schrimm, M., Buscail, R., Adjeroud, M.: Spatial variability of the biogeochemical composition of surface sediments in an insular coral reef ecosystem: Moorea, French Polynesia. Estuar. Coast. Shelf Sci. **60**, 515–528 (2004)
59. Loya, Y.: Plotless and transect methods. In: Stoddart, D.R., Johannes, R.F. (eds.) Coral Reefs Res methods, pp. 197–217. UNESCO, Paris (1978)
60. Core Team, R.: A Language and Environment for Statistical Computing. R Foundation for Statistical Computing, Vienna, Austria (2012)
61. Oksanen, J.: Multivariate Analysis of Ecological Communities in R, pp. 1–40 (2015)
62. Bray, J.R., Curtis, J.T.: An ordination of the upland forest communities of southern Wisconsin. Ecol. Monogr. **37**, 325–349 (1957)
63. Clarke, K.R.R.: Non-parametric multivariate analyses of changes in community structure. Aust. J. Ecol. **18**, 117–143 (1993)
64. McClanahan, T., Graham, N.: Recovery trajectories of coral reef fish assemblages within Kenyan marine protected areas. Mar. Ecol. Prog. Ser. **294**, 241–248 (2005)
65. Hoegh-Guldberg, O., Mumby, P.J., Hooten, A.J., et al.: Coral reefs under rapid climate change and ocean acidification. Science **80–318**, 1737–1742 (2007)
66. Hughes, T.P., Rodrigues, M.J., Bellwood, D.R., et al.: Phase shifts, herbivory, and the resilience of coral reefs to climate change. Curr. Biol. **17**, 360–365 (2007)
67. Mumby, P.J., Harborne, A.R., Williams, J., et al.: Trophic cascade facilitates coral recruitment in a marine reserve. Proc. Natl. Acad. Sci. **104**, 8362–8367 (2007)
68. Estes, J.A., Duggins, D.O.: Sea otters and kelp forests in Alaska: generality and variation in a community ecological paradigm. Ecol. Monogr. **65**, 75–100 (1995)
69. Sinclair, A.R.E., Olsen, P.D., Redhead, T.D.: Can predators regulate small mammal populations? Evidence from house mouse outbreaks in Australia. Oikos **59**, 382–392 (1990)
70. Dublin, H.T., Sinclaire, A.R., McGlade, J.: Elephants and fire as causes of multiple stable states in the Seregenti-Mara woodlands. J. Anim. Ecol. **59**, 1147–1164 (1990)
71. Cheal, A.J., Emslie, M., MacNeil, M.A., et al.: Spatial variation in the functional characteristics of herbivorous fish communities and the resilience of coral reefs. Ecol. Appl. **23**, 174–188 (2013)
72. Garpe, K.C., Yahya, S.A.S., Lindahl, U., Ohman, M.C.: Long-term effects of the 1998 coral bleaching event on reef fish assemblages. Mar. Ecol. Prog. Ser. **315**, 237–247 (2006)
73. Thomas, C.D., Thomas, C.D., Cameron, A., et al.: Extinction risk from climate change. Nature **427**, 145–148 (2004)
74. Graham, N.A.J., Wilson, S.K., Jennings, S., et al.: Dynamic fragility of oceanic coral reef ecosystems. Proc. Natl. Acad. Sci. USA **103**, 8425–8429 (2006)
75. Hiddink, J.G., ter Hofstede, R.: Climate induced increases in species richness of marine fishes. Glob. Change Biol. **14**, 453–460 (2008)
76. Knowlton, N., Jackson, J.: Shifting baselines, local impacts, and global change on coral reefs. PLoS Biol (2008)
77. Wilson, S.K., Graham, N.A.J., Pratchett, M.S., et al.: Multiple disturbances and the global degradation of coral reefs: are reef fishes at risk or resilient? Glob. Change Biol. **12**, 2220–2234 (2006)
78. Horn, M.H.: Biology of marine herbivorous fishes. Oceanogr. Mar. Biol. **27**, 167–272 (1989)
79. Bellwood, D.R., Choat, J.H.: A functional analysis of grazing in parrotfishes (family Scaridae): the ecological implications. Environ. Biol. Fishes **28**, 189–214 (1990)
80. Williams, I.D., Polunin, N.V.C.: Large-scale associations between macroalgal cover and grazer biomass on mid-depth reefs in the Caribbean. Coral Reefs **19**, 358–366 (2001)

81. Walker, B., Kinzig, A., Langridge, J.: Plant attribute diversity, resilience, and ecosystem function: the nature and significance of dominant and minor species. Ecosystems **2**, 95–113 (1999)
82. Adam, T.C., Schmitt, R.J., Holbrook, S.J., et al.: Herbivory, connectivity, and ecosystem resilience: response of a coral reef to a large-scale perturbation. PLoS One (2011)
83. Lamy, T., Legendre, P., Chancerelle, Y., et al.: Understanding the spatio-temporal response of coral reef fish communities to natural disturbances: insights from beta-diversity decomposition. PLoS One **10**, 1–18 (2015)
84. Hay, M.E.: Patterns of fish and urchin grazing on caribbean coral reefs: are previous results typical? Ecology **65**, 446–454 (1984)
85. McClanahan, T.R., Shafir, S.H.: Causes and consequences of sea urchin abundance and diversity in Kenyan coral reef lagoons. Oecologia **83**, 362–370 (1990)
86. Hughes, T.P., Reed, D.C., Boyle, M.J.: Herbivory on coral reefs: community structure following mass mortalities of sea urchins. J. Exp. Mar. Biol. Ecol. **113**, 39–59 (1987)
87. Carpenter, R.C.: Mass mortality of Diadema antillarum. I. Long-term effects on sea urchin population-dynamics and coral reef algal communities. Mar. Biol. **104**, 67–77 (1990)
88. Moreau, F., Chancerelle, Y., Galzin, R., et al.: Les aires marines protégées de Moorea, 10 années de suivi (2004–2014) 159 p. RA-204. AMP Moorea (2014)
89. Carpenter, R.C.: Mass mortality of Diadema antillarum. II. Effects on population densities and grazing intensity of parrotfishes and surgeonfishes. Mar. Biol. **104**, 79–86 (1990)
90. Mumby, P.J., Hastings, A., Edwards, H.J.: Thresholds and the resilience of Caribbean coral reefs. Nature **450**, 98–101 (2007)

Using Time Series Analysis for Estimating the Time Stamp of a Text

Costin-Gabriel Chiru and Madalina Toia

Abstract Language is constantly changing, with words being created or disappearing over time. Moreover, the usage of different words tends to fluctuate due to influences from different fields, such as historical events, cultural movements or scientific discoveries. These changes are reflected in the written texts and thus, by tracking them, one can determine the moment when these texts were written. In this paper, we present an application based on time series analysis built on top of the Google Books N-gram corpus to determine the time stamp of different written texts. The application is using two heuristics: words' fingerprinting, to find the time interval when they were most probable used, and words' importance for the given text, to weight the influence of words' fingerprinting for estimating the text time stamp. Combining these two heuristics allows time stamping of that text.

Keywords Time series analysis • Time stamping • Peak detection • Google corpus

1 Introduction

The usefulness of automatic text time stamping is evident, no matter how new or old the text is. Firstly, it might help authenticate old books or letters. Secondly, for unsigned manuscripts, it could assist putting them inside a context. Thirdly, for more recent publications, it could improve the search engines by allowing the researchers to choose the most up to date information by sorting through the data chronologically.

C.-G. Chiru (✉) · M. Toia
Department of Computer Science and Engineering, Politehnica University from Bucharest, 313 Splaiul Independetei, Bucharest, Romania
e-mail: costin.chiru@cs.pub.ro

M. Toia
e-mail: mada.toia@gmail.com

I. Rojas et al. (eds.), *Advances in Time Series Analysis and Forecasting*, Contributions to Statistics, DOI 10.1007/978-3-319-55789-2_3

Finding out when a recent book was written is quite easy. One may look at the copyright page and find when was it published, which edition it is and when was the first one written. For old books, however, these standards do not apply. Antiquarians might search for the author's bibliography or might use the other books advertised at the end of the current one to time stamp it. Details about the publisher, such as its name or address might be also useful in determining an approximate time stamp for the book. But when the text to be dated is not from a particular book, or the book itself is very old and copyright information or even publishing details do not exist, then it becomes more complicated.

Fortunately, there are solutions even for these cases. English nowadays is different than the one spoken by Shakespeare's characters. Over the centuries, words tend to be simplified in order to help pronunciation. Even though the change is not abrupt, it is more likely to find the older form in the past than after the newer form got popular. Grammar also changes over time. Hence, sentences with words in a stranger order are more likely to be written in a more distant time. Considering these small hints, linguists can give a rough estimate about the publishing time of a text.

Since language is continuously changing, with new words being added while others being forgotten, the aim of this paper is to use these language variations in order to determine when a text was written. Thus, we built a model that uses time series to fingerprint the words from a document, along with the importance of those words for the given document.

The paper continues with the presentation of the background for this work before delving into the details of the application. Section 4 highlights how this work can be used by presenting a case study and what outcomes may be expected, by showing our results. The paper ends with our conclusions and possible research directions for extending the work done so far.

2 Background and Related Work

2.1 *Language Modeling*

A language model is a probabilistic way of defining a language. It is built from samples of text and it describes word probabilities inside a distribution. The changes inside its corpus can be visible in this distribution. For example, in a model describing old English the word "*thou*" will have a higher probability of appearing, than its current version, "*you*". In order to time stamp the texts, we intended to capture these changes. Therefore, we needed language models for every period of time so that, based on them, to figure out to which of them the text belonged.

To obtain the needed language models we used the N-gram approach. An N-gram is a continuous group of N letters, syllables or words that appear in sequences of text or speech [1]. Each of these sequences has assigned a probability based on the number of occurrences inside a training corpus. More precisely, an

N-gram model can be used to predict the most probable n-th element from a sequence after seeing the previous n-1 elements and knowing the sequence's probability. This model is not entirely accurate, the results depending heavily on the training set.

2.2 *Google Books N-Gram Corpus*

The corpus was obtained by scanning over 5 million books from the Google Books Library, resulting in around 5 billion words. Even though the texts represent only about 4% of all the published books [2], they can generate a good model for our case. The resulted corpus contains the words used between 1500 and 2008 (in multiple languages) along with their frequency of use inside those books. The available N-grams range from 1 to 5 words.

For this implementation, we created a database containing only the unigrams (N = 1) from the English corpus. We chose to do so because of the smaller size of this database, which translates in a much shorter response time of the application. Even though we cannot model grammatical changes with this data, we can still follow changes inside the lexicon. However, a downside of this database is the fact that there are not that many texts between 1500 and 1800, so the precision in those centuries is smaller.

2.3 *Related Work*

There are multiple theories on how or why a language is constantly changing. Some say this is because people tend to simplify pronunciation [3] or slightly change letters inside words because they write them as they hear them (e.g. color and colour). Another change might be due to the fact that people are social. They tend to mimic one another and this is why people living in the same region have different sayings and accents than others from other regions from the same country. Also, because of this tendency they end up misusing the same words and changing their meaning [4, 5].

In [6], the authors used the Google N-gram corpus to model the language changes over time. They tried to find similarities in how often words are used, in each period of their lifetime, for multiple languages: English, Hebrew and Spanish. They studied how similar words replace one another because of the simplifying tendency and how long it takes to a normal word to reach his usage pike. They only chose the 1800–2008 time interval and normalized the words' usage to compensate for the vocabulary increase. The lifetime of a word was divided into 3 phases: infant, adolescent and mature. This allowed them to statistically discover that for the English corpus, the time it took a word to reach common usage was around 30–50 years.

In [7] the authors used articles from 5 French newspapers published between 1801 and 1944 to find out the decade on which a piece of text was written. They used a chronological method followed by classification. During the first stage, all the persons' names found in the text were gathered and searched for their birthday on Wikipedia, thus eliminating the time interval before the persons' birth. Another chronological information was related to neologisms and archaisms: the authors removed the time intervals when those words did not exist. Even though this approach was very efficient, with an error of only 3%, the resulting time intervals were larger than the desired granulation. The next step was to use classification for shrinking the outputted time frame. For this, they split the training data into N-grams and assigned them to each year. The closest year was found using cosine similarity.

Another attempt to classify a text based on time, was done in [8]. Here, the authors computed a statistical language model for each year inside the predefined time interval. The classification part returned the most similar model with a confidence score. They used two well-known Dutch newspapers with articles from January 1999 to February 2005 to build the training set and articles from another newspaper for test. The assumption was that if the granularity would be small (1 week), the main topics presented would be the same. In this case, the topics would be visible inside the language model built from that time frame, thus making classification easier.

Another attempt was done by Szymanski and Lynch [9], who won the "SemEval-2015 Task 7: Diachronic Text Evaluation" contest. For this task they used the provided training set (articles from The Spectator's archive published from 1500 to 2014) as well as the Google 1-gram corpus for English. For The Spectator training set, the authors split it into time periods with a granularity of 6, 12, 20 and 50 years. Then, they used feature extraction to select four types of features: character n-grams, part-of-speech tag, word and syntactic phrase-structure. For classifying, they used Naive Bayes and SVM. The obtained accuracy, depending on the chosen granularity, was 41.5 for 6 years, 45.9 for 12 years, 55.3 for 20 years and 73.3 for 50 years.

Zimmerman also proposed, using a Naive Bayes classifier, to obtain the publishing date of 19 Old English texts [10]. This language was spoken from the Anglo-Saxon settlement in Britain until the Norman invasion, and thus the evaluated books were published between 850 and 1230. The author divided this interval into five subgroups and used them as classes for the Naive Bayes classifier. He took advantage of the fact that the words order inside a sentence changed and used the apparition percentage of certain syntactic construction as features for classification. The classifier correctly determined the time frame for 36 texts out of 50. However, it performed much better for the extreme classes than for the ones in the transition period.

The other approaches concentrate on recent texts, for these being easier to gather proper training data. As stated before, there are not that may publications kept from before 1600. However, most of the undated books come from that time frame.

3 Implementation Details

3.1 Application's Design

The architecture of the built application is presented in Fig. 1. The user interacts with the user interface for selecting the text to be processed and view the results (the application's estimation of the time stamp of that text).

After the user inputs a text file, the application processes it and extracts the text vocabulary, in order to identify the distinct words. Each word is then processed inside the Word Processing module. This module receives a word as input and searches for its fingerprint, generating the moments of time when that word is most probable to have been used. For that, it looks for the word in the database and retrieves its time series. Afterwards, the obtained information is processed using the Simple Peak Detection algorithm to narrow down the time interval when it is most likely to have been used. The obtained intervals are then passed to the EMD module where they are filtered out to maintain only the ones with the highest probability for the usage of the given word (its fingerprint). The results are further passed to the Reducing Unit, where all the intervals resulted from the Word Processing module are overlapped, and the most likely period is shown to the user in the user interface.

Using the steps presented above, we reduce the problem of time stamping a document to the one of building its words' fingerprints (which in turn is based on identifying the words' spikes), and of combining these fingerprints by finding the weights corresponding to the importance of different kinds of words.

3.2 Building Words Time Series

As already stated, for this research we have used the 1-gram English Corpus extracted from Google Books. We believe this decision to be the right choice because of two reasons: first of all, it was built by digitizing writings from all areas, making it unbiased towards one domain or another, and secondly, its dimension ensures having a good coverage of the words from a specific language, being unlikely to find words in an analyzed writing that are not in the corpus. Moreover,

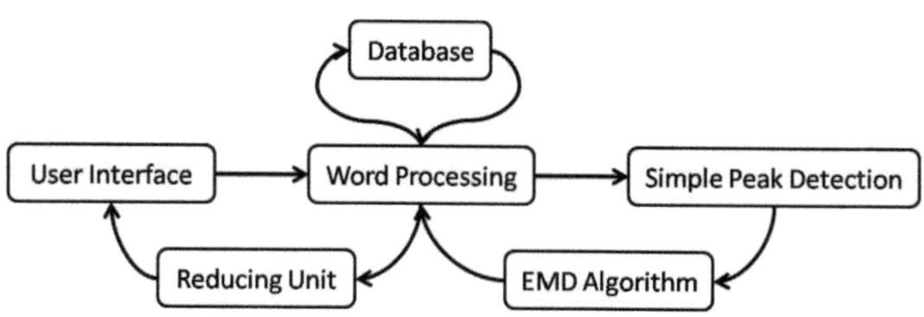

Fig. 1 Application architecture

having information about the words' usage across more than 300 years makes it a valuable resource for applying a time series approach.

Since the values from the corpus represent the number of times specific words have been found in the digitized documents written in a specific year, these values needed to be normalized according to the total number of words analyzed during that year. Thus, instead of using the number of times a word was used during an year, we actually used its frequency, computed as in (1), where n_i represents the total number of word's appearances on that particular year (i), while t_i represents the total number of words found in the digitized documents from that particular year;

$$p(w,i) = \frac{n_i}{t_i} \tag{1}$$

However, due to the low number of books published during the 16th and 17th centuries, there are some years having no information about words' usage. Moreover, this low number of publications from some years generates big variations in words' usage, negatively influencing the results of our approach. In order to decrease this noise, instead of using the words' frequencies from the Google corpus (computed using (1)), we used a smoothed version of it. This means that instead of using the word's frequency from a given year, we used the average value of the word's frequencies from a window of size k centered on that year, as in (2). Here, p_i is the word's frequency from year i, k is the size of the window and $i \in [1500, 2008]$.

$$v(i,k) = \frac{\sum_{j=-k}^{k} p_{i+j}}{2^{*}k + 1} \tag{2}$$

Thus, in order to build the time series needed for the next steps, for each word from the Google Books N-gram Corpus we first extracted its number of usages for each year in the available range ([1500, 2008]), then we normalized these values according to (1) and smoothed them using (2), so that to finally save the new data in an array that was further used by the application.

3.3 Words' Fingerprint

By fingerprinting a word we understand the automatic identification of the time period(s) in which that word is most probable to have been used. If we represent the word's time series graphically, these periods are highlighted by spikes in the graph and are caused by unusual uses of that word due to the fact that it is linked to some important events. This occurs in the case of people's names when they become famous, of countries in case a major event happens there, of different diseases if they become epidemic, etc.

The periods of time we were interested in are defined by the time intervals between the starting and ending points of different such spikes. Using this feature we were able to narrow down the time interval when a text could have been written to only those time periods when its words had spikes.

To identify these spikes, two different algorithms are used: simple peak detection (SPD), an algorithm based on the standard deviation of the time series and Earth Mover's Distance (EMD), which is used to further narrow down the results provided by SPD. We don't apply EMD directly, because this algorithm is much more expensive, computationally speaking, and thus we decided to first apply SPD as to reduce the interval on which EMD is applied and thus to reduce the required computation.

Simple Peak Detection (SPD)
The first step of this algorithm is to compute the time series mean (μ) and standard deviation (σ). Once these are computed, the algorithm considers a possible spike any interval [*start, stop*] for which the time series is initially ascending from the *start* position up to a point and then descending until the *stop* position, providing it respects the restriction $|x_i - x_{i+1}| > \varepsilon * \sigma$, where x_i is the word's frequency during year i, ε is a constant, and σ is the standard deviation of the time series. To exemplify how this algorithm is working, we give the example of *ice cream*, for which the time interval detected by SPD is [1905, 2008] (See Fig. 2).

Earth Mover's Distance (EMD)
The problem of the SPD algorithm is that it detects very large time intervals. In the case of *ice cream*, there are in fact two different intervals that are merged together, thus resulting a single large one. In order to obtain smaller intervals, that would allow a more precise time stamping of the analyzed document we employed a modified version of the EMD algorithm [11]. This algorithm is trying to select only those areas from the graph that resemble to a Gaussian distribution (as a spike is very similar to a Gaussian distribution).

Thus, we started from the intervals obtained in the previous step from SPD algorithm and applied whitening [12] in order to obtain a distribution with 0 mean and a standard deviation of 1, as needed by the EMD algorithm. If the initial time

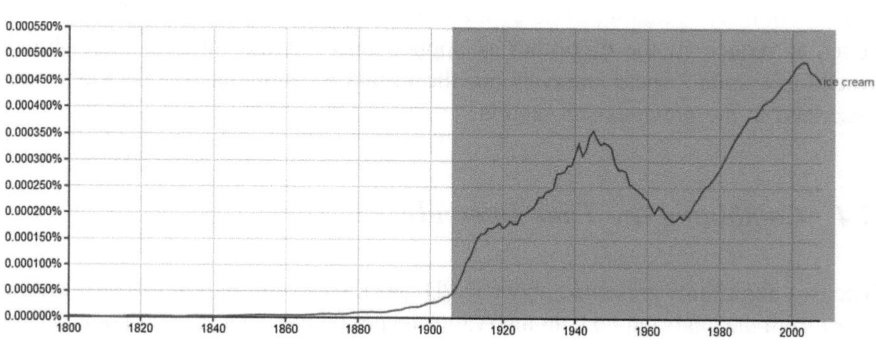

Fig. 2 The time interval detected by SPD as being the spike of "ice cream" expression

series for the word w_i was named *x(i, t)*, with t ∊ [l, r], where l and r represent the left and right margins of the interval, then the whitened series *g(i, t)* is given by (3).

$$g(i,t) = \frac{x(i,t) - \min_{p \in [l,r]} x(i,p)}{\sum_{q=l}^{r} [x(i,q) - \min_{p \in [l,r]} x(i,p)]} \tag{3}$$

Having the new values from the whitened time series, we compute μ and σ using (4) and (5) in order to be able to identify the normal distribution N(μ, σ) that approximates the time series.

$$\mu = \frac{\sum_{t=l}^{r} t^{*} g(i,t)}{r-l+1} \tag{4}$$

$$\sigma^2 = \frac{\sum_{t=l}^{r} (t-\mu)^{2*} g(i,t)}{r-l+1} \tag{5}$$

Using the built normal distribution N(μ, σ), we can generate its values for each year from our interval, thus generating how the time series should look like. After that, in order to decide whether our initial time series resembles a Gaussian distribution or not, we computed the distance between the whitened time series and the one obtained from N(μ, σ), using the following code:

```
acc = distance = 0;
for (t = l; t < r; t++){
    acc += g(i,t) - 1/(√(2π)σ) * e^(-(t-μ)^2/(2σ^2));
    distance += |acc|;
}
return distance
```

For the final results, we only choose those intervals with a distance smaller than a constant ε. Depending on how large the value of ε is, we may retrieve only intervals that look like a bell or be more permissive and accept some (small) variations.

Unfortunately, the EMD algorithm has a complexity of O(n^3), being too costly, computationally speaking, to be applied on the whole time interval. Therefore, in order to reduce it, the algorithm is applied only on the intervals that already resemble a spike. These intervals are the results of applying the SPD algorithm described in the previous sub-section.

3.4 Combining the Time Intervals

The two algorithms presented above (SPD and EMD) will output, for each distinct word from the analyzed text, an interval (time period having a starting point, an end and a peak—the maximum value) representing its fingerprint. However, in order to time stamp the document, these intervals must be combined so that the algorithm to return only the time period with the maximum probability.

Moreover, when computing the final time stamps, besides the information obtained from the words' fingerprinting, some additional information should be used: the words frequencies in the analyzed text and the words' properties. Some examples of features that could be used are the words' number of letters or if they were capitalized or not. We chose to increase the weights of capitalized words, because they are likely to express elements having stronger fingerprints (narrower intervals of high probability). Such words might represent people's names or places, which are more useful in time stamping the document. We also decided to decrease the weights of words with less than 4 letters, as most of them are prepositions, which are not very helpful for us.

Once all the intervals are computed, they are added to an array having a slot for each year from 1500 to 2008. The year with the highest sum is returned as the most probable year for the analyzed document.

3.5 Optimization

Both word extraction and peak detection are independent operations and thus we could use parallel computing to faster fulfill the task. For word extraction, the inputted documented is split into subparts, each going to be parsed and then its distinct words counted. This stage represents the mapping part inside the map reduce approach. The reduce phase implies summing up the distinct words in order to find the most relevant ones.

Computing each word's fingerprint and extracting the spikes is also highly parallel, this part also being executed by multiple mappers. At the end, the reducers would sum up the obtained intervals, considering each word's importance inside the text and the steepness of the spike.

4 Case Study and Results

After processing various texts with the help of this application, certain patterns had emerged. The program works best with books that add new words to the lexicon. For example, *Harry Potter* introducing words like *Quidditch* or *Hogwarts.* This happens because these specific words were introduced in this book series, and it is highly unlikely to appear in other texts. As it can be seen in Fig. 3, their frequency started to grow around 1998, when the first book was published, and started decreasing in 2007, after the last book was written.

This also happens in the case of fantasy books such as *A Game of Thrones* or *Lord of the Rings*, where words like *direwolf* or *hobbit* have a strong fingerprint around their publishing date. On the other hand, for books such as Jules Verne's

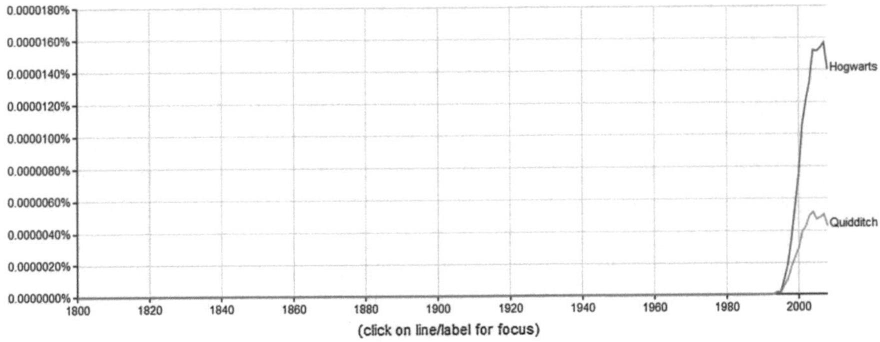

Fig. 3 Visualization of specific words from the Harry Potter book series

Travel to the Center of the Earth there are no such words and therefore the result is not very precise.

Historical events and famous peoples' names also have a good fingerprint, making history books or newspapers easier to date. Since the text tells something about a person or place, it means that these should have existed, helping time stamping. For example, we investigated Niccolo Machiavelli's *Il Principe* in which *Lorenzzo Medici*, *Cesare Borgia*, *Henry VIII* and *Pope Clement* are important characters. Although Machiavelli died in 1527, the book was published in 1532, while our application detects the year 1563 as the most likely for the text's publication. In Fig. 4, we present the time series of two of these names, *Medici* and *Cesare* (Cesare Borgia), having a strong fingerprint around 1565, while in Fig. 5 we present the time series of the other two characters (Clement and Henry VIII), having a smaller spike around 1520, besides the one between 1560 and 1570. The first spike is due to Henry VII creating his own religion, since the pope didn't allow him divorce Katherina de Aragon.

The delay between the publishing year and the one detected by the application is in this case mainly due to the fact that the book and characters were Italians, while English authors started talking about it after a while. Still, even if the book was

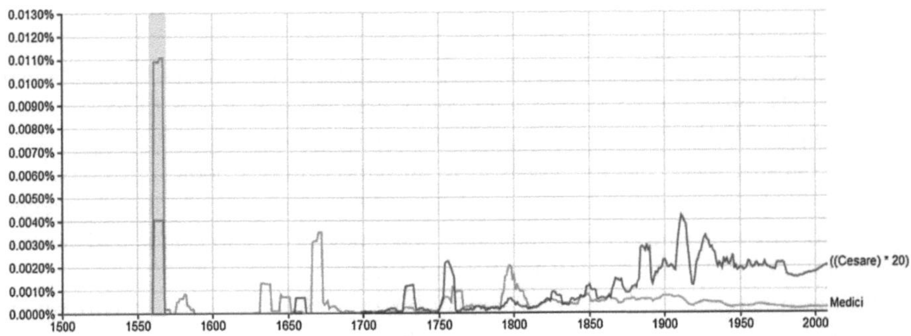

Fig. 4 The fingerprints of names *Medici* and *Cesare* for dating Machiavelli's *Il Principe*

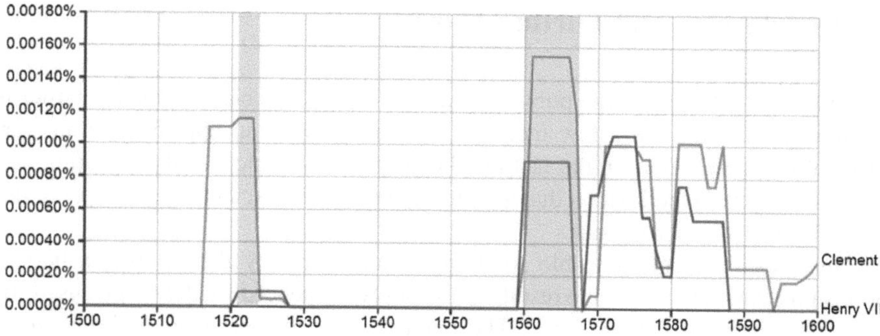

Fig. 5 The fingerprints of names *Clement* and *Henry VIII* for dating Machiavelli's *Il Principe*

written in English, such a delay is expected, as the people tend to write about past events a while after they happened.

As seen in the above example, people's names, places or countries tend to have a stronger influence when dating a document.

However, for documents containing few words with strong fingerprints, the time stamp obtained by the application will tend to be around 1500–1600. This happens because of the large number of words having spikes in that period: in the Google corpus, in the 16th century there are many years with 0 apparitions of different words, followed by periods with high frequency for those words (these frequencies being artificially high due to the smaller number of publications for those years). More test results can be found in Table 1.

Table 1 Test results for various books

Name of the book	Publish date	Estimated date
Fantasy books		
Harry Potter and the Sorcerer's Stone	1997	1998
A Game of Thrones	1996	2000
Lord of the Rings	1954	1978
A Journey to the Centre of the Earth	1864	1663
Engineering books		
Dark Energy, Dark Matter—NASA article	2013	1601
Elevator Systems of the Eiffel Tower	1889	1627
The Colored Inventor	1913	1591
The New York Subway	1904	1627
The Panama Canal	1913	1831
History books		
The Prince	1532 (Italian)	1583
Saint Austin	1900–1940	1780
Prince Henry the Navigator, the Hero of Portugal and of Modern Discovery, 1394–1460 A.D	(aprox) 1950	1594, 1595

5 Conclusions and Further Work

It is difficult to determine the publishing date of a text, especially when the writer is unknown, but there are a few methods that can provide an approximate answer. Most of them are based on the fact that language is dynamic, words appearing and disappearing from the lexicon due to cultural movements, scientific discoveries and the influence of other nations.

This paper presents a viable solution for automatically determining a book's publishing date. However, the results are strongly influenced by the corpus used for extracting the words' time series. The main issue is the fact that the books used aren't evenly distributed over the time frame.

The tested method has both advantages and disadvantages. As presented in the previous section's example, the spike overlapping method works best for documents containing words strongly linked to a time period. For example, historical books, newspaper articles or science fiction publication are easier to time stamp. However, for texts containing few to no words with strong fingerprints, the results are far from satisfying. The obtained results are in concordance with those reported in [7].

During the application testing we were able to identify some important sources of errors. Probably the biggest one was related to the small number of publications from the 16th and 17th centuries. This lead to spikes in words time series, strongly influencing the obtained results. Another issue related to the corpus is represented by the wrongly dated books inside the corpus. Another source of errors is the fact that authors tend to write about past events after a while since they happened, fact that introduces some delays in identifying the publishing year.

Besides these errors that are not easily corrected, there were some errors introduced by our implementation decisions. The most important one was related to only using the unigrams from the corpus for obtaining a good response time from the application. The downside is that unigrams are not good for capturing the context of their usage. If larger n-grams were used, the estimated time stamps would have been closer to the real publishing dates. To support this claim, we provide the example of the NASA's article about *dark matter*. Taken individually, as unigrams, these words provide inconclusive results, as they don't have strong fingerprints. However, if they are considered as forming a bigram, the result would be more accurate, the *Dark matter* bigram having a spike after 1980. Some other important sources of errors that might be corrected in the future are the way the words are fingerprinted and the way the relevant intervals are combined.

Once these sources of errors were identified, some possible improvements have also emerged: the use of larger n-grams (bigrams or trigrams) instead of unigrams, ignoring the time period before 1800, weighting the words differently according to their part-of-speech, using the neologisms and archaisms when fingerprinting the words (which might help decide that the text could not have been written before or after a time period) and finally changing the fingerprinting method to also accept plateaus, not only spikes—this might be leading to obtaining acceptable results for more documents, but with larger errors.

Acknowledgements This work has been funded by University Politehnica of Bucharest, through the "Excellence Research Grants" Program, UPB–GEX. Identifier: UPB–EXCELENȚĂ–2016 Aplicarea metodelor de învățare automată în analiza seriilor de timp (Applying machine learning techniques in time series analysis), Contract number 09/26.09.2016.

References

1. Jurafsky, D., Martin, J.: Speech and Language Processing. Prentice Hall (2000)
2. Michel, J.-B., Shen, Y.K., Aiden, A.P., Veres, A., Gray, M.K., The Google Books Team, Pickett, J.P., Hoiberg, D., Clancy, D., Norvig, P., Orwant, J., Pinker, S., Nowak, M.A., Aiden, E.L.: Quantitative analysis of culture using millions of digitized books. Science **331**(6014), 176–182 (2011)
3. Fromkin, V., Robert, R., Hyams, N.: An Introduction to Language, 7th edn. Thomson Wadswor (2003)
4. Wijaya, D.T., Yeniterzi, R.: Understanding semantic change of words over centuries. In: DETECT'11, pp. 35–40 (2011)
5. Mitra, S., Mitra, R., Riedl, M., Biemann, C., Mukherjee, A., Goyal, P.: That's sick dude!: automatic identification of word sense change across different timescales. In: 52nd ACL, pp. 1020–1029 (2014)
6. Petersen, A.M., Tenenbaum, J., Havlin, S., Stanley, H.E.: Statistical laws governing fluctuations in word use from word birth to word death. Sci. Rep. **2**, 313 (2012)
7. Garcia-Fernandez, A., Ligozat, A.-L., Dinarelli, M., Bernhard, D.: When was it written? Automatically determining publication dates. In: String Processing and Information Retrieval, pp. 221–236 (2011)
8. de Jong, F., Rode, H., Hiemstra, D.: Temporal language models for the disclosure of historical text. In: Proceedings of the AHC'05, pp. 161–168 (2005)
9. Szymanski, T., Lynch, G.: UCD: diachronic text classification with character, word, and syntactic N-grams. In: SemEval 2015, 879–883 (2015)
10. Zimmermann, R.: Dating hitherto undated old English texts based on text-internal criteria. http://www.old-engli.sh/my-research.php
11. Rubner, Y., Tomasi, C., Guibas, L. J.: A metric for distributions with applications to image databases. In: Computer Vision and Image Understanding, pp. 86–109 (2004)
12. Bishop, C.M.: Pattern Recognition and Machine Learning. Springer, New York (2001)

Using LDA and Time Series Analysis for Timestamping Documents

Costin-Gabriel Chiru and Bishnu Sarker

Abstract Identifying the moment of time when a book was published is an important problem that might help solving the problem of authorship identification and could also shed some light into identifying the realities of the human society during different periods of time. In this paper, we present an attempt to estimate the publication date of books based on the time series analysis of their content. The main assumption of this experiment is that the subject of a book is often specific to a time period. Therefore, it is likely to use topic modeling to learn a model that might be used to timestamp different books, given for training many books from similar periods of time. To validate the assumption, we built a corpus of 10 thousand books and used LDA to extract the topics from them. Then, we extracted the time series of particular terms from each topic using Google Books N-gram Corpus. By heuristically combining the words' time series and the topics from a document, we have built that document's time series. Finally, we applied peak detection algorithms to timestamp the document.

Keywords Natural Language Processing · Time Series Analysis · Google Books N-gram Corpus · Topic Modeling · LDA

1 Introduction

Language is continuously changing. If we analyze it from a temporal point of view, we can see that new words are invented while others are forgotten over time. Moreover, the use of specific words fluctuates a lot due to important historical events, social needs or technological breakthroughs. Most of the time, these

C.-G. Chiru (✉) · B. Sarker
Department of Computer Science and Engineering, University Politehnica from Bucharest, 313 Splaiul Independetei, Bucharest, Romania
e-mail: costin.chiru@cs.pub.ro

B. Sarker
e-mail: bishnukuet@gmail.com

I. Rojas et al. (eds.), *Advances in Time Series Analysis and Forecasting*, Contributions to Statistics, DOI 10.1007/978-3-319-55789-2_4

changes are reflected in language and thus they can be tracked and used to timestamp documents written in different periods of time. Nowadays, this information is manually used by librarians to authenticate old books or to identify the broader context in which they were written. However, this is a tedious operation, requiring very large amounts of time for analyzing the content of the books and the elements specific to each period of time. If such an operation could be automated, it could assist the librarians in doing their work faster and more reliable.

In this paper, we present an attempt to estimate the publication date of books starting from the topics that are present in each of them. To do that, we start from the assumption that the writings from different periods of time reflect the concerns of the citizens from that time. For example, during or shortly after periods of war there might be published more books about heroic deeds in battles or about the struggle of people on the battlefield. During rainy periods, the published books might talk about floods and their catastrophic effects. When Neil Armstrong reached the Moon, there were a lot of articles talking about space travel. During the 90s, when AIDS was a problem for humanity, there were many articles talking about medicine and the efforts to find cures to this disease and so on. Thus, we hope that by identifying the topics that are present in different books, we will be able to estimate their timestamp.

Identifying the topics from a book can be done with the help of topic modeling, by employing Latent Dirichlet Allocation (LDA) [1–3]. LDA receives a corpus of texts at input and outputs three things: the words belonging to the topics, along with their probabilities of belonging to these topics and the degree of membership of the documents from the corpus to each of these topics. Having this information, we can estimate, for a new document, what are the topics that are present in it, along with their distribution. However, for deriving the required information, we first need a corpus to train LDA on, that should be as balanced as possible in terms of the contained topics and the moments of time when the text was written. Therefore, we used a database of about 10 thousand books published at different moments in history and debating about various topics, from which we extracted 300 topics. In this way, every new document that is analyzed can be expressed as a mixture of the 300 topics extracted by the LDA.

Still, knowing these topics (along with their mixture coefficients) for a document is not enough to timestamp it. Another piece of information is still missing from the puzzle: the periods of time when these topics were popular or not. In other words, we need to know when it is probable to use one topic or another. In order to derive this information, we started from the information provided by the LDA in order to select the most representative 20 words for each topic. Then, using Google Books N-gram Corpus [4], we extracted the time series of the usage of these words over the history and then combined them in order to obtain the time series for the whole topic. Once this information is also available, the new document may be timestamped based on the distribution of the contained topics along with their distribution in time.

In the next section, we will present other attempts at timestamping documents. Afterwards we will describe the steps that were undertaken to reach our goal and we will exemplify them through a case study. In Sect. 4 we provide an interpretation of the obtained results, explaining what didn't work as expected. Finally, we conclude this paper with our observations and suggestions of improving the current approach.

2 Related Work

Google Books N-gram Corpus [4] was released in 2010 and since then it was used for various purposes, people from various domains trying to extract interesting trends from this data. Thus, it has proven its utility in various fields, ranging from Scientific Referencing and History of Science [5], to medicine [6], culture [7], Linguistics and Digital Humanities [8], etc.

In medicine, it was used to highlight the changes in the popularity of specific drugs over time, historical epidemiology of drug use and misuse, and adoption and regulation of drug technologies [6]. Koplenig [8] used Google Books N-gram Corpus to demonstrates the importance of metadata that is a powerful source of information for digital humanities.

For the task of timestamping, Islam et al. [9] used Google Books N-gram Corpus and an unsupervised approach to map books to the moment of time when they were written. However, the dataset they considered was for 36 books written between 1551 and 1969 which is considerably smaller in size. Moreover, the work presented here is significantly different than the one from [9] in that it is based on topic modeling using LDA and it has a different approach of estimating the time series.

Szymanski and Lynch [10], used Google Books N-gram Corpus to augment the data provided for the SemEval-2015 Task 7 contest. They won the contest by applying Naive Bayes and SVM for classifying the text with different granularities. The features they used for training the classifiers were character n-grams, part-of-speech tags, words and syntactic phrase-structures.

Other researchers also used different resources for timestamping documents. For example, Garcia-Fernandez et al. [11] used articles from 5 French newspapers published between 1801 and 1944 to find out the decade on which a piece of text was written. They used features such as people's names to eliminate the time interval before they were born, or archaisms and neologisms found in the text for eliminating the time intervals when those words did not exist.

The research having the largest influence on our paper was intended to analyze the correlation between significant historic periods (events) and the text written during that period [12]. The authors also presented several peak detection algorithms that we found to be helpful for timestamping the documents.

3 Experiment Outline

The experiment was design so that to follow a series of 6 major steps: build the corpus (1), preprocess the dataset (2), apply topic modeling using LDA (3), extract the years and the corresponding frequencies for topic specific terms (4), combine the terms' frequencies from different years to find the potential timestamp for the considered document (5) and finally plot the results: the document's time series along with the peaks representing the timestamp of the document. These steps will be further detailed in this section, starting with the corpus description.

3.1 Corpus Description

For this experiment, we started from a large dataset of around 10 thousand books on different subjects, such as *science fiction* (e.g. "Doctor Who", "Star Trek", "Legions of fire", etc.); *speculative fiction* (e.g. "Adventure of the Lady on the Embankment", etc.); *fantasy* (e.g. "Bitten", etc.); *autobiographical novels* from Charles Bukowski; *philosophy* from Immanuel Kant, Ludwig Wittgenstein, Francis Bacon, etc.; *science* from Albert Einstein, Stephen Hawking, Greg Egan, etc. We tried to cover topics from as many domains as possible, hoping that this will help LDA learn better topics. Although most of the books were in English, there were still a few written in other languages, and thus they had to be eliminated from the corpus.

3.2 Dataset Preprocessing

Before applying LDA, the dataset had to go through certain preprocessing to make it useable. First of all, the books were in Portable Document Format (.pdf), as this is one of the popular formats for electronic documents nowadays. Therefore, the first step was to extract the text from the .pdf files using PDFMiner [13]. As soon as we had the plain text, we had to clean the data by applying 4 filters. The first one removed the stop words based on a list with 540 stop words. Next, we removed the words having less than 4 characters, as most of these words were functional words (prepositions, conjunctions, etc.) that did not provide an important information gain and thus we decided to ignore them. The third filter removed words that occurred less than 20 times and more than 10,000 times. We considered that the ones occurring too rarely might have been misspells, thus having high chances to not being present in the Google Books N-gram Corpus, while the ones occurring too often could be common words that are present in all the documents, and thus not helping in discriminating between them. Finally, we removed the words that did not contained only by letters: words containing numbers, punctuations, or special characters have been also removed from the text.

3.3 Training the Topic Model Using LDA

One major step of our experiment was the training of a Latent Dirichlet Allocation model to discover the topics from the dataset and find the association between the documents and various topics. For this purpose, we used the steps described in [14] to train an LDA model using MALLET. We asked MALLET to discover 300 topics and for each of these topics, we only considered the top 20 keywords with the highest probabilities to belong to that topic. To illustrate the results of the training, in Table 1 we present 10 of the topics learned from the dataset. MALLET also outputs another set of results containing the documents and their membership to different topics with the associated probabilities. An example for the book *A Fistful of Sky* written by Nina Kiriki Hoffman is provided in Table 2.

Table 1 10 Topics discovered by MALLET from the dataset

Topic	Dirichlet parameter	Keywords
0	0.00122	Lydia asher rhion falcon vampire ysidro eye hand time light voice dark tara jasek julian silver satruff harlow night
1	0.00135	Ethan roman kickaha rome gaius wolff eilan gate september time hunnar vibulenus lord tran caillean wall emperor avalon tribune
2	0.00203	Project ryson dwarf delver lief taverik goblin time gutenberg holli human electronic looked foundation hand head term conor town
3	0.00239	Jennifer judge lieutenant chief time vaughn officer head looked commander worsel lensman cloudd carr hand captain maury welkin nodded
4	0.00084	Kerrick city time harrison hunter chentelle jonah looked herilak murgu creature josan life fargi enge death speak hand eistaa
5	0.00154	Sara tobin doris tucker arkoniel april korin tamir time blue tharin looked wizard hand minotaur child niryn lhel kieran
6	0.00082	Annja sebastian mile looked rolf time garin head hand eye nodded gregor smiled people baneen shook frowned sword left
7	0.00159	Kelly tarja manuel damin brak harshini adrina joan mantle carl defender fisk pfeiffer child demon looked citadel sister
8	0.00172	Wendy gail clive colt cora whale savn brion time bringas vlad quentin hand runciter looked moment eye carey nicole
9	0.00086	Valeria kerrec slim euan purple tristran rider shoogar time denton dance magic honorius stallion power nadine briana gothard nate
10	0.0448	Hand eye head body hair woman arm mouth voice lip finger breath heart skin blood shoulder feel love smile

Table 2 Document-topic associations for the book A Fistful of Sky written by Nina Kiriki Hoffman

Topic	Probability of associations
196	0.48557740634883567
10	0.10388047142519742
70	0.0888011283699817
113	0.0767591931418019
216	0.0671712937355634
18	0.054321100502322306
208	0.04242948339815659
162	0.023406143620481785
32	0.02125307825973484

3.4 Retrieving of Year Wise Frequency for Keywords

As the main objective of this experiment was to determine the time series of each book, we first needed to determine the time series of the words appearing in that book. Therefore, we had to obtain the frequencies of terms occurrences in different years and we identified in Google Books N-gram Corpus a very good resource that fitted this purpose.

Google Books N-gram Corpus repository contains around 5 billion words collected from the digitized version of 5,195,769 books that were written starting from 1500, representing about 4% of all the books that were ever written in the world [4].

The dataset shows how many times in each year a specific *n*-gram (from unigrams to 5-grams) was used, where a unigram is a string uninterrupted by spaces (i.e. a word, a number, a typo, etc.), while an *n*-gram is a sequence of n unigrams. In this experiment, we have only used the unigram dataset, which is provided in the following format: *Term <tab> Year <tab> Word Counts <tab> Volume Count <tab> Newline*. For example, one of the lines from the corpus is: *Time <tab> 1931 <tab> 23 <tab> 12*, which means that in 1931 the word “Time” occurred 23 times in 12 books that were published during that particular year.

Google Books N-gram Corpus is case-sensitive, making a difference between the same word written in different ways. For example, the words “time” and “tIme“ are distinctly presented in the dataset. Therefore, for our task, the dataset had to be pre-processed and then certain cleaning tasks needed to be carried out.

Thus, first of all we modified all the words to be in lowercase and then added together the frequencies of the same word in the same years to get the year wise frequency values for each term. For a better understanding, we provide the following example represented by two lines from the Google Books N-gram Corpus:

Time	1931	23	12
TiMe	1931	45	41

After adding the frequencies, we will have the following result:

time	1931	68	–

We did not consider the volume count for any purposes and thus we have ignored it in our work.

The next step was to apply on the obtained data the same cleaning filters that we used for the books' content to get the data in the desired format.

As soon as the topics were available, with the keywords that were maintained for each of them and with the documents-topics mappings, we were ready to query Google Books N-gram Corpus for retrieving the words' frequencies for each year in order to build their time series. Thus, for each document, for each topic it was associated with, we have considered the keywords corresponding to those topics and retrieved their time series from Google Books N-gram Corpus (the merged frequencies for each year).

3.5 Determining the Document Timestamp and Plotting the Result

Once the time series for all the words from the document were extracted, we could start combining them in order to detect the document time series. However, instead of simply combining the words' time series based on their frequency in the document, we decided to take an extra step and to build first the time series of the topics extracted using LDA and then to use these time series to build the document time series. We resorted to this extra step hoping that, by using the hidden topics, the results will be more accurate than without using them.

Nevertheless, since the data that extracted from Google Books N-gram Corpus is represented by the total number of occurrences of a term during each year (not its frequency), these numbers must be normalized, by dividing them to the total count of occurrences per year. In Figs. 1 and 2, we present the time series of the words from a specific topic before and after applying normalization.

The two figures clearly show the anomaly that arises if the data is not normalized. As it is can be seen in Fig. 1, the recent years have much higher peaks, but this is due to the fact that for these years, the Google Books project had much more data (digitized books) than for the past years. Therefore, normalization was required in order to obtain the proper usage distribution of these words (as shown in Fig. 2).

In the next step, a weighted sum formula was applied to obtain the topics' time series. The individual time series of different words from a specific topic were combined based on the weights corresponding to the words' probability to belong

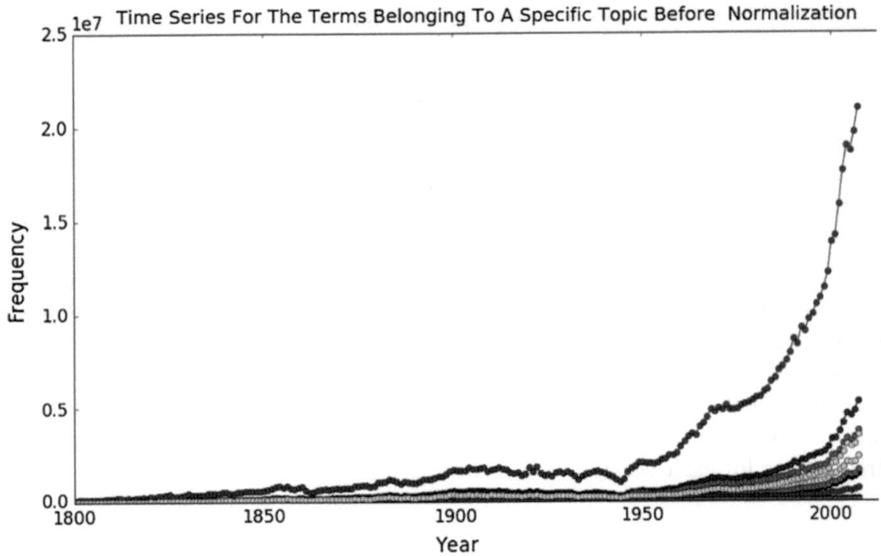

Fig. 1 Time series for the words belonging to a specific topic before normalization

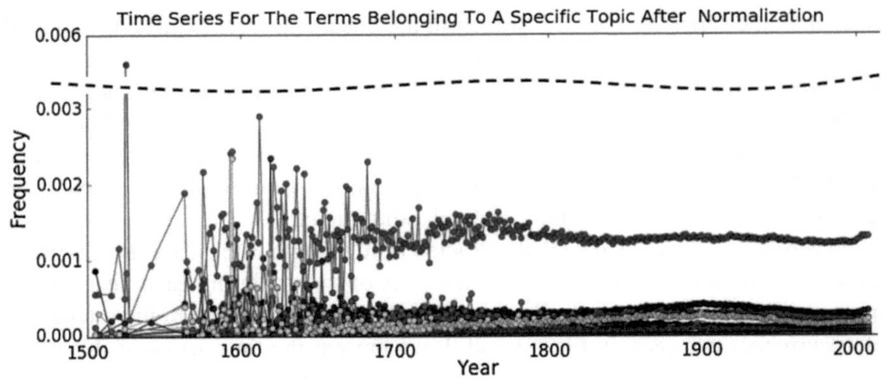

Fig. 2 The time series for the words from Fig. 1 after normalization

to that topic. Fig. 3 presents how the time series of the individual words from Fig. 2 were combined in order to build the topic's time series.

The process continued by computing the combined time series for a document using the time series of the topics that were associated with that document. Fig. 4 shows the time series for all the topics from the book used to generate Table 2 (A

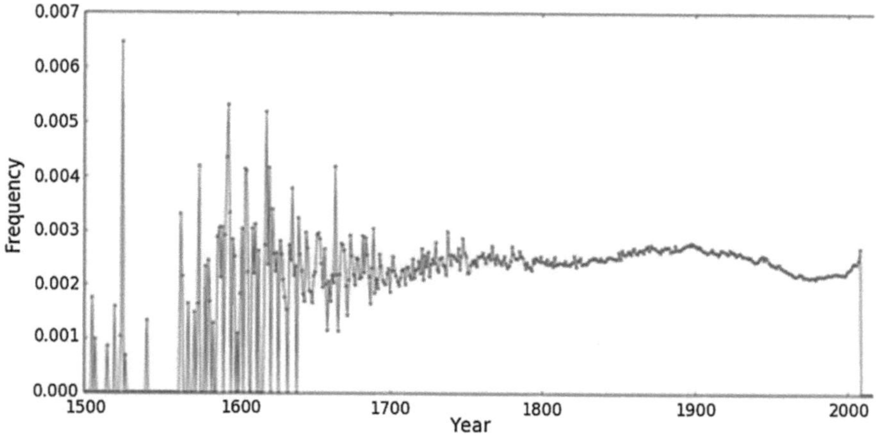

Fig. 3 The topic time series obtained by combining the time series of the words from Fig. 2

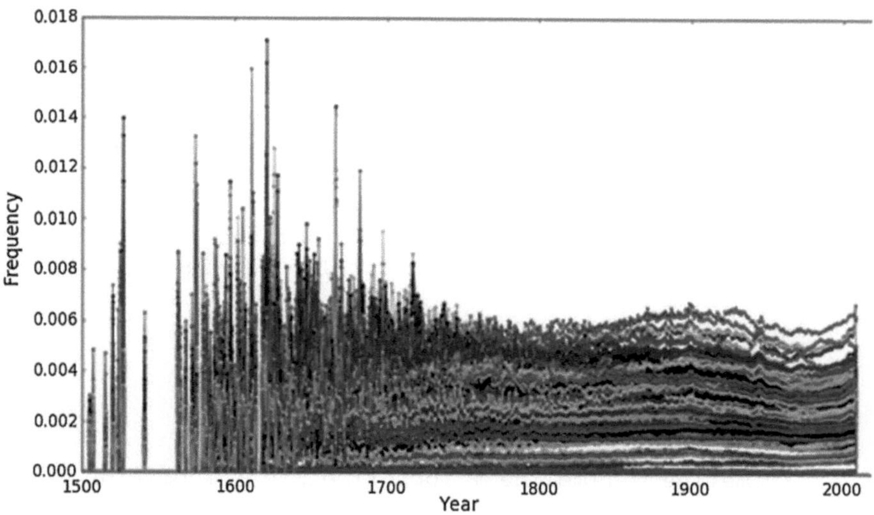

Fig. 4 Time series for the topics from the book *A Fistful of Sky* by Nina Kiriki Hoffman

Fistful of Sky by Nina Kiriki Hoffman), while in Fig. 5 we present the books' time series obtained by combining the topics from Fig. 4. Again, we are using a weighted sum in which the weights are represented by the associations probabilities provided by the LDA algorithm, as shown in the next algorithm for computing the document time series.

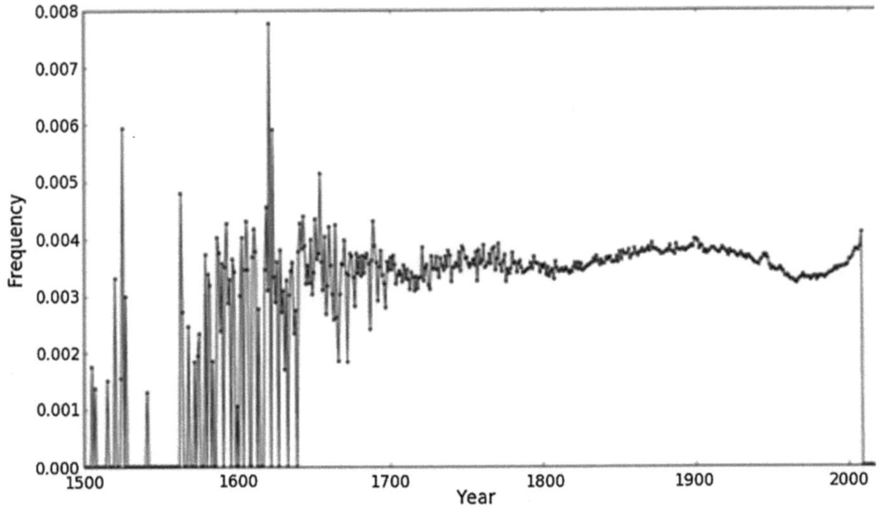

Fig. 5 The document's time series for the book *A Fistful of Sky* obtained by combining the time series of the topics from Fig. 4

```
Input:
  TopicTermsMapping: Returned from Mallet,
  TermFrequencyPerYear: Obtained from Google Ngram
  DocumentTopicMapping: Returned from Mallet
Output:
  Timeline: Year wise combined  normalized frequencies.
Algorithm: Computing the document Time Series
For a document d:
  TopicProbMapping = DocumentTopicMapping[d]
  For each topic t in TopicProbMapping:
    TopicProb = TopicProbMapping[t]
    TopicTerms = TopicTermsMapping[t]
    TopicTimeline = CombineTerms(TopicTerms)
    Timeline = Timeline + TopicTimeline * TopicProb
Return Timeline
Function CombineTerms(TopicTerms):
  For year in range (1500,2016):
      For each term in TopicTerms:
        YearWiseFrequencies[year] +=
                                  TermFrequencyPerYear[year]
  Return YearWiseFrequencies
```

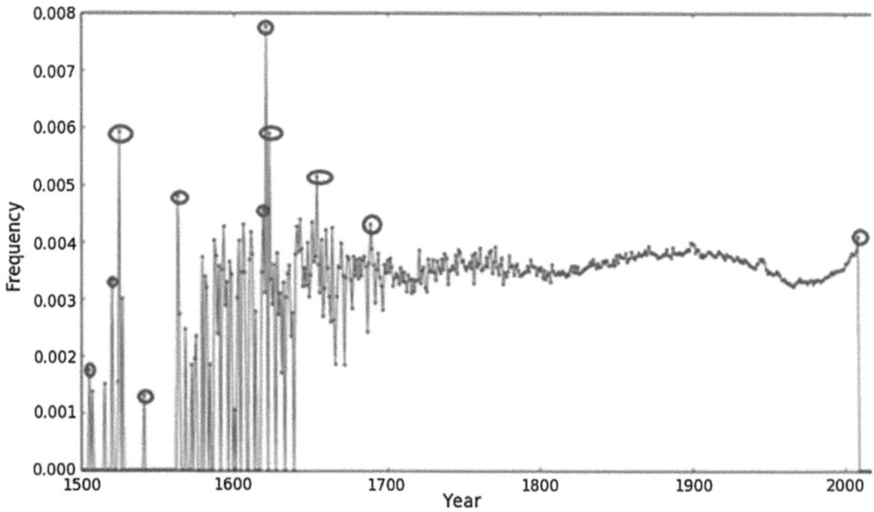

Fig. 6 The peaks from the document's time series for the book *A Fistful of Sky*

The final step of our approach consisted in using peak detection algorithms to predict what are the most likely years for timestamping the book. For this task, we used the algorithms presented in [12]. For the consideed book, the identified peaks are described in Fig. 6.

4 Results Interpretation

As it can be seen in Fig. 6, the time series of the analyzed book resulted in multiple peaks, grouped in different areas of the graph, which was a little puzzling, as we would have expected that all the peaks to be in the same area.

Moreover, the highest of the peaks is around the year 1620, while the book was written in 2002. Besides, most of the peaks are grouped between 1500 and 1700. However, there is also a peak after 2002, this being in fact the real date for this book and being identified close to the publishing date.

The peaks that can be seen between 1500 and 1700 might be explained by the topic of this book - a family that is capable of doing magic and the particular case of Gypsum LaZelle who could cast only curses. As this period of time was the classical period of witch-hunts, resulting in thousands of executions [15], it is expected that most of the books from that time to have dealt with this subject. Since then, the topic of witchery was only present in fiction books, thus having much lower peaks than during that time interval. Therefore, the high frequency of peaks during that time interval, along with the fact that the highest ones are also there may be explained by this event. However, the presence of these peaks greatly complicates the problem of timestamping the book because it introduces the need to find ways of identifying the real peak when multiple ones are detected.

5 Conclusions and Future Work

In this paper we presented an attempt to estimate the publication year of a book starting from the topics that are debated between its covers. Harnessing the great power of Google Books N-gram Corpus and a reasonably large corpus of data at hand, we have endeavored to predict the publication year of a book using LDA topic modeling techniques as a core machine learning method.

The obtained results showed the need for additional knowledge in order to discern the good from the bad ones. While the results for the presented example were influenced by an event that happened between 16th and 17th centuries, it is interesting to investigate the results obtained using this methodology in the case of other books and to see whether the methodology works better for specific categories of books.

In the future, we plan to do an extensive evaluation of the presented methodology to assess its error rate. We also intend to compare the obtained results with the ones presented in [9] with the purpose of identifying whether topic modeling helped or damaged the documents timestamping.

Two different directions could be further investigated in order to have a full-fledged analysis and perhaps much better results. First of all, instead of only using the unigram dataset from Google Books N-gram Corpus, one could also use the bi-grams and tri-grams. However, the accuracy improvement that might be obtained comes with the price of much larger needs in terms of computational power, storage space and time required for analysis. The second possibility is to change the weights of different words when computing the topics' (and documents') time series, based on their part-of-speech, considering that some parts-of-speech (nouns, verbs) have larger importance for a text than the others.

Moreover, since we observed the anomalies from the results, two possibilities exist: either to use the data that is reliable (in other words, to restrict the analysis to the years 1800-2008) or to use additional sources of information that could limit the analysis time span (such as to limit the analysis to the interval when the author was alive, if the author is known). However, both directions are difficult to follow. While for the second case it is not trivial to find such sources of information, the problem that arises in the first case is a little trickier: after 1800, the document time series becomes rather sinusoidal, with very small spikes, which might become impossible to be detected using the current spike detection algorithms. Thus, this might trigger the need for finding alternative ways of determining the spikes of the graph and consequently the time stamp of the document.

Finally, another source of errors might be the dataset that we used. While we tried to gather documents debating about topics from as many domains as possible, most of the documents came from the science fiction domain. This means that it is possible that some of the 300 topics that we extracted with LDA to actually contain almost the same words (with different probabilities) and thus to be really difficult to discriminate between them. This fact can be seen in the right half of the graph from Fig. 4, where the topics almost have a linear distribution after 1800. A more comprehensive corpus, with more topics covered, that could also consider the actual distribution of documents, might alleviate this problem. However, such a corpus is very difficult to find.

Acknowledgements This work has been funded by University Politehnica of Bucharest, through the "Excellence Research Grants" Program, UPB – GEX. Identifier: UPB–EXCELENȚĂ–2016 Aplicarea metodelor de învățare automată în analiza seriilor de timp (Applying machine learning techniques in time series analysis), Contract number 09/26.09.2016.

References

1. Blei, D.M., Ng, A.Y., Jordan, M.I.: Latent Dirichlet allocation. J. Mach. Learn. Res. **3**, 993–1022 (2003)
2. Chen, E.: Introduction to Latent Dirichlet Allocation. http://blog.echen.me/2011/08/22/introduction-to-latent-dirichlet-allocation/22 Aug 2011
3. AlSumait, L., Barbará, D., Domeniconi, C.: On-line lda: adaptive topic models for mining text streams with applications to topic detection and tracking. In: Data Mining, 2008. ICDM'08, pp. 3–12 (2008)
4. Michel, J.-B., Shen, Y.K., Aiden, A.P., Veres, A., Gray, M.K., The Google Books Team, Pickett, J.P., Hoiberg, D., Clancy, D., Norvig, P., Orwant, J., Pinker, S., Nowak, M.A., Aiden, E.L.: Quantitative analysis of culture using millions of digitized books. Science **331**(6014), 176–182 (2011)
5. Sparavigna, A.C., Marazzato, R.: Using Google Ngram viewer for scientific referencing and history of science. arXiv preprint arXiv:1512.01364 (2015)
6. Montagne, M., Morgan, M.: Drugs on the internet, part IV: Google's Ngram viewer analytic tool applied to drug literature. Subst. Use Misuse **48**(5), 415–419 (2013)
7. Patrick, J.: Using the Google N-Gram corpus to measure cultural complexity. Literary Linguist. Comput. **28**(4), 668–675 (2013)
8. Koplenig, A.: The impact of lacking metadata for the measurement of cultural and linguistic change using the Google ngram data set—reconstructing the composition of the german corpus in times of WWII. In: Digital Scholarship in the Humanities, fqv037 (2015)
9. Islam, A., Mei, J., Milios, E.E., Keselj, V.: When was macbeth written? mapping book to time. In: Computational Linguistics and Intelligent Text Processing. Springer International Publishing, pp. 73–84 (2015)
10. Szymanski, T., Lynch, G.: UCD: Diachronic Text Classification with Character, Word, and Syntactic N-grams. SemEval **2015**, 879–883 (2015)
11. Garcia-Fernandez, A., Ligozat, A.-L., Dinarelli, M., Bernhard, D.: When was it written? automatically determining publication dates. In: String Processing and Information Retrieval, pp. 221–236 (2011)
12. Popa, T., Rebedea, T., Chiru, C.: Detecting and describing historical periods in a large corpora. ICTAI **2014**, 764–770 (2014)
13. Yusuke, S.: PDFMiner. http://euske.github.io/pdfminer/index.html (2008)
14. Digital Research Infrastructure for the Arts and Humanities: Topic modeling with MALLET. https://de.dariah.eu/tatom/topic_model_mallet.html#topic-model-mallet (2015)
15. Ankarloo, B., Clark, S., Monter, W.: Witchcraft and magic in Europe. The Athlone Press (2002)

Part II
Multi-scale Analysis of Univariate and Multivariate Time Series

Fractal Complexity of the Spanish Index IBEX 35

M.A. Navascués, M.V. Sebastián, M. Latorre, C. Campos, C. Ruiz and J.M. Iso

Abstract We study and compare the reference of the Spanish stock market IBEX 35 with other international indices from consolidated as well as emerging economies. We look for similarities and differences between the Spanish index and the markets chosen, from a self-affine perspective. For it we compute fractal parameters which provide an indication of the erraticity of the data. We perform inference statistical tests, in order to elucidate if the computed parameters are significantly different in the Spanish selective. Beginning from the daily closing values of the IBEX of more than one decade, we investigate the stability in mean and variance, and test the necessity of the transformation of the record in order to improve its normality or stabilize and minimize the deviation. We use appropriate statistical methodologies, as ARIMA and ARCH, to obtain good explicative models of the series considered, and estimate its parameters of interest.

Keywords Stock market indices · Fractional brownian motions · Fractal dimension · Arch models

M.A. Navascués (✉) · M. Latorre · C. Campos
Escuela de Ingeniería y Arquitectura, Universidad de Zaragoza, Zaragoza, Spain
e-mail: manavas@unizar.es

M. Latorre
e-mail: mario.latorrepellegero@gmail.com

C. Campos
e-mail: ccampos@unizar.es

M.V. Sebastián · C. Ruiz
Centro Universitario de la Defensa de Zaragoza, Zaragoza, Spain
e-mail: msebasti@unizar.es

C. Ruiz
e-mail: cruizl@unizar.es

J.M. Iso
Inspección General Del Ejército, Barcelona, Spain
e-mail: jisopere@et.mde.es

I. Rojas et al. (eds.), *Advances in Time Series Analysis and Forecasting*,
Contributions to Statistics, DOI 10.1007/978-3-319-55789-2_5

1 Introduction

A clear reflection of the performance of a national economy is the evolution of its stock exchange. It is an accepted fact that the stock data follow self-similar patterns, and our wish in this paper is to quantify this kind of geometric properties of the charts. We consider the time series composed of the daily closing prices of several international stock indices, from a fractal point of view. In this way we want to show that the new mathematical methodologies may contribute both qualitatively and quantitatively to the knowledge and forecasting of the share prices.
Fractality is closely related to chaos theory. One of the main characteristics of the chaotic systems is the sensitivity with respect to initial conditions. This feature is frequently observed in the markets, where very similar conditions display completely different performances. For instance a rumor may cause an increase or decline of the shares or an apparently ordered behavior to become unstable and erratic. We notice that the complexity of the economic data is well fitted by this type of models.
In Sect. 2 we compute two different parameters: the first is related to a framework of fractal noise, and the second is the fractal dimension. We use these quantifyers to study the Spanish stock index IBEX 35 (Figs. 7 and 8), and compare it with other international markets. We have chosen two indices of consolidated economies (German DAX (Fig. 1) and Dow Jones (Fig. 2)), and two markets of the so-called MINT (Mexico, Indonesia, Nigeria and Turkey) group (Indonesian IDX (Fig. 3) and Mexican IPC (Fig. 4)) [7], in order to unveil their mutual correlations, analogies and divergences.

In Sect. 3 we apply models of time series in order to analyze the Spanish selective over more than one decade. In particular we investigate the stability in mean and variance, and obtain good explicative models.

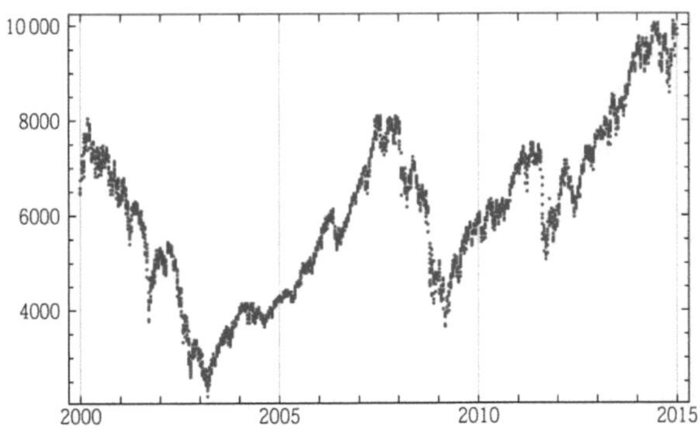

Fig. 1 Chart of DAX in the period 2000–2014

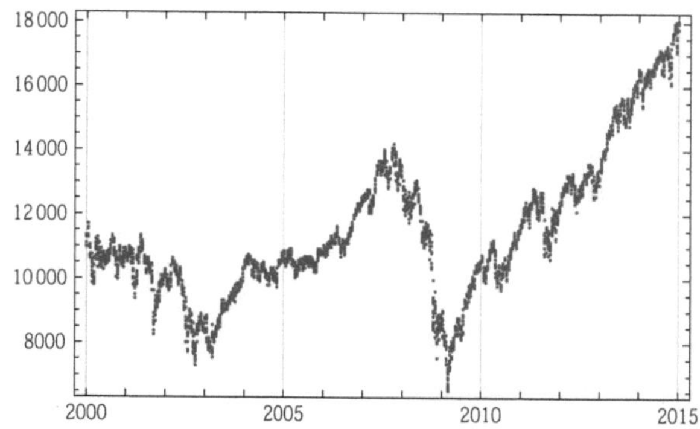

Fig. 2 Chart of Dow Jones in the period 2000–2014

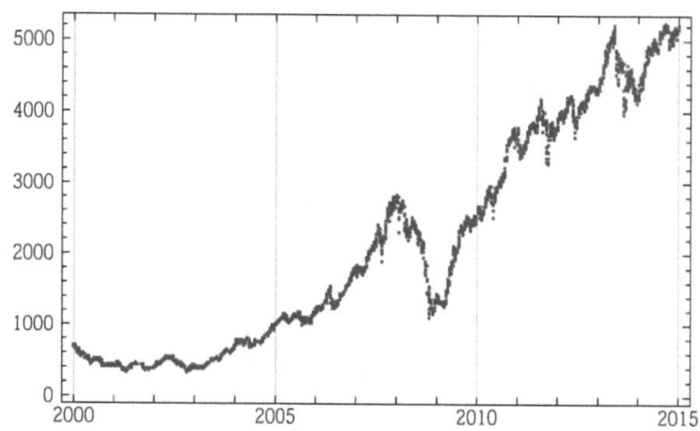

Fig. 3 Chart of IDX in the period 2000–2014

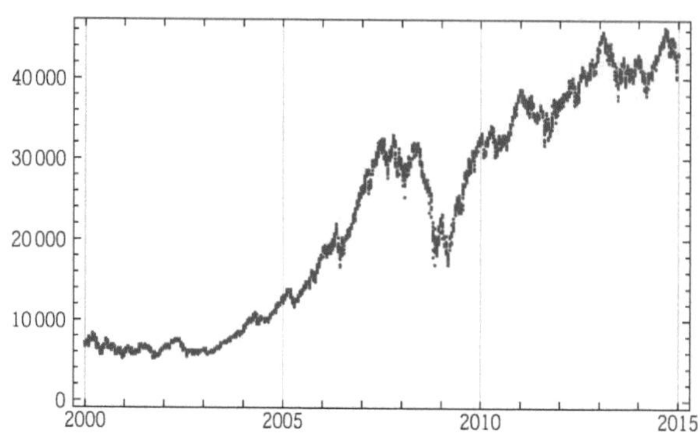

Fig. 4 Chart of IPC in the period 2000–2014

2 Fractal Complexity

In this Section we approach the study of the geometric complexity of the data from two different points of view. In the first case we study the existence of a power law in the Fourier spectrum of the series. If so, the variables may admit a model of fractal noise (or colored noise). In the second case, we compute the fractal dimension of the records, testing the convenience of a fractional Brownian motion scenario.

2.1 Fractal Noise

In general it is said that the economical series are well represented by colored noises, and we wished to test this hypothesis. A variable of this type satisfies an exponential power law:

$$S(f) \simeq kf^{-exp},$$

where f is the frequency, and $S(f)$ is the spectral power. In our case we compute discrete powers corresponding to discrete frequencies (mf_0), where $f_0 = 2\pi/T$ is the fundamental frequency and $m = 1, 2, \ldots,$ (T is the length of the recording). A logarithmical regression of the variables provides the exponent (in negative) as slope of the fitting.

For it we construct first a truncated trigonometric series in order to fit the data [6]. Since the Fourier methods are suitable for variables of stationary type, we subtracted previously the values on the regression line of the record. We analyzed the index year by year and obtained an analytical formula for every period, as sum of the linear part and the spectral series. We compute in this way a discrete power spectrum which describes numerically the great cycles of the index. We perform a graphical test in order to choose the number of terms for the truncate sum. We find that 52 summands are enough for the representation of the yearly data, that corresponds to the inclusion of the cycles of weekly length. The formula is almost interpolatory. The harmonics of the series allow us to obtain the spectral powers. These quantities enable a numerical comparison between different indicators, for instance. The numerical procedure is described in the reference [6]. Concerning this methodology, we consider that the test performed is merely exploratory if one wants to measure the geometric complexity. The necessary truncation of the series defined to compute the powers collects the macroscopic behavior of the variables, but omits the fine self-affine oscillations.

In Table 1 the exponents computed for the years listed are presented. The mean values obtained on the period are: 1.97, 1.98, 1.98, 2.01 and 2.03 for DAX (Germany), Dow Jones (USA), IBEX (Spain), IDX (Indonesia) and IPC (Mexico), respectively, with standard deviations 0.21, 0.15, 0.23, 0.14 and 0.31. The results suggest fairly a structure close to a red noise, whose exponent is 2. The values for a white noise would be 0. The correlations obtained in their computation are about 0.8. Table 2 shows the averages for every index in the pre-crisis (2000–2007) and crisis (2008–2014) periods, respectively.

2.2 Fractional Brownian Motion

We perform now a comparative numerical study of the mentioned international stock indicators from a self-affine point of view. In this case we have computed annual dimensions, whose values are included in Table 3. The fractal dimension (D) is related to the Hurst parameter (H) through the relation: $D = 2 - H$. The scalar H

Table 1 Exponent of power law for every index and year

Year	DAX	DJ	IBEX	IDX	IPC
2000	1.741	1.808	1.466	2.220	1.772
2001	2.062	2.235	1.818	2.257	1.871
2002	2.200	2.135	1.998	1.800	1.992
2003	2.049	2.191	2.247	2.072	2.156
2004	2.283	2.027	2.074	1.938	1.943
2005	1.926	1.897	2.135	2.020	2.071
2006	2.190	1.971	2.301	2.038	1.991
2007	2.200	1.925	2.028	2.125	2.970
2008	1.861	1.714	1.933	2.161	2.005
2009	1.692	1.881	1.814	1.990	1.634
2010	1.695	2.125	1.873	1.981	2.343
2011	1.919	1.940	1.811	1.882	1.833
2012	2.051	1.815	2.001	1.781	1.924
2013	1.603	2.038	2.352	1.977	1.771
2014	2.012	2.073	1.896	1.914	2.128

Table 2 Mean and standard deviations of the exponent in the periods pre-crisis (2000–2007) and crisis (2008–2014)

Year	DAX	DJ	IBEX	IDX	IPC
Mean 2000–2007	1.966	2.024	2.008	2.059	2.096
SD 2000–2007	0.212	0.151	0.266	0.148	0.372
Mean 2008–2014	1.833	1.941	1.954	1.955	1.948
SD 2008–2014	0.173	0.148	0.188	0.117	0.236

is associated with the concept of fractional Brownian motion [4, 8, 9]. This type of variable (let us denote it by $B_H(t, \omega)$) satisfies the self-similar equation

$$\{B_H(t_0 + T, \omega) - B_H(t_0, \omega)\} \approx \{h^{-H}(B_H(t_0 + hT, \omega) - B_H(t_0, \omega))\},$$

where $\approx$ means that they have the same probability distribution. The increment variables are Gaussian with mean zero and variance proportional to T^{2H} ([4], Corollary 3.4).

We take advantage of this property to compute H, and consequent dimension D. The correlation in the computation of the parameters is in all cases very close to 1.

We summarize now the results obtained for the fractal dimension:

Table 3 Annual fractal dimensions

Year	DAX	DJ	IBEX	IDX	IPC
2000	1.541	1.543	1.781	1.466	1.483
2001	1.465	1.502	1.524	1.444	1.432
2002	1.573	1.548	1.580	1.440	1.518
2003	1.542	1.523	1.526	1.455	1.534
2004	1.492	1.499	1.477	1.460	1.498
2005	1.550	1.550	1.515	1.422	1.506
2006	1.522	1.558	1.480	1.508	1.509
2007	1.524	1.563	1.560	1.529	1.501
2008	1.559	1.653	1.619	1.476	1.557
2009	1.480	1.493	1.500	1.514	1.478
2010	1.565	1.529	1.535	1.573	1.505
2011	1.487	1.565	1.548	1.575	1.576
2012	1.591	1.565	1.541	1.616	1.514
2013	1.532	1.481	1.469	1.522	1.528
2014	1.473	1.470	1.524	1.611	1.512

Regarding DAX index, the maximum value (1.59) is recorded in 2012 and the minimum (1.46) in 2001. Thus, the range variation is 0.13 which represents a 8% with respect to the maximum value.

In the American Dow, the highest dimension is reached in 2008, with value 1.65. The minimum occurs in the year 2014, with a value of 1.47. The parameter varies over the period 0.18 points, representing a 11% of the peak value.

In the Spanish index the absolute extremes occur in 2000 and 2013 with values 1.78 and 1.47 respectively. The second minimum (1.62) is recorded in 2008. The range of variation is 0.31 (17%).

For the Indonesian index, the maximum is reached in 2012 with value 1.62, and in 2005 there is a minimum of 1.42. The range of variation in the scalar is 0.19 (12%).

Regarding the Mexican IPC, the maximum is set to 1.58 in 2011. The second maximum is given in 2008. A minimum of the dimension is found in 2001 with value 1.43. The range of absolute variation is 0.14, a 9% with respect to the peak value of the period.

Table 4 shows the averages and typical deviation of the fractal dimensions in the periods pre-crisis, crisis and total.

The results obtained display a fair uniformity. However the index IBEX presents the greatest variability, ranging from 1.47 up to 1.78, the highest dimension obtained. The greatest variance corresponds to the Spanish index as well, specially in the first period. The Spanish selective and Dow Jones register local maxima during the year 2008 (outbreak of the crisis) that, in the American case, is global as well. Figures 5 and 6 depict the time evolution of the parameter computed for the IBEX and IDX respectively over the years studied.

Table 4 Mean and standard deviations of the fractal dimensions classified in periods

Year	DAX	DJ	IBEX	IDX	IPC
Mean 2000–2007	1.526	1.536	1.555	1.466	1.498
SD 2000–2007	0.039	0.025	0.098	0.039	0.030
Mean 2008–2014	1.527	1.537	1.534	1.555	1.524
SD 2008–2014	0.047	0.064	0.046	0.053	0.033
Mean 2000–2014	1.526	1.536	1.545	1.507	1.510
SD 2000–2014	0.039	0.045	0.076	0.063	0.033

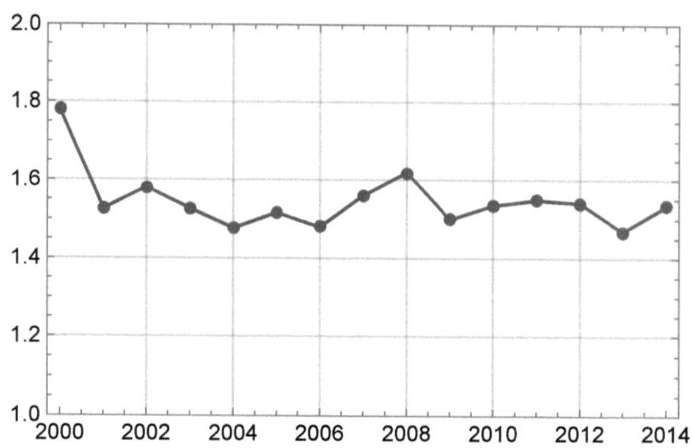

Fig. 5 Evolution of the fractal dimension of IBEX 35 over the period considered

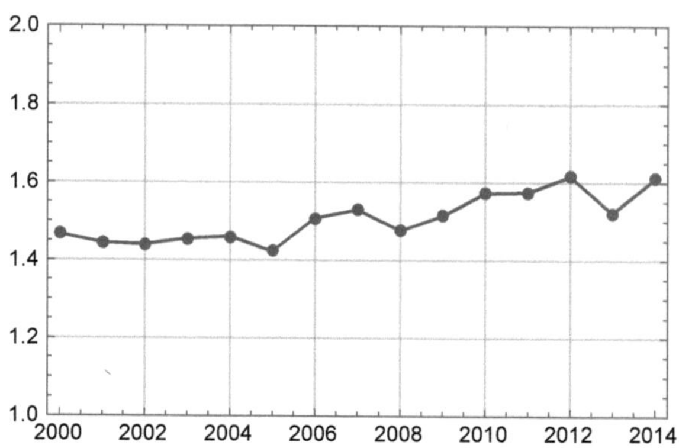

Fig. 6 Evolution of the fractal dimension of IDX over the period considered

The values obtained are in all instances very close to 1.5, typical scalar of a Brownian motion, but we believe that the models described by Mandelbrot [4, 5] are more accurate for this type of series.

2.3 Statistical Tests

We have performed a non-parametric Mann-Whitney test to the parameter obtained for the five indices. The objective was to find (if any) significant differences in the values with respect to the market considered. The samples were here the annual fractal dimensions of each selective. The parametric tests require some hypotheses on the variables like, for instance, normality, equality of variances, etc. In our case we cannot assume the normality of the distribution because is unknown. Commonly this condition may be acceptable for large samples, but the size is small here, and for this reason we chose a non-parametric test, being Mann-Whitney a valid alternative. We provide the results of the test applied to the outcomes of the five indices and fifteen years. The p-values are shown in Table 5. It can be observed that in all cases the p-value is greater than 0.05, showing no statistically significant evidence to reject that the fractal dimensions of these indices come from similar distributions.

However the Spanish index records the minimum p-values (Table 5, digits in bold), pointing to a disparity in the IBEX with respect to IDX and IPC. It is followed by Dow Jones, with respect to IDX and IPC as well.

2.4 Linear Correlations

We have computed now the mutual linear correlations of the parameter between the different indices. The results are shown in Tables 6 and 7. We divided the time in two periods as before: years 2000–2007 and years 2008–2014, in order to check the variations due to the recent economic crisis.

Table 5 p-values of the test of fractal dimension differences

p-values	DAX	DJ	IBEX	IDX	IPC
DAX	*X*	0.595	0.838	0.233	0.267
DJ	0.595	*X*	0.775	0.161	0.148
IBEX	0.838	0.775	*X*	**0.098**	**0.137**
IDX	0.233	0.161	0.098	*X*	0.713
IPC	0.267	0.148	0.137	0.713	*X*

Table 6 Correlation matrix for the dimensions of the period 2000–2007

Hurst exp.	DAX	DJ	IBEX	IDX	IPC
DAX	1	0.67	0.34	−0.14	0.75
DJ	0.67	1	0.24	0.45	0.44
IBEX	0.34	0.24	1	−0.01	−0.14
IDX	−0.14	0.45	−0.01	1	0.11
IPC	0.75	0.44	−0.14	0.11	1

Table 7 Correlation matrix for the dimensions of the period 2008–2014

Hurst exp.	DAX	DJ	IBEX	IDX	IPC
DAX	1	0.52	0.34	0.00	0.05
DJ	0.52	1	0.91	−0.35	0.61
IBEX	0.34	0.91	1	−0.20	0.53
IDX	0.00	−0.35	−0.20	1	−0.14
IPC	0.05	0.61	0.53	−0.14	1

The correlation in the fractal behavior between the values of IBEX and the indices IDX and IPC is negative in the first period. It becomes positive in the second segment in the Mexican case, though remaining poor.

The relation of the Indonesian index with respect to the rest is very low, even negative, and the coefficient decreases in the second period in most of the cases.

The Mexican market increases its relation with Dow Jones and IBEX in the second period.

2.5 Conclusions

The fractal tests in the stock records analyzed in the period 2000–2014 provide a variety of outcomes which we summarize now.

- The numerical results present a great uniformity. Nevertheless, the p-values obtained in the statistical test point to greater numerical differences in IBEX and Dow Jones with respect to IDX and IPC.
- The disparity between consolidated and emerging economies is better appreciated in the first period (see Table 4). The fractal dimensions of Indonesia and Mexico are slightly lower than the rest of the indices. However these magnitudes increase in the second period, tending to match the complexity of the rest of the values.
- We can observe an index IBEX highly influenced by the 2008 financial crisis (Fig. 5).

- This indicator records the greatest variability due to high values in 2000 and 2008 (Table 3).
- The correlation in the fractal behavior between the values of IBEX and the indices IDX and IPC is negative in the first period. It becomes positive in the second segment in the Mexican case, though remaining poor.
- The mean average of the fractal dimensions is 1.52. The standard deviations are around 0.05.
- In most cases, the dimensions tend to rise in the second period, pointing to a higher complexity and erraticity during the crisis.
- In general, we observe a mild anti-persistent behavior in the markets ($D > 1.5$), except in Indonesia and Mexico in the first period. However, it is likely that the globalization process will lead to a greater uniformity.
- The results obtained from the power law exponent are around 1.99, very close to the characteristic value of a red noise or Brownian motion (2), with a typical deviation of 0.21.
- The stock records may admit a representation by means of colored noises, in particular of red noise, but refined by a model of fractional Brownian motion quite strict. The numerical results suggest that the fractal dimension may be a predictor of changes in the market.

The study of an increasingly complex economy requires the use of the most sophisticated scientific tools. The numerical analysis of the market might constitute a third procedure for the prognosis of the economic behaviors, acting in conjunction with the fundamental and technical analyses. According to the results, we infer that the fractal dimension is suitable for the numerical description of this type of economical signals, providing a measure of the persistence of the trends in the stock data.

3 ARCH Model

Volatility is a feature characteristic of the economic time series and its study (and subsequent control) is a challenge for the financial community. A good statistical tool for this purpose is the methodology of time series. Chatfield considers, in the reference [2], a wide range of models for this technique. In general, in the econometric models the variance is not constant and consequently the traditional procedures as ARIMA models of Box et al. [1] are not suitable for the processing of financial time series. Engle [3] considers a class of stochastic processes called ARCH models, where the variance conditioned to past information is not constant. We analyze, from that perspective, the data of closing daily values of the reference index of the Spanish stock market IBEX 35, from January 1992 to October 2010. The chart of the series is shown in Fig. 7.

The variable is clearly non-stationary, there are great oscillations in mean, that changes with no apparent pattern. Furthermore, the record presents a high variability.

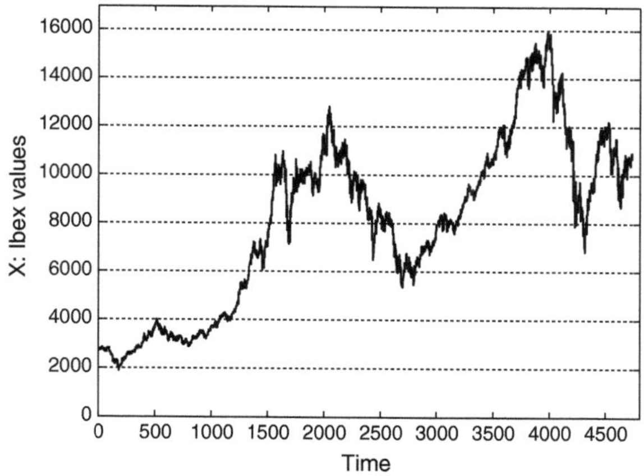

Fig. 7 Daily closing values of IBEX 35 from 1992 to 2010

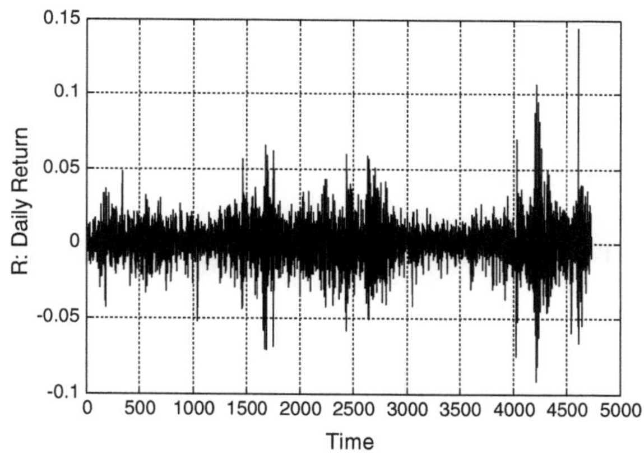

Fig. 8 Values of daily returns of IBEX 35 from 1992 to 2010

Let us consider the daily return of the series X defined in terms of its samples as:

$$R_t = \frac{X_t - X_{t-1}}{X_{t-1}} = \frac{\Delta X_t}{X_{t-1}}$$

where $\Delta X_t = X_t - X_{t-1}$ and $t = 1, 2, \ldots$, Δ is called difference operator. The time graph of the series of relative differences is shown in Fig. 8.

Although the value of the mean of daily return series is stabilized, a great variance is observed at some points. This fact is an indication of the existence of cores of variability: great changes tend to follow great changes and small changes tend to shadow small changes. This pattern aims to sign unpredictability (where a positive sign in the return points to a profit and a negative sign to loss). High variability of the daily return series tends to be the response to fewer benefits than expected and vice versa, low variability is the answer to high expected profits. That is to say, small (large) variations of the series of daily return of IBEX 35 with respect to its mean are followed by great (small) variations, and consequently the variance (or volatility) is conditioned by its recent past, that is to say, presents some type of inertia. Furthermore, the autocorrelation of daily return series is virtually non-existent (and the same applies to the partial autocorrelation), unlike the case of the squared series, indicating that the variances have autocorrelation, suggesting that the variability of the process is non-stationary. The effects of cores of variability can be studied with GARCH (Generalized AutoRegressive Conditional Heteroskedasticity).

The GARCH (1, 1) model considered for the daily return series of IBEX 35 is

$$R_t = 0.0007951 + Z_t;$$

$$\sigma_t^2 = 2.7274 \times 10^{-6} + 0.88528\sigma_{t-1}^2 + 0.10231Z_t^2$$

where Z_t is called innovation process, being stationary, usually a normal random variable with mean ν and variance σ^2.

Similar models GARCH (1, 1) could be obtained for other indices like DAX or DJ, for instance.

References

1. Box, G.E., Jenkins, G.M., Reinsel, G.C.: Time Series Analysis: Forecasting and Control. Wiley (2013)
2. Chatfield, C.: The Analysis of Time Series: An Introduction, 5th edn. Chapman & Hall (1996)
3. Engle, F.R.: Autoregressive conditional heterocedasticity whit estimates of the variance of United Kingdom inflation. Econometrica **50**(4), 987–1008 (1982)
4. Mandelbrot, B.B., Ness, J.V.: Fractional brownian motion, fractional noises and applications. SIAM Rev. **10**, 422–437 (1968)
5. Mandelbrot, B.B., Hudson, R.L.: The (Mis)Behaviour of the Markets: A Fractal View of Risk. Profile Books, Ruin and Reward (2004)
6. Navascués, M.A., Sebastián, M.V., Ruiz, C., Iso, J.M.: A numerical power spectrum for electroencephalographic processing. Math. Methods Appl. Sci. (2015). doi:10.1002/mma.3343
7. O'Neill, J.: Who you calling a BRIC?. Bloomberg View (web), November 12 (2013)
8. Peters, E.E.: Chaos and Order in the Capital Market: a New View of Cycles. Prices and Market Volatility. Wiley, New York (1994)
9. Peters, E.E., Peters, D.: Fractal Market Analysis: Applying Chaos Theory to Investment and Economics. Wiley, New York (1994)

Fractional Brownian Motion in OHLC Crude Oil Prices

Mária Bohdalová and Michal Greguš

Abstract Widespread use of information and communication technologies has caused that the decisions made in financial markets by investors are influenced by the use of techniques like fundamental analysis and technical analysis, and the methods used are from all branches of mathematical sciences. Recently the fractional Brownian motion has found its way to many applications. In this paper fractional Brownian motion is studied in connection with financial time series. We analyze open, high, low and close prices as a selfsimilar processes that are strongly correlated. We study their basic properties explained in Hurst exponent exponent, and we use them as a measure of predictability of time series.

Keywords Fractional brownian motion · Hurst exponent · Time series

1 Introduction

The theory of Brownian motion (Bm) focuses on non-stationary Gaussian processes and their generalization to the fractional Brownian motion (fBm) [1–4] and the corresponding fractional Gaussian noise (fGn). Fractional Brownian motion is a centered self-similar Gaussian process with stationary increments, which depend on a parameter $H \in (0, 1)$ called the Hurst exponent. This proces includes a wide class of self-similar stochastic processes whose variances are scaled by the power law N^{2H}, where N is the number of increments in fBm and $0 < H < 1$. If $H = 0.5$, the process corresponds to the classical Brownian motion with independent increments. The term fractional was proposed by Mandelbrot [5] in connection with fractional integration and differentiation. In general, increments in fBm are strongly correlated and have long memory. Hence, fBm has become a powerful mathematical model for studying

M. Bohdalová (✉) · M. Greguš
Faculty of Management, Comenius University in Bratislava, Odbojárov 10, Bratislava, Slovakia
e-mail: maria.bohdalova@fm.uniba.sk
URL: http://www.fm.uniba.sk

M. Greguš
e-mail: michal.gregus@fm.uniba.sk

I. Rojas et al. (eds.), *Advances in Time Series Analysis and Forecasting*, Contributions to Statistics, DOI 10.1007/978-3-319-55789-2_6

correlated random motion with wide application in physics, hydrology, mathematical finance, etc. [6–8]. Fractal Brownian motion is a process without independent increments, and possessing long-range dependence and self-similarity properties. Long-range dependence in a stationary time series occurs when the covariances tend to zero like a power function, and they do so slowly so that their sums diverge. The self-similarity property means invariance in distribution under a suitable change of scale. One of the simplest stochastic processes is the Gaussian process, and a similar process with stationary increments is the fractional Brownian motion. This process is a generalization of the classical Brownian motion [6, 8].

Financial markets are complex systems and the emergence of complexity could be spontaneous. Peters [9] proposed a fractal market hypothesis based on fractal theory. Fractal theory takes into account complexity in financial markets, and it can be described by pure mathematical formulas. The efficient market hypothesis tells us market prices follow a random walk (random walk is necessary for an application of statistical analysis to a time series of price changes). It also tells us that any predictable trend can be elliminated by an arbitrage in a small time period [10].

Fractal market hypothesis is a theory describing capital markets. It combines fractals and other concepts from chaos theory with commonly used quantitative methods to explain and predict the behavior of markets. Fractal market hypothesis pays attention to the randomness of daily events occurring in the market, and it also includes outliers in the form of rapid movements and market crashes.

Fractal market hypothesis proposes the following [11]:

- The market is stable when it consists of investor covering a large number of investment horizons. This ensures there is enough liquidity for traders.
- The information set is more related to market sentiment and technical factors in the short run rather than in the long run. As investment horizons increase, long run fundamental information becomes much more significant.
- If there is an event which makes the validity of basic information questionable, investors focusing on the long run either stop participating in the market or start trading based on the short run information set. When the over all investment horizon of the market shrinks to a uniform level, the market stops being stable.
- Prices reflect a combination of short run technical trading and long run fundamental valuation.
- If a security has no connection with the economic cycle, it is very likely that in such a case there will be no long run trend. Trading, liquidity, and short term information will become much more important than before, and they will dominate over the other elements in the market.

Fractal market hypothesis tells us that information is valued according to the investment horizon of the particular investor. Since different investment horizons value information in a different way, the diffusion of information will also be uneven. At any given point of time, prices may not reflect all available information, but only the information important and available to that investment horizon. Fractal market hypothesis offers an economic and mathematical structure to fractal market analysis.

Using the fractal market hypothesis, we are able to understand the behavior of the markets better.

The effective market analysis is heavily dependent on the rationality of investors. Rationality is defined as the ability to evaluate financial instruments on the basis of all available informations and to price them accordingly [11]. The basic assumption of effective markets hypothezis is that all investment decisions are rational and based on a fair game, in which players try to maximize their gains.

This paper considers the problem of identifying, measuring and characterizing long-range dependence in the OHLC crude oil prices. Problem known as detecting the long memory is being solved in many papers, for example, in [10, 12, 13]. In this paper we have identified the long memory and the length of the business cycle that may be used for predicting future prices, and we verify the hypothesis about efficient markets.

This paper is organized as follows. Next section introduces fractional Brownian motion. Section 3 describes our data analysis. Conclusion describes our findings.

2 Fractional Brownian Motion

A Gaussian process $B^H = \{B_t^H, t \geq 0\}$ is called a fractional Brownian motion (fBm) of Hurst exponent $H \in (0, 1)$ if it has zero mean and the covariance function:

$$E\left(B_t^H B_s^H\right) = R_H(t, s) = \frac{1}{2}\left(s^{2H} + t^{2H} - |t - s|^{2H}\right). \tag{1}$$

The fractional Brownian motion has the following properties:

1. Self-similarity: For any constant $a > 0$, the processes $\{a^{-H} B_{at}^H, t \geq 0\}$ and $\{B_t^H, t \geq 0\}$ have the same probability distribution.
2. Stationary increments: From (1) follows that the increment of the process in an interval $[s, t]$ has a normal distribution with zero mean and variance

$$E\left(\left(B_t^H - B_s^H\right)^2\right) = |t - s|^{2H}. \tag{2}$$

Hurst exponent is widely used as a measure of the long-term memory of time series. There are known some statistical techniques for estimation of the Hurst exponent. Some of them were described in [2, 9–11], etc. In this paper we have used R/S analysis because this method enables detecting cycles in analysed time series.

2.1 R/S Analysis for Estimation of the Hurst Exponent

To estimate the Hurst exponent H for time series $\{R_t\}_{t=1}^n$ we take the following steps (see [9–11, 14, 15]):

1. We divide period n into m continuous subperiods of length q, such that $mq = n$. The rescaled range will be calculated for first rescaling or normalizing data by subtracting the sample mean R_m. We create the new series $\{Z_t\}_{t=1}^{n}$

$$Z_r = R_r - \bar{R}_m, r = 1, 2, \dots, n \quad (3)$$

2. We create a cummulative time series $\{Y_r\}_{r=1}^{n}$:

$$Y_1 = Z_1 + Z_r, r = 2, 3, \dots, n \quad (4)$$

Note that, by definition, the last value of Y (Y_n) will always be zero because Z has a mean of zero.
3. The adjusted range, R_n, is the series obtained as the maximum minus the minimum value of the Y_r:

$$r_n = max(Y_1, Y_2, \dots, Y_n) - min(Y_1, Y_2, \dots, Y_n). \quad (5)$$

The adjusted range r_n is always non-negative, it is the distance that the system travels for time index n.
4. A power-law behavior is expected, and H is the Hurst exponent:

$$(R/S)_n = c.n^H \quad (6)$$

The subscript, n, for $(R/S)_n$ refers to the R/S value for $R_1, R_2, \dots, R_n$ and c is a constant.
5. The R/S value of Eq. (6) is referred to as the rescaled range because it has zero mean, and it is expressed in terms of the local standard deviation. In general, the R/S value scales as we increase the time increment, n, by a power? law value equal to H. Generally, the Hurst exponent H is expressed as

$$log(R/S)_n = log(c) + H.log(n) \quad (7)$$

For $H = 0.50$ the system is independently distributed. When H differs from 0.50, the observations are not independent. Each observation carries a memory of all the events that precede it. What happens today influences the future. The time is important.
6. The impact of the present on the future can be expressed as a correlation:

$$C = 2(2H - 1) - 1, \quad (8)$$

where C is a correlation measure and H is the Hurst exponent.

The Hurst exponent H is classified as [6, 11]

- If $H = 0.50$, the process $B^{0.5}$ is an ordinary Brownian motion, i.e. it is a random series. In this case the increments of the process in disjoint intervals are independent. The present does not influence the future. Its probability density function can be a normal curve, but it does not have to be.

- If $0 \leq H < 0.50$, the time series is antipersistent, or ergodic. Antipersistent time series are known as mean reverting series.
- If $0.50 < H < 1.00$, the time series have persistent or trend-reinforcing nature. Persistent time series are fBm or a biased random walk.

Hurst exponent can be used to verify the EMH (effective market hypothezis): The time series is a random walk. To verify EMH, we calculate the expected value of the adjusted range $E(R/S)_n$ and its variance $Var(E(R/S)_n)$ [9, 11]:

$$E(R/S)_n = \frac{n-0.5}{n} \left(\frac{\pi n}{2}\right)^{-0.5} \sum_{r=1}^{n-1} \sqrt{\frac{n-r}{r}} \tag{9}$$

$$Var(E(R/S)_n) = \left(\frac{\pi^2}{6} - \frac{\pi}{2}\right).n \tag{10}$$

The Eq. (9) can be used to generate the expected values of the Hurst exponent. The expected Hurst exponent is varying depending on the values of n. Any range will be appropriate as long as the system under study and the $E(R/S)_n$ series correspond to the same values of n. For financial purposes, we will begin with $n = 10$. The final value of n will depend on the system under study.

The R/S values are random variables, normally distributed, and therefore we would expect that the values of H would also be normally distributed:

$$Var(H_n) = \frac{1}{T}, \tag{11}$$

where T is the total number of observations in the sample. Note that $Var(H_n)$ does not depend on n or H, but it depends on the total sample size T. Then t-statistics is used to verify the significance of the null hypothesis. If the Hurst exponent H is approximately equal to its expected value $E(H)$, it means that the time series is independent and random in the analysed period (the Hurst exponent is insignificant). If the Hurst exponent H is greater (smaller) than its expected value $E(H)$, the time series is persistent (antipersistent) (the Hurst exponent is significant). If the series exhibits a persistent character, then the time series has a long memory and the ratios R/S_n are increasing. If the ratios R/S_n are decreasing, the time series is antipersistent. The breaks may indicate a periodic or nonperiodic component in the time series with some finite frequency. We calculated the V-statistics to estimate precisely where this break occurs [11]:

$$V_n = \frac{(R/S)_n}{\sqrt{n}}. \tag{12}$$

3 Data and Empirical Results

The aim of this paper is to analyse the fractal structure of daily OHLC prices of the NYMEX Crude oil over the daily period from March 30, 1983 to June 10, 2016. The data were collected on a daily basis from Investing.com [16]. Our sample covers 7561 data points, it contains a large number of observations and covers a long time period. Such a long period allows us to learn learn a lot about behaviour of the market. The period 1983–2016 is characterized by a remarkable crude oil cycle, for example in July 2008 and also the changes in oil prices incurred as a result of sanctions against Russia.

Figure 1 shows a strongly correlated time series of OHLC prices, (the correlation coefficient for each pair is close to one).

When analyzing the crude oil market, we use logarithmic returns R_t defined as:

$$R_t = \ln \frac{P_t}{P_{t-1}}, \tag{13}$$

where P_t is either open, high, low or close price known at time t. See Fig. 2 for open and high log returns, and Fig. 3 for low and close log returns.

Our analysis is based on Peter's books [9, 11]. We have prepared our own program code, written in SAS IML language for R/S analysis. The figures were prepared in Wolfram Mathematica by our own program code. We will use this methodology to validate the effective market hypothesis. We compute Hurst coefficient H and its expected value $E(H)$ using rescaled R/S analysis, and we will verify the null hypothezis: the time series is a random walk. If the Hurst exponent H and its expected value $E(H)$ are approximately equal then the time series is independent and random during the analysed period (the Hurst exponent is statisticaly insignificant).

Figure 4 and Table 1 show R/S analysis results for the open log returns during the analysed period. Hurst coeficient H is equal to 0.5445. The expected Hurst exponent is equal to $E(H) = 0.5408$. The standard deviation of $E(H)$ is 0.0115 for 7560 obser-

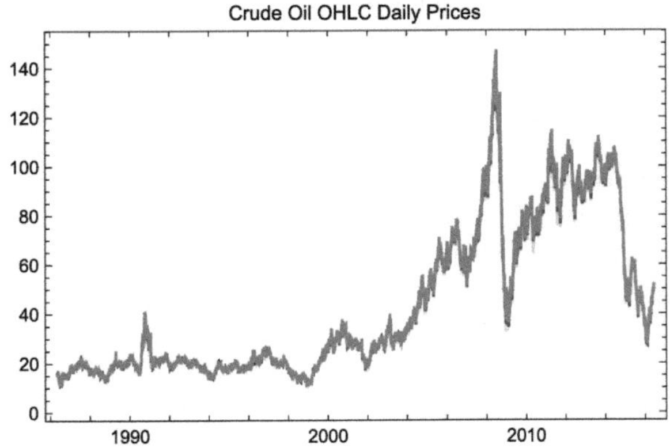

Fig. 1 Crude oil OHLC daily prices. Sample range: 3/30/1983–6/10/2016

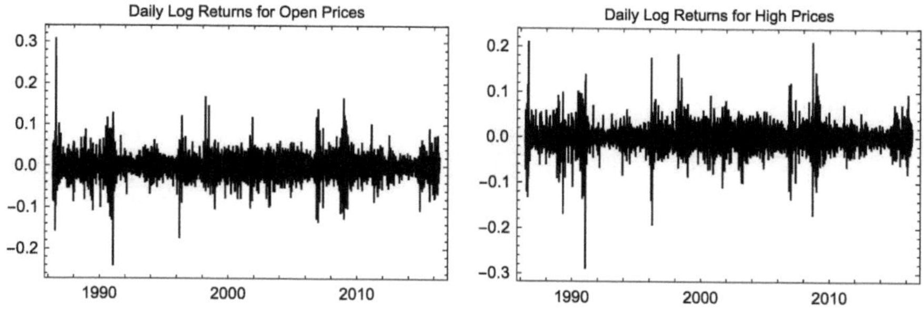

Fig. 2 Crude oil open, high daily log returns. Sample range: 3/30/1983–6/10/2016

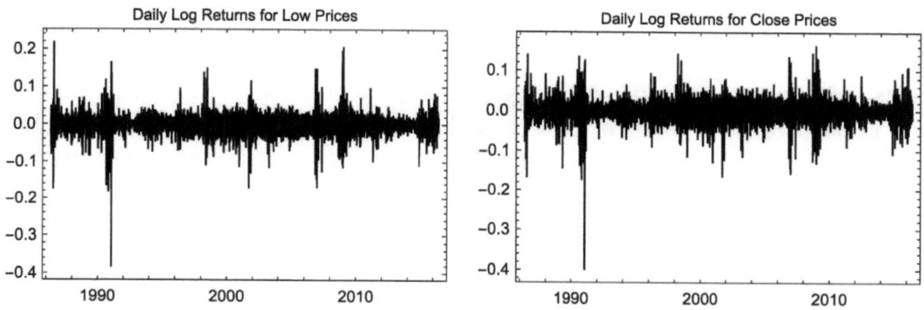

Fig. 3 Crude oil low, close daily log returns. Sample range: 3/30/1983–6/10/2016

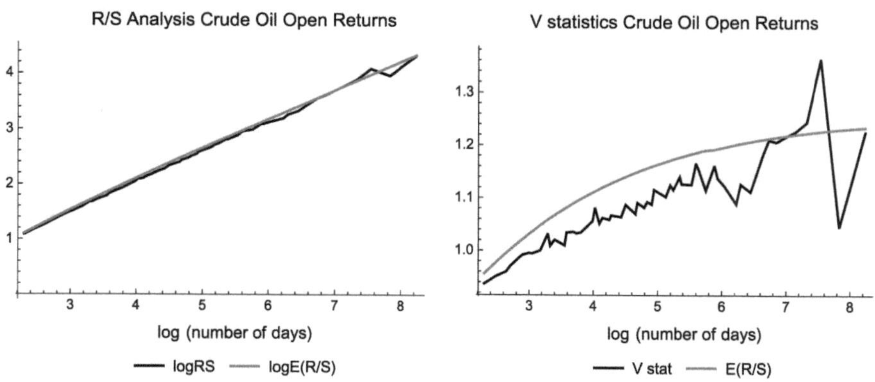

Fig. 4 R/S Analysis of crude oil daily open log returns. Sample range: 3/30/1983–6/10/2016

vations. The Hurst exponent for daily Crude oil open log returns is 0.31894 standard deviations away from its expected value. This is not significant result at the 95% level, and we conclude that the time series is a random walk. This non-significant result is confirmed using a graph, (see Fig. 4) where R/S (black line) and $E(R/S)$ (gray line) or V-statistics are plotted. However, breaks in R/S graph (Fig. 4) appear at 27 and 1890 observations ($\log 27 = 3.2958$, $\log 1890 = 7.5443$). Both graphs clearly confirm stops groving at $n = 27$, $n = 1890$ observations. These breaks may

Table 1 Estimation of the Hurst exponent, regression results, open log returns

Time period	Hurst exp. H	$E(H)$	n	t-stat.
Whole period	0.5445	0.5408	54	0.3189
$10 < n < 27$	0.5885	0.6017	9	−1.1436
$27 \leq n < 1890$	0.5488	0.5338	43	1.3113
$1890 \leq n < 3780$	0.8936	0.5085	2	33.4855

be a signal of a periodic or nonperiodic component in the time series with frequency of approximately 27 or 1890 periods. We run regression to estimate Hurst exponent for R/S values in the next subperiods: $n < 27, 27 \leq n < 1890, 1890 \leq n < 3780$. Table 1 shows the results of the regression analysis for estimating the Hurst exponents and their expected values during the analyzed periods. During the periods for $n < 27$ and $27 \leq n < 1890$, the time series has a random character (t-statistics is less than 1.96 (0.975 quantile of the Normal distribution), the Hurst exponent is insignificant. We have found that only for period $1890 \leq n < 3780$ H is significant. H is equal to 0.8936, the time series has persistent character. It means, that the earlier history had a random character and the recent history exhibits a long memory effect.

When we analyse high log returns using R/S analysis (Fig. 5 and Table 2) for the whole analysed period, we have obtained similar results as for the open log returns. H is equal to 0.5446, and this value is not significant at the 95% level. We conclude that the time series is a random walk. These non-significant results are confirmed using graphs, see Fig. 5. However, this time series exhibits four breaks (Fig. 5). The breaks appear for 27, 210, 360 and 1890 observations. Both graphs clearly confirm stops groving four times. Only the first two breaks give insignificant results. During periods $n < 27$ and $27 \leq n < 210$, the Hurst exponent is insignificant, the time series has a random character. Breaks at 360 and 1890 are significant. Table 2 shows these results. We have found that H is significant for periods $210 \leq n < 360$, $360 \leq n <$

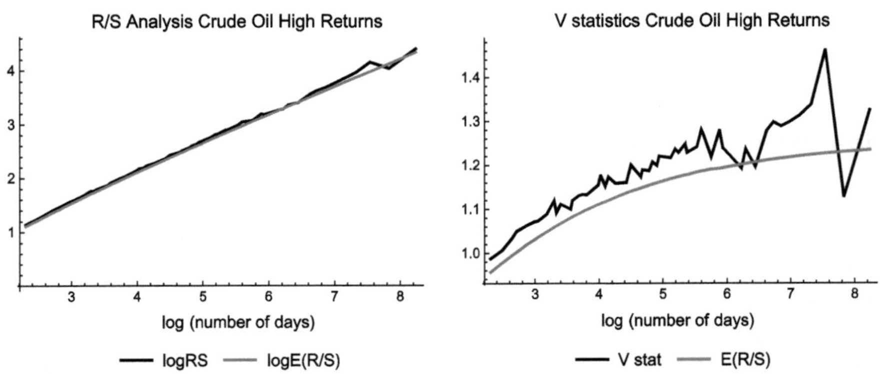

Fig. 5 RS analysis of the crude oil daily hight log returns. Sample range: 3/30/1983–6/10/2016

Table 2 Estimation of the Hurst exponent, regression results, high log return

Time period	Hurst exp. H	$E(H)$	n	t-stat.
Whole period	0.5446	0.5408	54	0.3282
$10 < n < 27$	0.6170	0.6017	9	1.3315
$27 \leq n < 210$	0.5580	0.5507	25	0.6414
$210 \leq n < 360$	0.5473	0.5233	6	2.0864
$360 \leq n < 1890$	0.5968	0.5170	12	6.9310
$1890 \leq n < 3780$	0.9017	0.5085	2	34.193

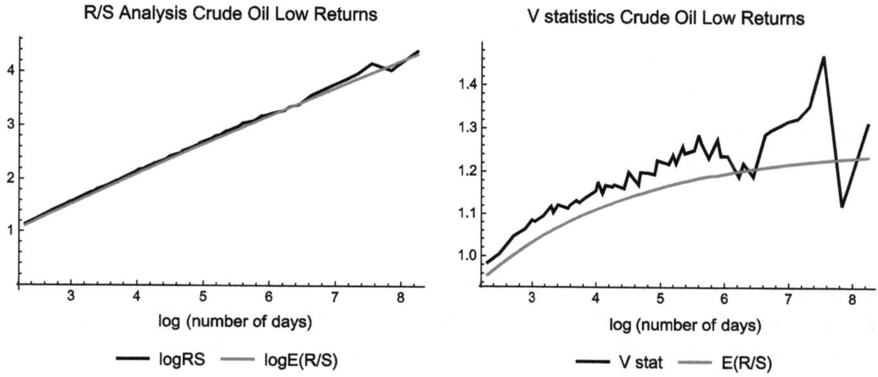

Fig. 6 RS analysis of crude oil daily low log returns. Sample range: 3/30/1983–6/10/2016

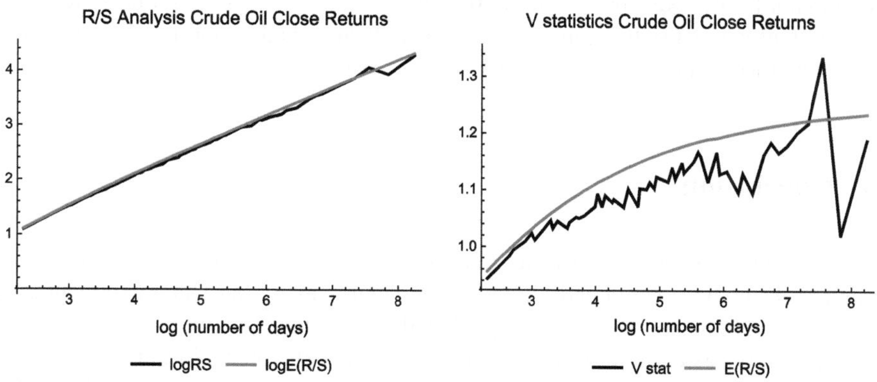

Fig. 7 RS analysis of crude oil daily close log returns. Sample range: 3/30/1983–6/10/2016

1890, $1890 \leq n < 3780$ H is significant and the time series has a persistent character. It means that the earlier history had random character, and recent history exhibits a long memory effect in three cycles.

R/S analysis of the low and close log returns (Figs. 6 and 7; Tables 3 and 4) gives similar results. H is equal to 0.5446 or 0.5354 for the whole analysed period, again

Table 3 Estimation of the Hurst exponent, regression results, low log return

Time period	Hurst exp. H	$E(H)$	n	t-stat.
Whole period	0.5446	0.5408	54	0.3308
$10 < n < 360$	0.5633	0.5578	40	0.4761
$360 \leq n < 1890$	0.6046	0.5170	12	7.6151
$1890 \leq n < 3780$	0.8868	0.5085	2	32.8976

Table 4 Estimation of the Hurst exponent, regression results, close log return

Time period	Hurst exp. H	$E(H)$	n	t-stat.
Whole period	0.5354	0.5408	54	−0.4749
$10 < n < 210$	0.5524	0.5644	34	−1.0400
$210 \leq n < 360$	0.5251	0.5234	6	0.1517
$360 \leq n < 1890$	0.5913	0.5171	12	6.4494
$1890 \leq n < 3780$	0.8771	0.5085	2	32.0488

these values are not significant at the 95% level. The time series are random walks. These results of non-significance are confirmed using graphs, see Figs. 6 and 7. Both time series exhibit two breaks. Breaks appear after 360 and 1890 observations. Both graphs clearly confirm stops growing two times. The Hurst exponent is insignificant during $n < 360$, and the time series have a random character. We have found H is significant for periods $360 \leq n < 1890$, $1890 \leq n < 3780$ and both time series have a persistent character. It means that the earlier history had a random character, and recent history exhibits a long memory effect in two cycles.

4 Conclusion

Information obtained from fractal analysis can be used for investor decisions. In this paper, we have examined the market efficiency using the Hurst exponent as a market trend indicator. We have analysed OHLC log returns to give the answer on the question of persistent long-term memory and market efficiency. Empirical evidence from NYMEX Crude oil data illustrates that during the analysed period 3/30/1983–6/10/2016 all OHLC log returns exhibit a random walk at significance level 0.05. After detailed analysis, we have found there are some cycles in the analysed data. All analysed log returns time series exhibit one cycle for the number of observation from 1890 to 3780 (approximately 7.5 years). The same fractal structure with two cycles exhibits low and close log returns. Only high log returns exhibit 3 cycles with a different size. Investors are advised to follow the major developments of high prices and take advantage of detected cycles.

We also ask how stable are our findings? Market reacts to new information, and the way it reacts is not much different from the way it reacted in 1983, even though the type of information is different. Therefore the underlying dynamics and, in particular, the statistics of the market have not changed. This would be especially true for fractal statistics. Our result ascertains that the fractal Brownian motion is present in NYMEX crude oil prices. Our findings are in line with the ones found in [10, 12] [13]. Our empirical results indicate that the crude oil market is consistent with the effective market hypothesis and that the market exhibits at various times inefficient behaviour over short time.

References

1. Bayraktar, E., Poor, H.V., Sicar, K.R.: Efficient estimation of the hurst parameter in high frequency financial data with seasonalities using wavelets. In: Proceedings of 2003 International Conference on Computational Intelligence for Financial Engineering (CIFEr2003), Hong-Kong, 21–25 Mar 2003
2. Beran, J.: Statistics for Long-memory Processes. Chapman and Hall, New York (1994)
3. Di Matteo, T., Aste, T., Dacorogna, M.M.: Long-term memories of developed and emerging markets: Using the scaling analysis to characterize their stage of development. J. Bank. Financ. **29**, 827–851 (2005). doi:10.1016/j.jbankfin.2004.08.004
4. Kahane, J.P.: Some Random Series of Functions, 2nd edn. Camridge University Press, London (1985)
5. Mandelbrot, B.B., Van Ness, J.W.: Fractional Brownian motions, fractional noises and applications. SIAM Rev. **10**, 422–437 (1968)
6. Nualart, D.: Fractional Brownian motion: stochastic calculus and applications, In: Proceedings of the International Congress of Mathematicians, pp. 1541–1562. European Mathematical Society, Madrid, Spain (2006)
7. Van Ness: Fractional Brownian motion, fractional noises and applications. SIAM Rev. **10**, 422–437 (1968)
8. Qian, H., Raymond, G.M., Bassingthwaighte, J.B.: On two-dimensional fractional Brownian motion and fractional Brownian random field. J. Phys. A: Math. Gen. **31**, L527–L535 (1998)
9. Peters, E.E.: Fractal Market Analysis. Wiley, New York (1994)
10. Li, D.–Y., Nishimura, Y., Men, M.: Why the long-term auto-correlation has not been eliminated by arbitragers: evidences from NYMEX. Energy Econ. **59**, 167–178 (2016). doi:10.1016/j.eneco.2016.08.006
11. Peters, E.E.: Chaos and Ordered in the Capital Markets. Wiley, New York (1996)
12. Li, D.–Y., Nishimura, Y., Men, M.: Fractal markets: liquidity and investors on different time horizons. Phys. A **407**, 144–151 (2014). doi:10.1016/j.physa.2014.03.073
13. Power, G.J., Turvey, C.G.: Long-range dependence in the volatility of commodity futures prices: Wavelet-based evidence. Phys. A **389**, 79–90 (2010). doi:10.1016/j.physa.2009.08.037
14. Bohdalová, M., Greguš, M.: Fractal analysis of forward exchange rates. Acta Polytech. Hung. **7**(4) 57–69 (2010)
15. Bohdalová, M., Greguš, M.: Markets, Information and Their Fractal Analysis. E-Leader, New York: CASA, 1–8, (2010)
16. Investing Ltd.: Crude-oil. http://www.investing.com/commodities/crude-oil-historical-data (n.d.)

Time-Frequency Representations as Phase Space Reconstruction in Symbolic Recurrence Structure Analysis

Mariia Fedotenkova, Peter beim Graben, Jamie W. Sleigh and Axel Hutt

Abstract Recurrence structures in univariate time series are challenging to detect. We propose a combination of symbolic and recurrence analysis in order to identify recurrence domains in the signal. This method allows to obtain a symbolic representation of the data. Recurrence analysis produces valid results for multidimensional data, however, in the case of univariate time series one should perform phase space reconstruction first. In this chapter, we propose a new method of phase space reconstruction based on the signal's time-frequency representation and compare it to the delay embedding method. We argue that the proposed method outperforms the delay embedding reconstruction in the case of oscillatory signals. We also propose to use recurrence complexity as a quantitative feature of a signal. We evaluate our method on synthetic data and show its application to experimental EEG signals.

Keywords Recurrence analysis · Symbolic dynamics · Time-frequency representation · Lempel-Ziv complexity · EEG

M. Fedotenkova (✉)
NEUROSYS team, INRIA, F-54600 Villers-lès-Nancy, France
e-mail: maria.fedotenkova@gmail.com

M. Fedotenkova
UMR no, 7503, CNRS, Loria, 54500 Vanduvre-lès-Nancy, France

M. Fedotenkova
Université de Lorraine, 54600 Villers-lès-Nancy, France

P.b. Graben
Bernstein Center for Computational Neuroscience, Berlin, Germany

J.W. Sleigh
Waikato Clinical School of the University of Auckland, Auckland, New Zealand

A. Hutt
Deutscher Wetterdienst, Offenbach Am Main, Germany

I. Rojas et al. (eds.), *Advances in Time Series Analysis and Forecasting*,
Contributions to Statistics, DOI 10.1007/978-3-319-55789-2_7

1 Introduction

Recurrent temporal dynamics is a phenomenon frequently observed in time series measured in biological systems. For instance, bird songs exhibit certain temporal structures, that recur in time [28]. Other examples are returning epileptic seizures [2], recurrent brain microstates in language processing [4] and in early auditory neural processing [13]. All these latter phenomena are observed in electroencephalographic data (EEG). To detect such temporal recurrent structures, typically one applies recurrence analysis [7, 21] based on Poincaré's theorem [24]. This approach allows the detection of recurrence structures in multivariate time series. To retrieve recurrence structures from univariate time, several methods have been suggested, such as delay embedding techniques.

However, most existing methods do not take into account specifically the oscillatory nature of the signals as observed in biological systems. To this end, we propose a technique to embed the univariate time series in a multidimensional space to better consider oscillatory activity. The approach is based on the signals time-frequency representation. In a previous work we have sketched this approach [27] already but without discussing its performance subject to different time-frequency representations. The present work shows this detailed discussion and suggests a new method to classify signals according to their recurrence complexity. Applications to artificial data permits evaluating the method and compare it to results gained from the conventional delay embedding technique. Final applications to experimental EEG data indicates the method's future application.

2 Analysis Methods and Data

2.1 Symbolic Recurrence Structure Analysis

Recurrence is a fundamental property of nonlinear dynamical systems, which was first formulated by Poincaré in [24]. It was further illustrated in the recurrence plot (RP) technique proposed by Eckmann et al. [7]. This relatively simple method allows to visualize multidimensional trajectories on a two-dimensional graphical representation. The RP can be obtained by plotting the recurrence matrix:

$$R_{ij} = \Theta\left(\varepsilon - ||\boldsymbol{x}_i - \boldsymbol{x}_j||\right), \qquad i,j = 1, 2, \dots, N, \tag{1}$$

where $\boldsymbol{x}_i \in \mathbb{R}^d$ is the state of the complex system in the phase space of dimension d at a time instance i; $|| \cdot ||$ denotes a metric, Θ is the Heaviside step function, and ε is a threshold distance.

It can be seen from (1), that if two points in the phase space are relatively close, the corresponding element of the recurrence matrix $R_{ij} = 1$, which would be represented by a black dot on the RP.

Instead of analyzing RPs point-wise we concentrate our attention on recurrence domains, labeling each domain with a symbol, thus obtaining recurrence plots of symbolic dynamics. The RP from symbols were successfully used in several studies (see, for instance, [6, 8, 17]). Here, we use the symbolic recurrence structure analysis (SRSA) proposed in [3], this technique allows to obtain symbolic representations of the signal from the RP, the latter being interpreted as a set of rewriting rules. According to these rules, large time indices are substituted with smaller ones when two states, occurring at these times, are recurrent. The process starts by initializing a symbolic sequence with discrete time at which the signal is computed, i.e., $s_i = i$. Next, this sequence is recursively rewritten based on the elements in the RP, namely, $s_i \rightarrow s_j$ if $i > j$ and $R_{ij} = 1$. Afterwards, the sequence is scanned for monotonically increasing indices and each of them is mapped to one symbol $s_i = 0$, which labels transient states. This is done to differentiate metastable states from transitions between them. More detailed description of the method and examples can be found in [3, 5].

By examining (1) one can see that the resulting recurrence matrix and, thus, symbolic sequence strongly depend on the distance threshold parameter ε. Several techniques for optimal ε estimation exist [22], most of which are heuristic. SRSA aims to obtain an optimal value of ε from the data.

Here, we propose two approaches to estimate ε optimally, based on (i) the principle of maximal entropy and (ii) Markov chain modeling of the system. The former implies that the system spends an equal amount of time in each recurrence domain [3], while the latter takes into account the probabilities of the system's transition from one recurrence state to another [5]. Each of these approaches assumes a certain model for the system's dynamics, hence for each ε value we can calculate the value of a utility function, which describes how well an obtained symbolic sequence fits to the proposed model. The optimal value of the threshold distance ε^* will then be the one maximizing the value of the utility $u(\varepsilon)$ function:

$$\varepsilon^* = \arg\max_{\varepsilon} u(\varepsilon) \ . \tag{2}$$

The utility function is different for both models. In the first case, the utility function is presented with the normalized symbolic entropy:

$$u(\varepsilon) = -\frac{\sum_{k=0}^{n-1} p_k(\varepsilon) \log p_k(\varepsilon)}{n(\varepsilon)} \ , \tag{3}$$

where $p_k(\varepsilon)$ is the relative frequency of the symbol k, $n(\varepsilon)$ is the cardinality of the alphabet the (number of states). Here, we divide the entropy by the cardinality of the alphabet in order to compensate for the influence of the alphabet size.

The second model rests upon the following assumptions about the ideal system's dynamics. (i) The system's metastable states exhibit mainly self-transitions, i.e., transition probabilities p_{ii} are larger than the probabilities of other transitions. (ii) There are no direct transitions from one metastable state to another one without passing through a transient state, i.e., $p_{ij} = 0$ when $i \neq j$ for $i, j > 0$. (iii) Probabilities of transitions from and to transient states, p_{0i} and p_{i0}, respectively, are distributed according to the principle of maximum entropy. We can now construct a transition matrix corresponding to the desired dynamics:

$$\boldsymbol{P} = \begin{bmatrix} 1-(n-1)q & r & r & \cdots & r \\ q & 1-r & 0 & \cdots & 0 \\ q & 0 & 1-r & \cdots & 0 \\ \vdots & \vdots & \vdots & \ddots & \vdots \\ q & 0 & 0 & \cdots & 1-r \end{bmatrix}, \tag{4}$$

here, the total number of states is n and the number of recurrence states is $n-1$, diagonal elements correspond to the probabilities of self-transitions, $q = p_{i0}$ and $r = p_{0i}$ for $i, j > 0$ are transition probabilities to and from a transient state $s_0 = 0$.

Keeping in mind the three criteria of the optimal dynamics, we can achieve the desired utility function by: (i) maximizing the trace of the transition matrix $\operatorname{tr} P = 1 + (n-1)(1-q-r)$; (ii) maximizing the normalized entropy of transition probabilities of the first row and the first column of $\boldsymbol{P}$ after neglecting p_{00}, i.e., $p'_{0i} = p_{0i} / \sum_{i=1}^{n-1} p_{0i}$ for the first row and $p'_{i0} = p_{i0} / \sum_{i=1}^{n-1} p_{i0}$ for the first column. (iii) suppressing transitions between recurrence states by simultaneously maximizing the trace and the entropies of the first row and column of $\boldsymbol{P}$, due to normalization condition $\sum_{i=0}^{n-1} p_{ij} = 1$. Then the utility function is given by:

$$u(\varepsilon) = \frac{1}{n-2}\Big(\operatorname{tr} \boldsymbol{P}(\varepsilon) + h_r(\varepsilon) + h_c(\varepsilon)\Big), \tag{5}$$

where h_r and h_c are the entropies of the first row and column of $\boldsymbol{P}$ (see [5] for more details).

2.2 Phase Space Reconstruction

A dynamical system is defined by an evolution law in a phase space. This space is d-dimensional, where each dimension corresponds to a certain property of a system (for instance, position, and velocity). Each point of the phase space refers to a possible state of the system. An evolution law, which is normally given by a set of differential equations, defines the system's dynamics yields a trajectory in phase space.

In certain cases only discrete measurements of a single observable are available, in this situation a phase space should be reconstructed according to Takens's theorem [26], which states that phase space presented with a d-dimensional manifold can be embedded into a $2d+1$-dimensional Euclidean space preserving the dynamics of the system. Several method of phase space reconstruction exist: delay embedding [26], numerical derivatives [23] and others (see for instance [16]).

In this work we propose a new method of phase space reconstruction based on the time-frequency representation of a signal. A time-frequency representation (TFR) is a distribution of the power of the signal over time and frequency. Here, the power in each frequency band contributes to a dimension of the reconstructed phase space. This approach is well-adapted for non-stationary and, especially, for oscillatory data, allowing better detection of oscillatory components rather than creating RPs point-wise from the signal. In this article we compare performance of the SRSA with different reconstruction methods, delay embedding and two different TFRs: spectrogram and scalogram.

2.2.1 Delay Embedding

Assume, we have a time series which represents scalar measurements of a system's observable in discrete time:

$$x_n = x(n\Delta t), \quad n = 1, \ldots, N \ , \tag{6}$$

where Δt is measurement sampling time. Then the reconstructed phase space is given by:

$$s_n = \left[x_n, x_{n+\tau}, x_{n+2\tau}, \ldots, x_{n+(m-1)\tau}\right], \quad n = 1, \ldots, N - (m-1)\tau \ , \tag{7}$$

where m is the embedding dimension and τ is the time delay.

These parameters play an important role in correct reconstruction and should be estimated appropriately. Optimal time delay τ should be chosen such that delay vectors from (7) are sufficiently independent. The most common technique to correctly estimate the τ parameter is based on the average mutual information [9, 19]. Moreover, the main attribute of appropriately chosen dimension m is that the original d-dimensional manifold will be embedded into an m-dimensional space without ambiguity, i.e., self-crossing and intersections. We apply the method of false nearest neighbors [14, 15], which permits the estimation of the minimal embedding dimension.

2.2.2 Time-Frequency Representation

Time-frequency representation of a signal shows the signal's energy distribution in time and frequency. In this work we analyze two different types of TFR: the spectrogram and the scalogram (based on continuous wavelet transform).

The spectrogram $S^h(t, \omega)$ of a signal $x(t)$ is the square magnitude of its short-time Fourier transform (STFT):

$$X_h(t,\omega) = \int_{-\infty}^{+\infty} x(\tau)h^*(t-\tau)\mathrm{e}^{-i\omega\tau}\,\mathrm{d}\tau \ , \tag{8}$$

where $h(t)$ is a smoothing window and $*$ denotes the complex conjugate. This yields $S_h(t,\omega) = |X_h(t,\omega)|^2$.

The continuous wavelet transform (CWT) [1] is obtained by convolving the signal with a set of functions $\psi_{ab}(t)$ obtained by translation and dilation of a mother wavelet function $\psi_0(t)$:

$$T_\psi(b,a) = \frac{1}{\sqrt{a}} \int_{-\infty}^{+\infty} x(t)\psi_0^*\left(\frac{t-b}{a}\right)\mathrm{d}t \ , \tag{9}$$

then, by analogy with the spectrogram, the squared magnitude of the CWT is called scalogram: $W_\psi(b,a) = \left|T_\psi(b,a)\right|^2$. In practice, the scale a can be mapped to a pseudo-frequency f and the dilation b represents a time instance and hence the time-frequency distribution can be written as $W_\psi(t,f)$.

The scalogram was computed using analytical Morlet wavelet, and a Hamming window with 80% overlap was chosen for the spectrogram. In all the methods the window length and scale locations were chosen such as to achieve a frequency resolution of 0.2 Hz for synthetic data and 1 Hz for experimental data.

2.3 Complexity Measure

To quantitatively assess the obtained symbolic sequences we propose to measure its complexity. We present here three different complexity measures. These are the cardinality of the sequence's alphabet and the number of distinct words obtained from the sequence [12], where a word is a unique group of the same symbols. In addition, we compute the well-known Lempel-Ziv (LZ) complexity [18], which is related to the number of distinct substrings and the rate of their occurrence along the symbolic sequence. All of the complexity measures have in common the notion of complexity, that is the number of distinct elements required to encode the symbolic string. The more complex the sequence is, the more of such elements are needed to present it without redundancy.

To demonstrate these measures we generated 100 artificial signals of two kinds (see below) with random initial conditions and random noise.

2.4 Synthetic Data

2.4.1 Transient Oscillations

The time series is a linear superposition of three signals, which exhibit sequences of noisy transient oscillations at a specific frequency [27]. These frequencies are 1.0, 2.25 and 6.3 Hz, cf. Fig. 1a. The sampling frequency is 50 Hz and the signal has a duration of 70 s. Figure 1 shows the three different transient oscillations whose sum represents the signal under study.

2.4.2 Lorenz System

The solution of the chaotic Lorenz system [3, 20] exhibits two wings which are approached in a unpredictable sequence. These wings represent metastable signal states. Figure 1b shows the time series of the z-component of the model.

2.5 Experimental Data

We examine electroencephalographic data (EEG) obtained during a surgery under general anesthesia [25]. The EEG data under investigation has been captured at frontal electrodes 2 min before (pre-incision phase) and 2 min after (post-incision phase) skin incision and last 30 s. The raw signal was digitized at a rate of 128 Hz and digitally band-pass filtered between 1 and 41 Hz using a 9th order Butterworth filter. The question in the corresponding previous study [25] was whether it is possible to distinguish the pre-incision from post-incision phase just on the basis of the captured EEG time series.

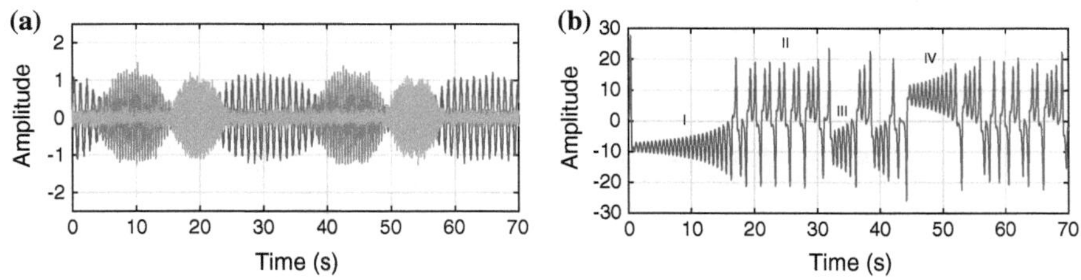

Fig. 1 Example signals of the synthetic data. **a** Three signals, whose sum represents the transient oscillation signal under study. **b** Solution of the Lorenz system along a single dimension

3 Results

3.1 Synthetic Data

3.1.1 Time-Frequency Embedding

To illustrate the method, Fig. 2 shows two different time-frequency representations of the transient oscillations signal. Spectrogram yields time-frequency intervals of high power at very good accordance with the underlying dynamics, cf. Sect. 2.4. In contrast, wavelet analysis smears out upper frequencies as a consequence of their intrinsic normalization of power. The symbolic sequences and the corresponding recurrence plots (middle and right-hand side of the panel) derived from the spectrogram fits perfectly with the underlying dynamics and are the same for both utility functions. They exhibit three different symbols in the symbolic sequence color-coded in blue, red and orange separated by transient states (color-coded in beige) in Fig. 2a and alternate in very good accordance to the three different transient oscillations. They are also visible as three rectangles of different size in the symbolic recurrence plot. Conversely, the scalogram yields only two recurrent signal features (entropy) and few recurrent states of brief duration (Markov), which do not reflect the underlying dynamics.

Typically, experimental neurophysiological signals exhibit a less regular temporal structure than given in the transient oscillations example. Solutions of the Lorenz system exhibit chaotic behavior, that is rather irregular and exhibits metastable oscillatory states. Since experimental EEG may exhibit chaotic behavior [10, 11], the

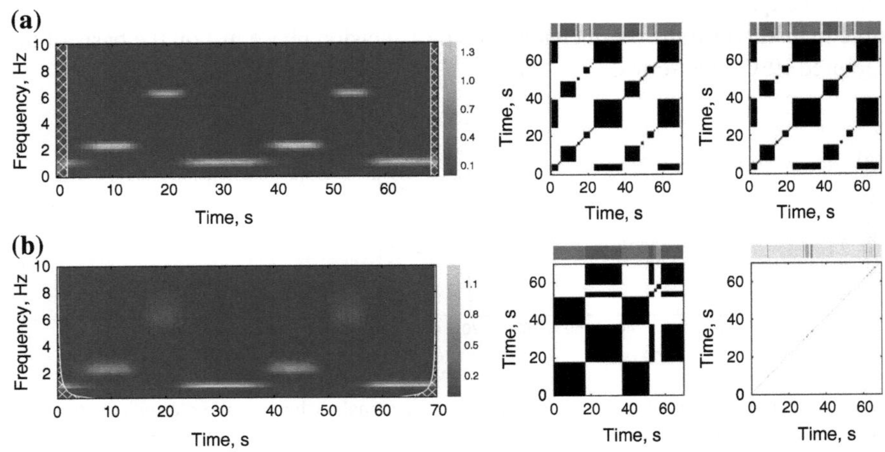

Fig. 2 Results for the transient oscillation signal. **a** Spectrogram; **b** scalogram. On each subfigure, *left* time-frequency representation, *middle* RPs with corresponding symbolic sequences above them (entropy utility function), *right* the same but with Markov utility function. In each symbolic sequence colors denote metastable states and transient states show in beige

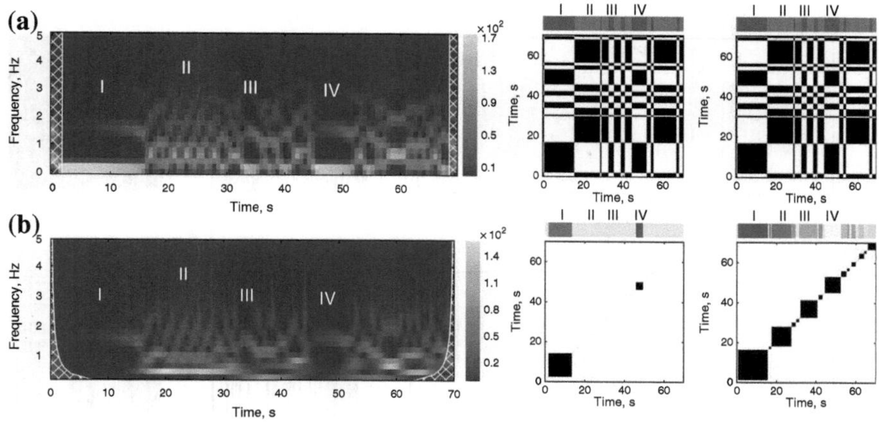

Fig. 3 Results for the Lorenz system. **a** Spectrogram; **b** scalogram. On each subfigure, *left* time-frequency representation, *middle* RPs with corresponding symbolic sequences above them (entropy utility function), *right* the same but with Markov utility function. In each symbolic sequence colors denote metastable states and transient states show in beige

Lorenz signal is tentatively closer to neurophysiological data. Figure 3 shows the TFR of the Lorenz signal. For both TFRs, one can well identify visually the four signal states I–IV marked in Fig. 1b. The color-coded symbolic sequences extracted from the spectrogram (seen in Fig. 3a) identify correctly the time windows of the signal states I–IV and are identical for both utility functions. The states I, II and IV are well captured, whereas the short state III is not well identified. The scalogram results are much worse in case of entropy utility function and only states I and IV are identified, while the Markov utility function captures all four states but no recurrence is present.

3.1.2 Delay Embedding

To illustrate the power of the method proposed, we compare our results to recurrence analysis results utilizing delay embedding, cf. Sect. 2.2. We consider the transient oscillations and the Lorenz signal, compute the optimal delay embedding parameters and apply the recurrence analysis technique to gain the symbolic sequences and the recurrence plots. Figure 4 reveals that the delay embedding essentially fails in detecting the recurrence domains in the transient oscillations compared to the time-frequency embedding (in case of both utility functions). In the Lorenz signal all states I-IV are captured in the symbolic sequence and visible in the recurrence plot, however the detection is much worse than with time-frequency embedding, cf. Fig. 3. Also the entropy utility function tends to produce few recurrent states with no transient states, whilst the usage of the Markov utility entails larger number of metastable and transient states.

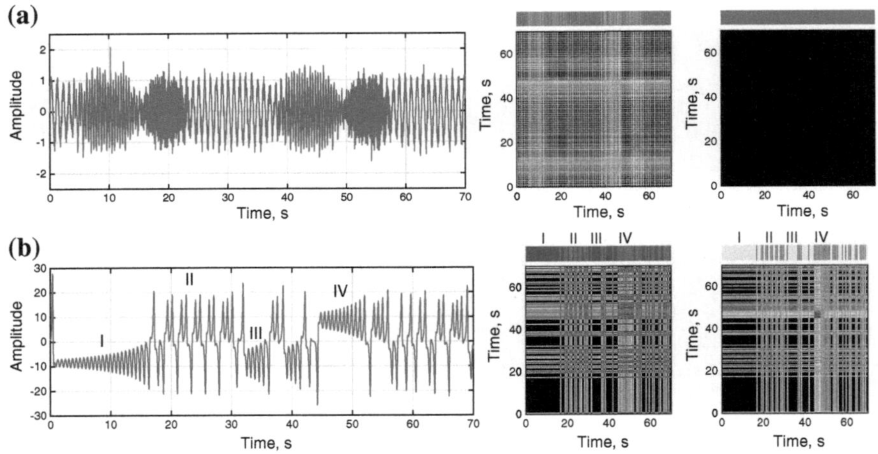

Fig. 4 Results obtained with delay embedding. **a** The transient oscillations, reconstruction parameters: $m = 5$, $\tau = 0.1$ s; **b** the Lorenz system, reconstruction parameters: $m = 3$ and $\tau = 0.16$ s

3.1.3 Complexity Measures

In order to quantify the intrinsic temporal structure, in addition we compute three complexity measures for each of the signals. To demonstrate the ability of complexity measures to distinguish temporal structures, Fig. 5 gives the distribution of complexity measures for both artificial datasets. We show results obtained with spectrogram, however the results for other embeddings are similar (not shown here for the sake of brevity). We observe that all complexity measures show significantly different distributions. Qualitatively, the largest difference between both signals is reflected in the LZ complexity measure. We also observe that in general

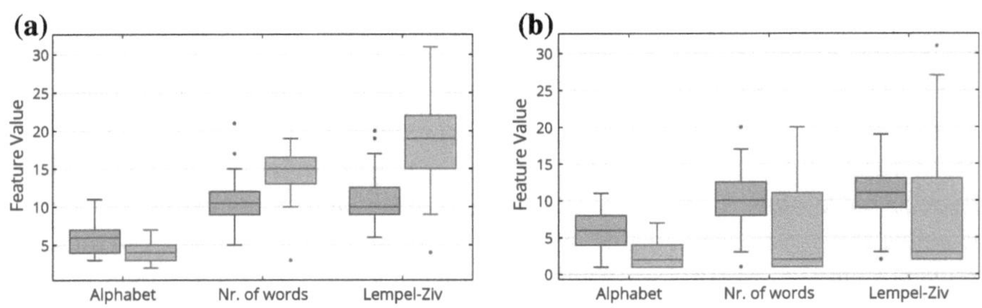

Fig. 5 Boxplots of three complexity measures for transient oscillations (*blue*) and Lorenz system (*red*) obtained with the spectrogram. **a** Entropy utility function; **b** Markov utility function. For each complexity measure, both distributions are significantly different (Kolmogorov-Smirnov test with $p < 0.001$)

complexities of the Lorenz system are larger than the ones of transient oscillations when obtained with Markov utility, it is the opposite for entropy utility function.

3.2 EEG Data

Finally, we study experimental EEG data. Figure 6 shows time-frequency plots (spectrogram) with corresponding symbolic sequences for two patients before and after incision during surgery. We observe activity in two frequency bands, namely strong power in the δ-band (1–5 Hz) and lower power in the α-range (8–12 Hz). This finding is in good accordance with previous findings in this EEG dataset [25]. The corresponding spectral power is transient in time in both frequency bands, whose temporal structure is well captured by the recurrence analysis with the entropy utility function as seen in the symbolic sequences. The symbolic analysis with the Markov utility function captures underlying dynamics well in case of patient #1099 (post-incision). In general Markov-based recurrence analysis tends to extract less recurrence domains separated by long transitions.

In order to characterize the temporal structure, we compute the symbolic sequences' recurrence complexity, which are shown in Table 1. We observe that the values of the various complexity measures are very similar in pre- and post-incision data and close between patients. However complexities obtained with the entropy utility function reveal larger differences between experimental conditions

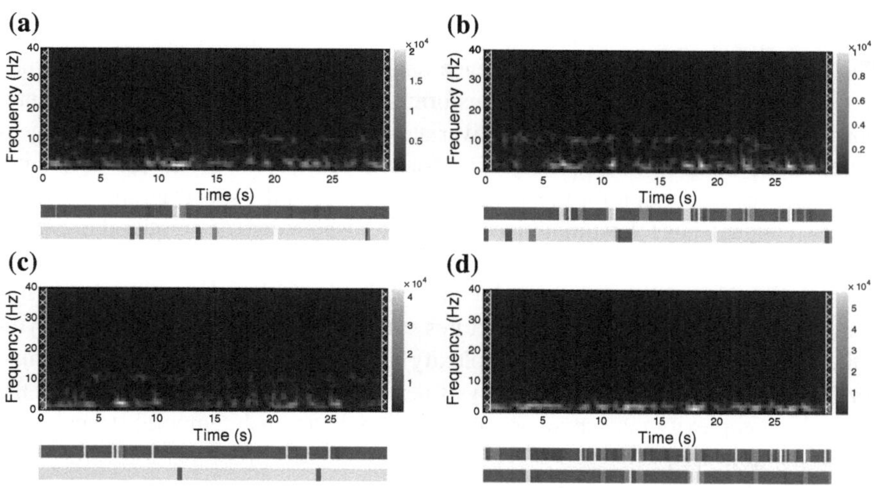

Fig. 6 Results for EEG signals obtained with spectrogram. Two colorbars below represent symbolic sequences obtained with entropy utility function (*top*) and Markov utility function (*bottom*). In each symbolic sequence colors denote metastable states and transient states show in beige. **a** Patient #1065 (pre-incision); **b** Patient #1065 (post-incision); **c** Patient #1099 (pre-incision); **d** Patient #1099 (post-incision)

Table 1 Complexity measures of EEG signals (spectrogram)

Complexity measure	Entropy		Markov	
	Pre-incision	Post-incision	Pre-incision	Post-incision
	Patient #1065			
Alphabet size	7	12	8	9
Nr. of words	19	25	15	12
Lempel-Ziv	22	27	13	13
	Patient #1099			
Alphabet size	5	13	3	8
Nr. of words	15	28	5	20
Lempel-Ziv	16	40	6	20

than between patients, whilst the Markov utility function demonstrates larger variation between patients than between the conditions. Since the time periods of pre- and post-incision data are captured several minutes apart and hence the corresponding data are uncorrelated, their similarity of complexity measures is remarkable pointing out to a constant degree of complexity in each patient. This is in line with the different complexity measures in both patients.

4 Discussion

The present work shows that recurrence analysis can be employed on univariate time series if, at first, the data is transformed into its time-frequency representation. This transform provides a multivariate time series whose dimensionality is equal to the number of frequency bins considered. We show that the best time-frequency representation for the synthetic time series is the spectrogram. We compare two approaches for estimation of optimal threshold distance required in SRSA. We demonstrate that a model of system's dynamics can be easily incorporated in the method through a utility function. However, if the model is not accurate the performance is worse. The recurrence structures extracted can be represented by a symbolic sequence whose symbolic complexity may serve as an indicator of the time series complexity. The EEG data analysis performed in this study indicates that the symbolic complexity may serve as a classifier to distinguish temporal structures in univariate time series.

References

1. Addison, P.S.: The Illustrated Wavelet Transform Handbook: Introductory Theory and Applications in Science, Engineering, Medicine and Finance. Institute of Physics Publishing, Bristol, UK, Philadelphia (2002)
2. Allefeld, C., Atmanspacher, H., Wackermann, J.: Mental states as macrostates emerging from EEG dynamics. Chaos **19**, 015102 (2009)
3. beim Graben, P., Hutt, A.: Detecting recurrence domains of dynamical systems by symbolic dynamics. Phys. Rev. Lett. **110**(15), 154101 (2013)
4. beim Graben, P., Hutt, A.: Detecting event-related recurrences by symbolic analysis: applications to human language processing. Philos. Trans. A Math. Phys. Eng. Sci. **373**(2034) (2015)
5. beim Graben, P., Sellers, K.K., Fröhlich, F., Hutt, A.: Optimal estimation of recurrence structures from time series. EPL **114**(3), 38003 (2016)
6. Donner, R., Hinrichs, U., Scholz-Reiter, B.: Symbolic recurrence plots: A new quantitative framework for performance analysis of manufacturing networks. Eur. Phys. J. Spec. Top. **164**(1), 85–104 (2008)
7. Eckmann, J.P., Kamphorst, S.O., Ruelle, D.: Recurrence Plots of Dynamical Systems. Europhys. Lett. EPL **4**(9), 973–977 (1987)
8. Faure, P., Lesne, A.: Recurrence plots for symbolic sequences. Int. J. Bifurc. Chaos **20**(06), 1731–1749 (2010)
9. Fraser, A.M., Swinney, H.L.: Independent coordinates for strange attractors from mutual information. Phys. Rev. A **33**(2), 1134–1140 (1986)
10. Freeman, W.J.: Evidence from human scalp EEG of global chaotic itinerancy. Chaos **13**(3), 1069 (2003)
11. Friedrich, R., Uhl, C.: Spatio-temporal analysis of human electroencephalograms: Petit-Mal epilepsy. Phys. D **98**, 171–182 (1996)
12. Hu, J., Gao, J., Principe, J.C.: Analysis of biomedical signals by the Lempel-Ziv Complexity: the effect of finite data size. IEEE Trans. Biomed. Eng. **53**(12), 2606–2609 (2006)
13. Hutt, A., Riedel, H.: Analysis and modeling of quasi-stationary multivariate time series and their application to middle latency auditory evoked potentials. Phys. D **177**, 203–232 (2003)
14. Kennel, M.B., Abarbanel, H.D.I.: False neighbors and false strands: A reliable minimum embedding dimension algorithm. Phys. Rev. E **66**(2), 026209 (2002)
15. Kennel, M.B., Brown, R., Abarbanel, H.D.I.: Determining embedding dimension for phase-space reconstruction using a geometrical construction. Phys. Rev. A **45**(6), 3403–3411 (1992)
16. Kugiumtzis, D., Christophersen, N.D.: State space reconstruction: method of delays vs singular spectrum approach. Res. Rep. Httpurn Nb NoURN NBN No-35645 (1997)
17. Larralde, H., Leyvraz, F.: Metastability for Markov processes with detailed balance. Phys. Rev. Lett. **94**(16), 160201 (2005)
18. Lempel, A., Ziv, J.: On the complexity of finite sequences. IEEE Trans. Inf. Theory **222**(1), 75–81 (1976)
19. Liebert, W., Schuster, H.G.: Proper choice of the time delay for the analysis of chaotic time series. Phys. Lett. A **142**(2), 107–111 (1989)
20. Lorenz, E.N.: Deterministic nonperiodic flow. J. Atmos. Sci. **20**(2), 130–141 (1963)
21. Marwan, N., Kurths, J.: Line structures in recurrence plots. Phys. Lett. A **336**, 349–357 (2005)
22. Marwan, N., Romano, M.C., Thiel, M., Kurths, J.: Recurrence plots for the analysis of complex systems. Phys. Rep. **438**(5–6), 237–329 (2007)
23. Packard, N.H., Crutchfield, J.P., Farmer, J.D., Shaw, R.S.: Geometry from a time series. Phys. Rev. Lett. **45**(9), 712 (1980)
24. Poincaré, H.: Sur le problème des trois corps et les équations de la dynamique. Acta Math. **13**(1), 3–270 (1890)
25. Sleigh, J.W., Leslie, K., Voss, L.: The effect of skin incision on the electroencephalogram during general anesthesia maintained with propofol or desflurane. J. Clin. Mon. Comput. **24**, 307–318 (2010)

26. Takens, F.: Detecting Strange Attractors in Turbulence. Springer (1981)
27. Tošić, T., Sellers, K.K., Fröhlich, F., Fedotenkova, M., beim Graben, P., Hutt, A.: Statistical frequency-dependent analysis of trial-to-trial variability in single time series by recurrence plots. Fronti. Syst. Neurosci. **9**(184) (2016)
28. Yildiz, I.B., Kiebel, S.J.: A hierarchical neuronal model for generation and online recognition of birdsongs. PLoS Comput. Biol. **7**, e1002303 (2011)

Analysis of Climate Dynamics Across a European Transect Using a Multifractal Method

Jaromir Krzyszczak, Piotr Baranowski, Holger Hoffmann, Monika Zubik and Cezary Sławiński

Abstract Climate dynamics were assessed using multifractal detrended fluctuation analysis (MF-DFA) for sites in Finland, Germany and Spain across a latitudinal transect. Meteorological time series were divided into the two subsets (1980–2001 and 2002–2010) and respective spectra of these subsets were compared to check whether changes in climate dynamics can be observed using MF-DFA. Additionally, corresponding shuffled and surrogate time series were investigated to evaluate the type of multifractality. All time series indicated underlying multifractal structures with considerable differences in dynamics and development between the studied locations. The source of multifractality of precipitation time series was two-fold, coming from the width of the probability function to a greater extent than for other time series. The multifractality of other analyzed meteorological series was mainly due to long-range correlations for small and large fluctuations. These results may be especially valuable for assessing the change of climate dynamics, as we found that larger changes in asymmetry and width parameters of multifractal spectra for divided datasets were observed for precipitation than for other time series. This suggests that precipitation is the most vulnerable meteorological quantity to change of climate dynamics.

Keywords Climate · Multifractal detrended fluctuation analysis · Time series · Meteorological quantities · European transect

J. Krzyszczak (✉) · P. Baranowski · M. Zubik · C. Sławiński
Institute of Agrophysics, Polish Academy of Sciences, ul. Doświadczalna 4, 20-290 Lublin, Poland
e-mail: jkrzyszczak@ipan.lublin.pl

H. Hoffmann
Institute of Crop Science and Resource Conservation (INRES), Katzenburgweg 5, 53115 Bonn, Germany

I. Rojas et al. (eds.), *Advances in Time Series Analysis and Forecasting*, Contributions to Statistics, DOI 10.1007/978-3-319-55789-2_8

1 Introduction

A typical way of analyzing climatic changes is to find out trends and oscillations of the relevant meteorological quantities [1]. However, in many cases this standard approach is not sufficient enough. Therefore more subtle methods are being developed and applied in order to project changes of meteorological parameters. These include fractal analysis and chaotic evolution analysis of the atmospheric system [2–4].

The multifractal nature of many geophysical systems and processes has been indicated in numerous papers. They include the distribution of clouds [5], the wind speed time series [6, 7], the air temperature time series [8–12], the ocean temperature time series [13], ground surface and soil temperature time series [14], the precipitation time series [15–19] or the ozone concentration time series [20]. Multifractal scaling was also reported for the Sun's magnetic field time series, stock market time series and heartbeat dynamics. Diminishing of multifractal nature was even observed in the arrangement of the street network in London [21].

The original version of the MF-DFA method [2] has been applied to the daily rainfall time series in the Pearl River basin and was compared with an universal multifractal model to show a relationship between topography and rainfall variability [22]. Applying MF-DFA on daily ground surface temperature records from four representative weather stations over China revealed considerable differences of the generalized Hurst exponents among sites [14]. These results indicated that the strength of the multifractal behavior of ground surface temperature is non-universal and depended on the geographical location of the station. Furthermore, it was possible to establish the multifractal properties of rainfall in space and time [23], to develop suitable models [24], and to use the models to predict rainfall extremes [25, 26].

Previous studies analyzing multifractal structure of meteorological time series used data from point-like sites, or at the most small areas (region). Usually, a specific meteorological quantity over a short time period was analyzed [7, 10, 14, 16, 17, 27]. Furthermore, climate shifts were indicated in literature, one around 1980, for which not only changes of mean temperature were observed, but also temperature variance changed in periods after and before 1980 [28] and another one in 2001/2002 [29]. Therefore, this work compares multifractal properties of various long-term meteorological time series coming from three meteorological stations located across a latitudinal European transect of varying climatic conditions divided into two separate periods in order to account for climate shifts. This will allow us to check whether changes in climate dynamics indicated in literature can be observed using MF-DFA analysis and to generalize the differences in the dynamics of meteorological processes. Thus, the aim of the present work is to analyze spatial and temporal variation of multifractal properties of daily meteorological time series.

2 Materials and Methods

2.1 Study Site and Meteorological Data

The analyzed study sites were located in three European countries with differing climatic conditions: Spain, Germany and Finland. The data used in the analysis came from meteorological stations located in Lleida (41°42′ N, 1°6′ E, 337 m a.s. l.), Spain, which represents semi-arid climate with Mediterranean-like precipitation patterns (annual average of 369 mm), foggy and mild winters and hot and dry summers (Köppen-Geiger classification: BSk), Dikopshof (50°48′ 29" N, 6°57′ 7′′ E, 60 m a.s.l.), Germany, which represents maritime temperate climate (Köppen-Geiger climate classification: Cfb) with significant precipitation throughout the year and Jokioinen (60°48′ N, 23°30′ E, 104 m a.s.l.), Finland, which is a subarctic climate with a strong seasonality of severe winters and cold short summers without dry season (Köppen-Geiger classification: Dfc).

Time series were measured daily from January 1st 1980 to December 31st 2010, using standard equipment comparable for all stations. Therefore, the time series were 11322 records long. Three variables were considered in the present study: air temperature [°C], precipitation [mm] and global radiation [MJ m^{-2} d^{-1}]. For Lleida, global radiation data had gaps of 48 days (11 days in September 1988 and 37 days in spring 1990). These gaps were filled by taking the absolute values of the associated grid cell in the ERA-interim dataset [30]. Series were divided into two subsets, one containing data from 1980 to 2001 (8035 records) and second one—2002–2010 (3287 records). Descriptive statistics of segmented meteorological time series are presented in Table 1.

2.2 MF-DFA Analysis

MF-DFA analysis of the nonstationary time series of x_k quantity having the length N requires performing of five steps [2]:

1. Creation of the "profile" $Y(i)$ by the subtraction of the mean value and integration of the time series to convert the noises into random walk:

$$\mathrm{Y(i)} = \sum_{k=1}^{i} [x_k - \langle x \rangle] \quad i=1, \ldots, N \tag{1}$$

2. Division of each "profile" into $N_s = int(N/s)$ non overlapping segments of equal length s. Since the length N of the series is often not a multiple of the considered time scale s, a short part at the end of the profile may remain. Therefore, the same procedure should be repeated starting from the opposite end to get $2N_s$ segments altogether.

Table 1 Descriptive statistics of the segmented meteorological time series from stations in Spain (Lleida), Germany (Dikopshof) and Finland (Jokioinen)

Variable	Precipitation (mm d^{-1})			Global radiation (MJ m^{-2} d^{-1})			Air temperature (°C)		
Site	Lleida	Dikopshof	Jokioinen	Lleida	Dikopshof	Jokioinen	Lleida	Dikopshof	Jokioinen
	1980–2001								
Mean	0.93	1.73	1.74	15.56	10.50	8.98	14.91	10.05	4.46
Min	0.00	0.00	0.00	2.90	<0.01	0.03	−8.30	−16.75	−33.35
Max	83.6	75.4	79.10	30.80	31.72	31.67	33.10	28.05	24.20
SD	3.86	3.78	3.88	8.10	7.56	8.33	7.51	6.74	9.27
Median	<0.01	0.07	0.10	15.20	8.92	6.20	14.50	10.30	4.50
	2002–2010								
Mean	0.93	1.69	1.66	15.88	10.94	9.26	15.38	10.71	5.09
Min	0.00	0.00	0.00	2.90	<0.01	0.03	−5.20	−11.75	−26.70
Max	47.4	50.52	50.00	30.80	31.33	30.72	30.65	28.85	25.00
SD	3.52	3.79	3.79	8.36	7.97	8.41	7.80	7.03	9.45
Median	<0.01	0.02	0.10	15.50	9.35	6.82	15.25	10.95	5.20

3. Calculation of the linear local trend using a least square fit for each of $2N_s$ segments and after that determination of the variance:

$$F^2(s,\nu) = \frac{1}{s}\sum_{i=1}^{s}\{Y[(\nu-1)s+i] - y_\nu(i)\}^2 \quad \nu = 1, \ldots, N_s \tag{2}$$

$$F^2(s,\nu) = \frac{1}{s}\sum_{i=1}^{s}\{Y[N-(\nu-N_s)s+i] - y_\nu(i)\}^2 \quad \nu = N_s+1, \ldots, 2N_s \tag{3}$$

where $y_\nu\ (i)$ is the fitting line in segment ν.
4. Obtaining the *q-th* order fluctuation function by averaging variance $F^2(s,\nu)$ over all $2N_s$ segments:

$$F_q(s) = \left\{\frac{1}{2N_s}\sum_{\nu=1}^{2N_s}\left[F^2(s,\nu)\right]^{q/2}\right\}^{1/q} \tag{4}$$

5. Determining the scaling behavior of the fluctuation functions by analyzing log-log plots $F_q(s)$ versus s for each value of q. For large values of s multifractal time series $F_q(s)$ increases as a power-law $F_q(s) \sim s^{h(q)}$ with generalized Hurst exponent $h(q)$ depending on q.
Subsequently, the multifractal spectrum is obtained by using the relationship $\tau(q) = qh(q) - 1$, and after that the Legendre transform:

$$\alpha = \frac{d\tau}{dq} \quad f(\alpha) = q\alpha - \tau(q) \tag{5}$$

The schematic representation of a multifractal spectrum with its most important parameters α_{max}, α_{min}, α_0, a_s and w marked is presented in Fig. 1.

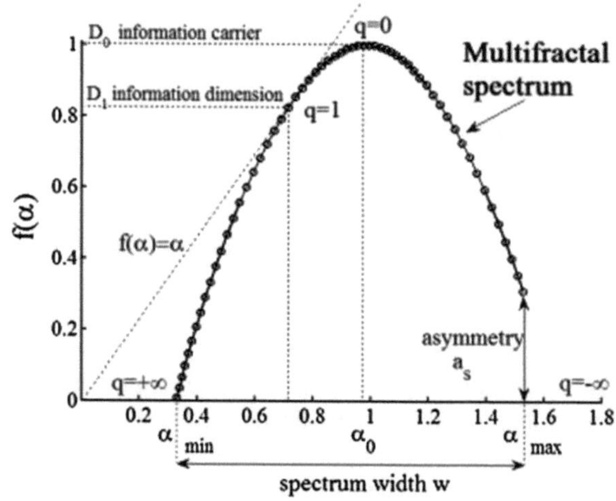

Fig. 1 Schematic presentation of main parameters of a multifractal spectrum

The α_{min} parameter indicates the most extreme and α_{max} the smoothest events in the studied process. A low value of α_0 indicates that the underlying process becomes correlated and loses fine structure, becoming more regular in appearance. The asymmetry parameter a_s achieves negative or positive values for a left- or right-skewed shape, respectively. The a_s is zero for symmetric shapes. A left-skewed spectrum means low fractal exponents of small weights, which correspond to dominance of extreme events [31]. A right-skewed spectrum denotes relatively strongly weighted high fractal exponents, which correspond to fine structures. The width of the spectrum w, which is difference between α_{max} and α_{min}, measures the length of the range of fractal exponents in the signal indicating the "richness" of the signal structure, i.e. more developed multifractality.

Kantelhardt [2] indicated that time series have two possible sources of multifractality. The first source is due to a broadness of the probability density function of the records contained in the studied time series, whereas the second source comes from different long-range correlations for small and large fluctuations. After obtaining the multifractal spectra, the main source of multifractality can be tested by randomly shuffling the series (what consists in generating a random permutation of the array elements of time series) to remove any temporal correlations. If spectra narrow significantly, long term correlations play the main role in the multifractality of the data. This is because shuffling of time series procedure destroys the long-range correlation. To check if the multifractality comes from broad distributions one need to analyze surrogate data. To obtain surrogates Amplitude Adjusted Fourier Transform (AAFT) has been applied in this paper [32]. If the multifractality in the time series is due to a broad probability density function only, the spectra obtained for the surrogate data indicate no multifractality [33, 34].

To calculate fluctuation function $F_q(s)$, scale s has to be determined. After several trials, we decided to use s ranging from 50 to 3000 events. The criterion for selection of the scale was the stability of obtained spectra. To prevent a potential distortion of the results by the so-called 'freezing' phenomenon [35], the range of q had to be limited to the [−4; 4]. This was due to the fact that the density distributions of all the studied meteorological time series had heavy tails.

3 Results and Discussion

Figures 2, 3 and 4 show the multifractal spectra of the segmented meteorological time series measured across European transect—for Jokioinen (Fig. 2), Dikopshof (Fig. 3) and Lleida (Fig. 4) stations. It results from those figures that all of the analyzed time series exhibit multifractal properties, both in first and second sub-period. After performing shuffling procedure obtained spectra are considerably narrower. Shuffled spectra suggest that long distance correlations play the main role for global radiation and air temperature time series and are important for the multifractality of precipitation time series, especially in Jokioinen and Dikopshof locations, since spectra for these sites become narrower than for Lleida. Surrogate

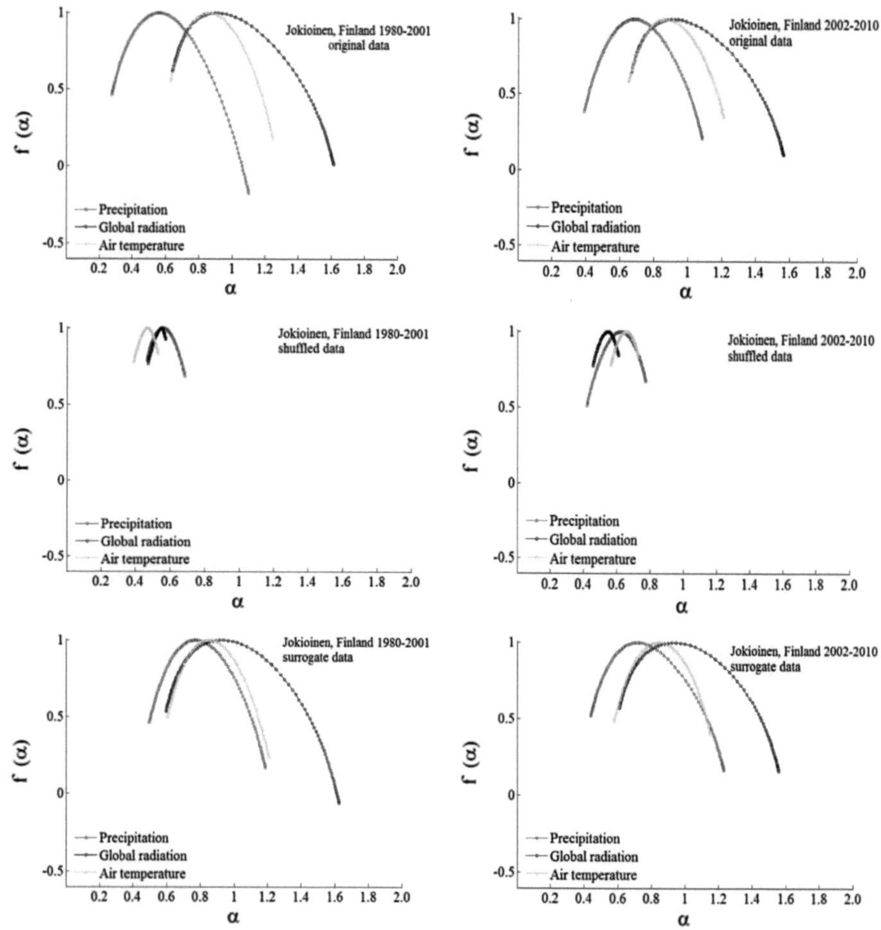

Fig. 2 Multifractal spectra of meteorological time series recorded at Jokioinen, Finland for 1980–2001 period (*left column*) and 2002–2010 period (*right column*). $f(\alpha)$ is singularity spectrum and α is singularity strength. Panels show original (*upper row*), shuffled (*middle row*) and surrogate data (*bottom row*)

spectra do not change vastly for global radiation and air temperature time series for all locations, what suggests that the broadness of the probability density function has almost no impact on the multifractality of those time series and that long range correlations are the main source of multifractality.

Contrary, for the precipitation time series, the change of surrogate spectra compared to spectra of original data is more visible, what suggests that broadness of probability density function plays more prominent role in multifractality of this time series. This is especially evident for Lleida, where surrogate spectra differ vastly from the spectra of original data.

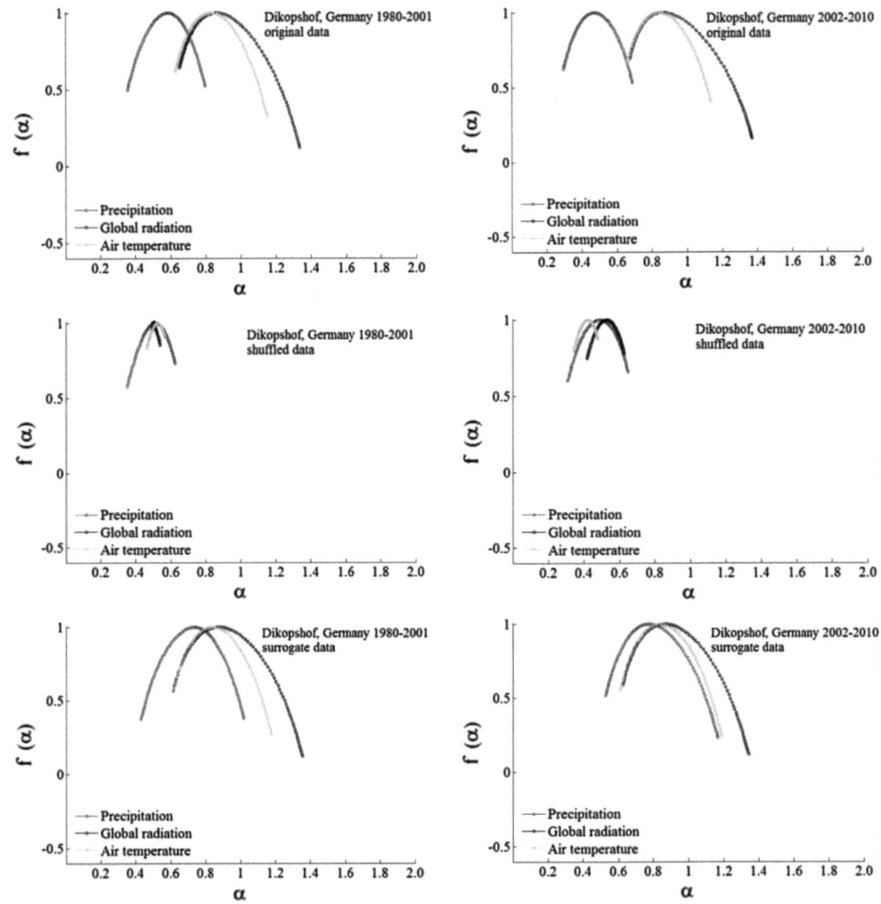

Fig. 3 Multifractal spectra of meteorological time series recorded at Dikopshof, Germany. $f(\alpha)$ is singularity spectrum and α is singularity strength. Panels show original (*upper row*), shuffled (*middle row*) and surrogate data (*bottom row*) for 1980–2001 (*left column*) and 2002–2010 (*right column*)

Those results are confirmed by the absolute differences found in the Hurst exponents (Fig. 5). As shown, the multifractality of global radiation and air temperature comes mainly from long range correlations whereas the broadness of the probability density function plays a minor role in multifractality, for all locations and both periods for the whole range of q. Contrary, for the precipitation time series the broadness of the probability density function is at least as important for the multifractality as long range correlations (Jokioinen site, both sub-periods and Dikopshof, 1980–2001 period) or even has a larger impact (Lleida site, both sub-periods and Dikopshof, 2002–2010 period).

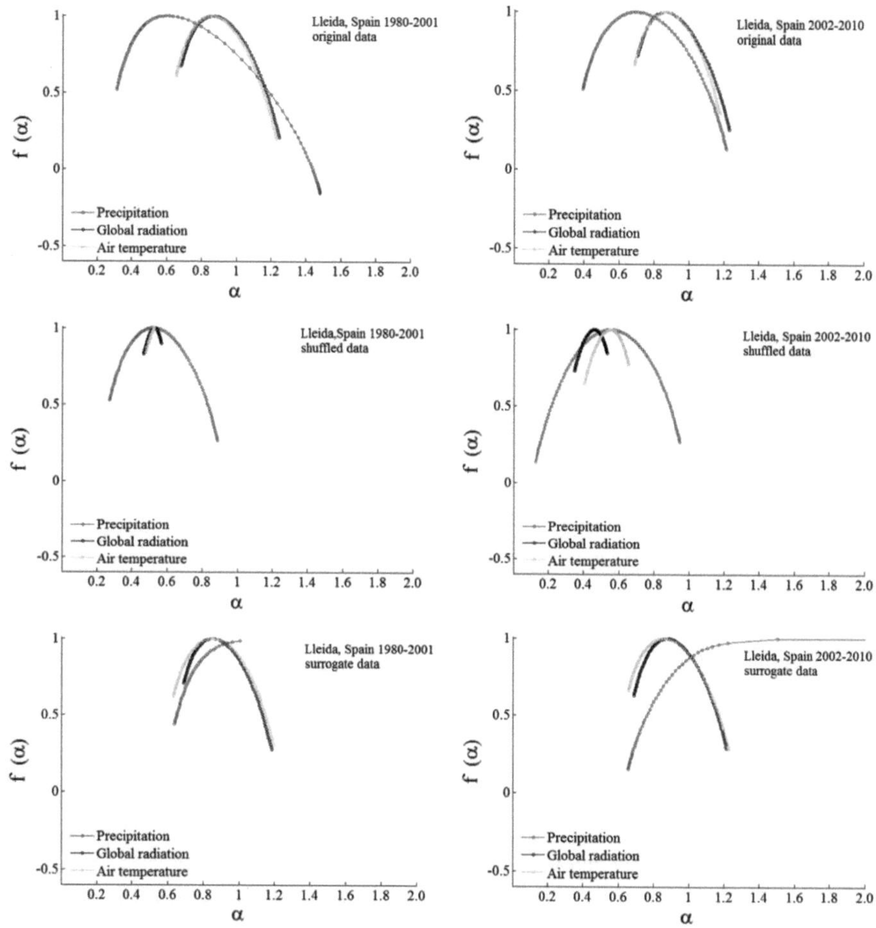

Fig. 4 Multifractal spectra of meteorological time series recorded at Lleida, Spain for 1980–2001 period (*left column*) and 2002–2010 period (*right column*). $f(\alpha)$ is singularity spectrum and α is singularity strength. Panels show original (*upper row*), shuffled (*middle row*) and surrogate data (*bottom row*)

Figure 6 presents a comparison of multifractal parameters describing the original spectra for both sub-periods. Almost all changes in spectra are consistent for all of the analyzed time series, namely we observe an increase of α_0 and a reduction of a_s (spectra becomes more symmetrical in shape, but stay right-skewed) and w parameters for the second sub-period (2002–2010), compared to the first one (1980–2001). Obtained results suggest that in second sub-period, after a climate shift indicated by Swanson and Tsonis [29], processes underlying the analyzed time series become less correlated and gain fine structure, lose "richness" of the signal structure, what means that multifractality of those time series is less developed and also extreme events are becoming more frequent. These characteristics are most

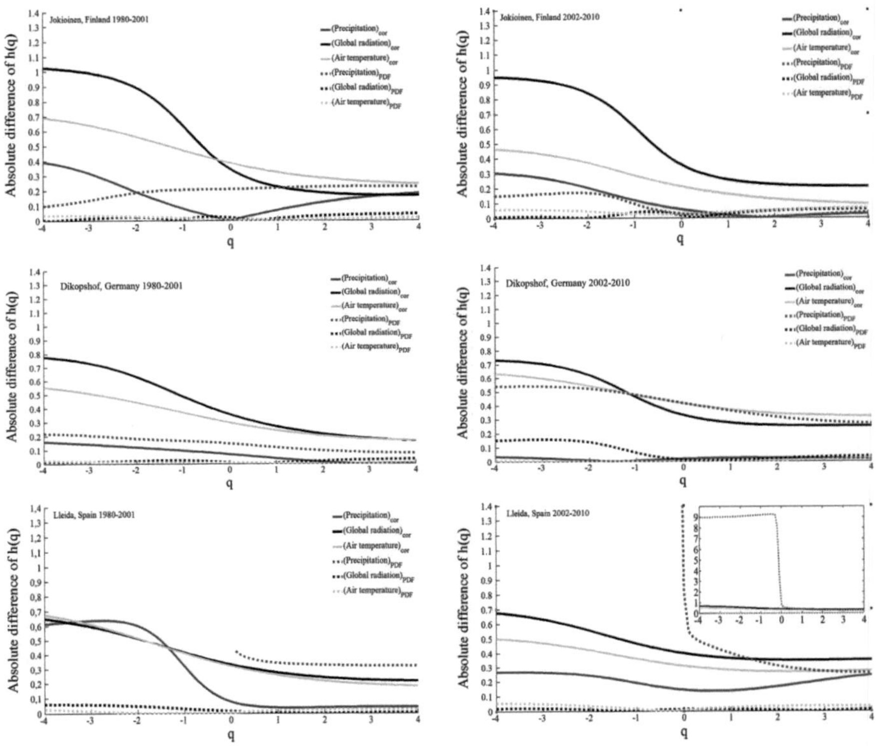

Fig. 5 Absolute differences of Hurst exponents for original and shuffled data $|h(q)-h_{shuf}(q)| = |h_{cor}(q)|$ and original and surrogate data $|h(q)-h_{sur}(q)| = |h_{PDF}(q)|$ as a function of q for studied meteorological time series, Jokioinen (*two upper plots*, *left panel* for 1980–2001 period and *right* for 2002–2010 period), Dikopshof (*two middle plots*) and Lleida (*two bottom plots*)

noticeable for precipitation, less for air temperature and hardly for global radiation time series. Only change of multifractal spectra parameters for Dikopshof station, especially precipitation time series are not in accordance to trend described above, as α_0 is slightly lower and spectra is changing from being slightly left-skewed to slightly right-skewed shape for the second sub-period (2002–2010), compared to the first one (1980–2001).

Above results show that the multifractal spectrum of precipitation time series deviates significantly from the spectra of other analyzed climate variables and it may be more vulnerable to changes in climate dynamics due to multifractality resulting mainly from the broad probability density function and not the long range correlations. This result is consistent with results from other papers [36, 38, 39].

Also, spatial differentiation of obtained spectra is clearly visible. The width of multifractal spectra for global radiation recorded at Jokioinen station, asymmetry of spectra of precipitation time series for Dikopshof or the width of the precipitation

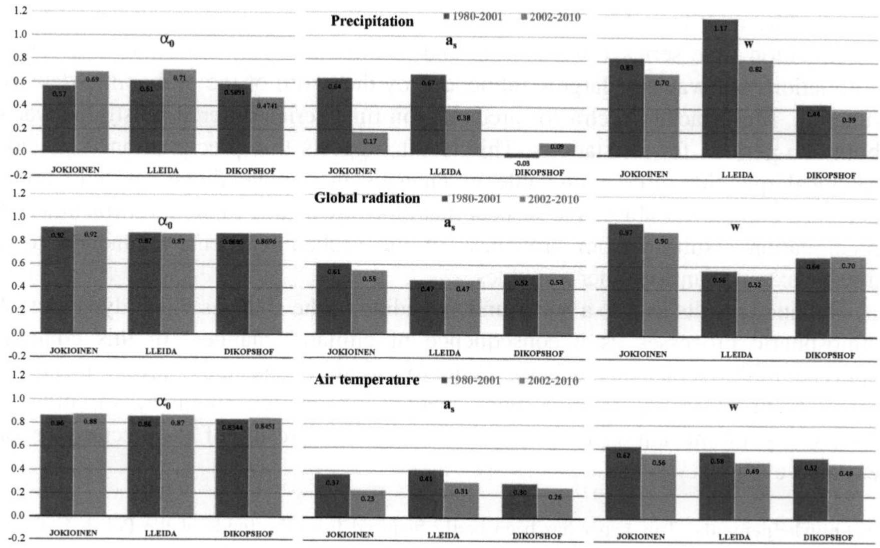

Fig. 6 Comparison of computed parameters (dimensionless) of multifractal spectra from the locations representing European transect and for data divided into 2 separate periods: 1980–2001 and 2002–2010. α_0 is a α-value corresponding to the maximum of the $f(\alpha)$ function; a_s is asymmetry parameter and w means the width of the multifractal spectrum

spectra for all stations are the most obvious examples. This result is clearly indicating that the studied meteorological quantities possess specific space dynamics, which can be attributed to climatic conditions.

4 Conclusions

We applied the MF-DFA technique to investigate the multifractal behavior of the chosen meteorological time series recorded at three locations in Europe lying across an European transect, including different climatic zones. We considered two periods 1980–2001 and 2002–2010 to check whether change in climate dynamics can be observed using the multifractal analysis. Sub-periods were chosen in order to account for climate shifts. The MF-DFA allowed to find not only quantitative information about the "complexity" of the studied series, but also to distinguish considerable differences in spectral properties between locations and analyzed periods. Except from precipitation, the multifractal spectra of the studied meteorological time series exhibit the typical single-humped shape that characterizes multifractal signals. We showed, by comparing the generalized Hurst exponent of the original time series with the shuffled and surrogate time series, that the multifractality of the air temperature and the global radiation is mainly due to the long range correlations. The broadness of the probability density function played a minor role for the multifractality of those time series. In contrast, the multifractality of precipitation time series was two-fold and resulted not only from the long range correlations but was also largely influenced by the width of the probability density function. Multifractal spectra for precipitation time series differed mostly between both sub-periods for all stations. This result suggests that precipitation is meteorological quantity more vulnerable to changes in climate dynamics than other analyzed climate variables. We suspect that this may be due to the multifractality of precipitation resulting from broadness of the probability density function to a greater extent than for other variables.

Obtained results lead to a better understanding of the changes in the dynamics of atmospheric processes as a consequence of climatic changes. In this context, applying multifractal analysis in the future to a more representative grid of locations seems to be promising to give more detailed information about the variation in complexity of the atmospheric processes (e.g. as a result of the occurrence of extreme events).

Acknowledgements This paper has been partly financed from the funds of the Polish National Centre for Research and Development in frame of the projects: LCAgri, contract number: BIOSTRATEG1/271322/3/NCBR/2015 and GyroScan, contract number: BIOSTRATEG2/298782/11/NCBR/2016. We acknowledge Finnish Meteorological Institute (FMI) for delivering us data for Jokioinen site [37]. HH was financially supported by the German Federal Ministry of Food and Agriculture (BMEL) through the Federal Office for Agriculture and Food (BLE), (2851ERA01J).

References

1. Balling, R.C., Vose, R.S., Weber, G.R.: Analysis of long-term European temperature records: 1751–1995. Clim. Res. **10**, 193–200 (1998)
2. Kantelhardt, J.W., Zschiegner, S.A., Koscielny-Bunde, E., Havlin, S., Bunde, A., Stanley, H. E.: Multifractal detrended fluctuation analysis of nonstationary time series. Phys. A **316**(1–4), 87–114 (2002)
3. Higuchi, T.: Approach to an irregular time series on the basis of the fractal theory. Physica D **31**, 277–283 (1988)
4. Kalauzi, A., Spasić, S., Ćulić, M., Grbić, G., Martać, Lj: Consecutive differences as a method of signal fractal analysis. Fractals **13**(4), 283–292 (2005)
5. Schertzer, D., Lovejoy, S.: Multifractal simulation and analysis of clouds by multiplicative process. Atmos. Res. **21**, 337–361 (1988)
6. Kavasseri, R.G., Nagarajan, R.: A multifractal description of wind speed records. Chaos Solitons Fractals **24**, 165–173 (2005)
7. Feng, T., Fu, Z., Deng, X., Mao, J.: A brief description to different multi-fractal behaviors of daily wind speed records over China. Phys. Lett. A **45**, 4134–4141 (2009)
8. Koscielny-Bunde, E., Roman, H.E., Bunde, A., Havlin, S., Schellnhuber, H.J.: Long-range power-law correlations in local daily temperature fluctuations. Philos. Mag. B **77**(5), 1331–1340 (1998)
9. Király, A., Jánosi, I.M.: Detrended fluctuation analysis of daily temperature records: Geographic dependence over Australia. Meteorol. Atmos. Phys. **88**, 119–128 (2005)
10. Bartos, I., Jánosi, I.M.: Nonlinear correlations of daily temperature records over land. Nonlinear Process. Geophys. **13**, 571–576 (2006)

11. Lin, G., Fu, Z.: A universal model to characterize different multi-fractal behaviors of daily temperature records over China. Phys. A **387**, 573–579 (2008)
12. Yuan, N., Fu, Z., Mao, J.: Different multifractal behaviors of diurnal temperature range over the north and the south of China. Theor. Appl. Climatol. **112**, 673–682 (2013)
13. Fraedrich, K., Blender, R.: Scaling of atmosphere and ocean temperature correlations in observations and climate models. Phys. Rev. Lett. **90**, 108501 (2003)
14. Jiang, L., Zhao, J., Li, N., Li, F., Guo, Z.: Different multifractal scaling of the 0 cm average ground surface temperature of four representative weather stations over China. Adv. Meteorol. **2013**, Article ID 341934 (2013)
15. Deidda, R.: Rainfall downscaling in a space-time multifractal framework. Water Resour. Res. **36**, 1779–1794 (2000)
16. García-Marín, A.P., Jiménez-Hornero, F.J., Ayuso, J.L.: Applying multifractality and the self-organised criticality theory to describe the temporal rainfall regimes in Andalusia (southern Spain). Hydrol. Process. **22**, 295–308 (2008)
17. De Lima, M.I.P., de Lima, J.L.M.P.: Investigating the multifractality of point precipitation in the Madeira archipelago. Nonlinear Process. Geophys. **16**, 299–311 (2009)
18. Gemmer, M., Fischer, T., Su, B., Liu, L.L.: Trends of precipitation extremes in the Zhujiang River Basin. South China J. Clim. **24**, 750–761 (2011)
19. Lovejoy, S., Pinel, J., Schertzer, D.: The global space—time cascade structure of precipitation: satellites, gridded gauges and reanalyses. Adv. Water Resour. **45**, 37–50 (2012)
20. Jimenez-Hornero, F.J., Jimenez-Hornero, J.E., de Rave, E.G., Pavon-Dominguez, P.: Exploring the relationship between nitrogen dioxide and ground-level ozone by applying the joint multifractal analysis. Environ. Monit. Assess. **167**, 675–684 (2010)
21. Murcio, R., Masucci, A.P., Arcaute, E., Batty, M.: Multifractal to monofractal evolution of the London street network. Phys. Rev. E **92**, 062130 (2015)
22. Yu, Z.-G., Leung, Y., Chen, Y.D., Zhang, Q., Anh, V., Zhou, Y.: Multifractal analyses of daily rainfall time series in Pearl River basin of China. Phys. A **405**, 193–202 (2014)
23. Valencia, J.L., Requejo, A.S., Gasco, J.M., Tarquis, A.M.: A universal multifractal description applied to precipitation patterns of the Ebro River Basin. Spain. Clim. Res. **44**, 17–25 (2010)
24. Veneziano, D., Langousis, A., Furcolo, P.: Multifractality and rainfall extremes: a review. Water Resour. Res.**42**, W06D15 (2006)
25. Venugopal, V., Roux, S.G., Foufoula-Georgiou, E., Arneodo, A.: Revisiting multifractality of high-resolution temporal rainfall using a wavelet-based formalism. Water Resour. Res. **42**, W06D14 (2006)
26. Yonghe, L., Kexin, Z., Wanchang, Z., Yuehong, S., Hongqin, P., Jinming, F.: Multifractal analysis of 1 min summer rainfall time series from a monsoonal watershed in eastern China. Theor. Appl. Climatol. **111**, 37–50 (2013)
27. Rodríguez, R., Casas, M.C., Redaño, A.: Multifractal analysis of the rainfall time distribution on the metropolitan area of Barcelona (Spain). Meteorol. Atmos. Phys. **121**, 181–187 (2013)
28. Huntingford, C., Jones, P.D., Livinia, V.N., Lenton, T.M., Cox, P.M.: No increase in global temperature variability despite changing regional patterns. Nature **500**, 327–330 (2013)
29. Swanson, K.L., Tsonis, A.A.: Has the climate recently shifted? Geophys. Res. Lett. **36**, L06711 (2009)
30. Dee, D., Uppala, S., et al.: The ERA-Interim reanalysis: configuration and performance of the data assimilation system. Q. J. R. Meteorol. Soc. **137**, 553–597 (2011)
31. Telesca, L., Lovallo, M.: Analysis of the time dynamics in wind records by means of multifractal detrended fluctuation analysis and the Fisher-Shannon information plane. J. Stat. Mech., P07001 (2011)
32. Theiler, J., Galdrikian, B., Longtin, A., Eubank, S., Farmer, D.J.: Using surrogate data to detect nonlinearity in time series. In: Nonlinear Modeling and Forecasting, pp. 163–188. Addison-Wesley (1992)
33. Min, L., Shuang-Xi, Y., Gang, Z., Gang, W.: Multifractal detrended fluctuation analysis of interevent time series in a modified OFC model. Commun. Theor. Phys. **59**, 1–6 (2013)

34. Mali, P.: Multifractal characterization of global temperature anomalies. Theor. Appl. Climatol. **121**(3), 641–648 (2014)
35. Kantelhardt, J.W., Koscielny-Bunde, E., Rybski, D., Braun, P., Bunde, A., Havlin, S.: Long-term persistence and multifractality of precipitation and river runoff records. J. Geophys. Res. **111**, D01106 (2006)
36. Baranowski, P., Krzyszczak, J., Slawinski, C., Hoffmann, H., Kozyra, J., Nierobca, A., Siwek, K., Gluza, A.: Multifractal analysis of meteorological time series to assess climate impacts. Clim. Res. **65**, 39–52 (2015)
37. Venäläinen, A., Tuomenvirta, H., Pirinen, P., Drebs, A.: A basic finnish climate data set 1961–2000—description and illustration. Finnish Meteorological Institute Reports, vol. 5. Finnish Meteorological Institute, Helsinki, Finland (2005)
38. Krzyszczak, J., Baranowski, P., Zubik, M., Hoffmann, H.: Temporal scale influence on multifractal properties of agro-meteorological time series. Agric. For. Meteorol. **239**, 223–235 (2017)
39. Hoffmann, H., Baranowski, P., Krzyszczak, J., Zubik, M., Sławiński, C., Gaiser, T., Ewert, F.: Temporal properties of spatially aggregated meteorological time series. Agric. For. Meteorol. **234–235**, 247–257 (2017)

Part III
Lineal and Non-linear Time Series Models (ARCH, GARCH, TARCH, EGARCH, FIGARCH, CGARCH etc.)

Comparative Analysis of ARMA and GARMA Models in Forecasting

Thulasyammal Ramiah Pillai and Murali Sambasivan

Abstract In this paper, two traditional Autoregressive Moving Average models and two different Generalised Autoregressive Moving Average models are considered to forecast financial time series. These time series models are fitted to the financial time series data namely Dow Jones Utilities Index data set, Daily Closing Value of the Dow Jones Average and Daily Returns of the Dow Jones Utilities Average Index. Three different estimation methods such as Hannan-Rissanen Algorithm, Whittle's Estimation and Maximum Likelihood Estimation are used to estimate the parameters of the models. Point forecasts have been done and the performance of all the models and the estimation methods are discussed.

Keywords Time series · Generalised autoregressive moving average · Hannan-Rissanen algorithm

1 Introduction

Time series is a set of well defined data items collected at successive points at uniform time intervals [1]. The goal of time series analysis is to predict a series that contains a random component. If this random component is stationary, then we can develop powerful techniques to forecast its future values [2]. Forecasting is important in the fields like finance, meteorology, industry and so forth [3].

It is known that the modelling of time series with changing frequency components is important in many applications. These type of time series cannot be identified using the existing standard time series techniques. However, one may propose the same classical model for all these cases. This may produce poor forecast values [4]. Due to that, Peiris introduced a new class of Autoregressive Moving Average

T.R. Pillai (✉)
School of Computing and Information Technology, Taylor's University, Subang Jaya, Malaysia
e-mail: thulasyammal.ramiahpillai@taylors.edu.my

M. Sambasivan
Taylor's Business School, Taylor's University, Subang Jaya, Malaysia

I. Rojas et al. (eds.), *Advances in Time Series Analysis and Forecasting*, Contributions to Statistics, DOI 10.1007/978-3-319-55789-2_9

(ARMA) type models with indices called Generalised ARMA (GARMA) to describe data with different frequency components [5]. This GARMA model can be described as parameter driven model.

Michael et al. extends the work of Zeger and Qadish (1988) and Li (1994) giving rise to the observation driven Generalised Autoregressive Moving Average (GARMA) model [6].

However, this article considers the parameter driven GARMA model which includes the additional parameter as an index. This GARMA model describes some hidden features of a time series. This approach leads to a significant improvement in the quality of forecasts of correlated data.

The property of variation in the density of crossings at a particular point in time series data sets is very common. These series display similar patterns in the autocorrelation function, partial autocorrelation function and the spectrum. The standard ARMA models cannot identify these variations and leads to misclassification problems in time series [5]. This encourages Peiris to introduce the new, generalized version of ARMA models with additional parameters or indices to control the degree of frequency or level of crossings [5]. These class of models cover the traditional ARMA family. Peiris had shown that this model can be utilized to model long memory or nearly long memory time series by suitably choosing the parameters [5].

Peiris had introduced Generalised Autoregressive (GAR(1)) model, defined as,

$$(1 - \alpha B)^{\delta} X_t = Z_t, \tag{1}$$

and the Generalised Moving Average (GMA(1)) model which is given as,

$$X_t = (1 - \beta B)^{\delta} Z_t. \tag{2}$$

These class of models cover the traditional first order Autoregressive (AR(1)) family and the traditional first order Moving Average (MA(1)) family when $\delta = 1$. Peiris had justified the importance of the GAR(1) model in practice using a set of real time series data. Peiris's results (estimates) were closer to the true values than in [5, 7]. Peiris et al. had justified the advantages of GMA(1) model in practice by using a set of real time series data given in series A of [8, 9]. The results obtained from GMA(1) were closer to true values than the traditional MA.

More recently, the GARMA(1, 1; 1, δ) model has been considered which is defined by,

$$(1 - \alpha B)X_t = (1 - \beta B)^{\delta} Z_t, \tag{3}$$

where $-1 < \alpha, \beta < 1$ and $\delta > 0$ [10]. This class of model covers the traditional Autoregressive Moving Average (ARMA(1, 1)) family when $\delta = 1$. The GARMA(1, 1; 1, δ) had been fitted to the forest area in Malaysia and Gross Domestic Product (GDP) in Malaysia [10, 11].

In addition, Shitan and Peiris studied the behaviour of the process GARMA(1, 1; δ, 1) [12]. The GARMA(1, 1; δ, 1) process is generated by,

$$(I - \alpha B)^{\delta} X_t = (I - \beta B) Z_t, \tag{4}$$

where $-1 < \alpha, \beta < 1$ and $\delta > 0$. Shitan had given two examples to illustrate the GARMA(1, 1; δ, 1) modelling [12].

The GARMA(1, 1; 1, δ) and GARMA(1, 1; δ, 1) models can be further generalised as follows:

$$(1 - \alpha B)^{\delta_1} X_t = (1 - \beta B)^{\delta_2} Z_t, \tag{5}$$

where $-1 < \alpha, \beta < 1$, $\delta_1 > 0$ and $\delta_2 > 0$. This model is denoted by GARMA(1, 1; δ_1, δ_2) and some properties of this model have been established [13].

It is interesting to note that the GARMA model can be further expanded to GARMA(1, 2; δ, 1) and it is given as below:

$$(1 - \alpha B)^{\delta} X_t = (1 - \beta_1 B - \beta_2 B^2) Z_t, \tag{6}$$

where $-1 < \alpha, \beta_1, \beta_2 < 1$ and $\delta > 0$. All these models have been shown to be useful in modelling time series data. GARMA(1, 2; δ, 1) performs better than ARMA(1, 1) for GDP data set of Malaysia [14]. Pillai had successfully illustrated the superiority, usefulness and applicability of the GARMA(1, 2; δ, 1) model using GDP data set of Malaysia [14].

The objective of this paper is to compare the performance of the ARMA and GARMA models and to compare the three estimation methods. The estimation methods are discussed in Sect. 2. In Sect. 3, we illustrate the applications of ARMA and GARMA modelling to a financial time series data namely Dow Jones Utilities Index data set (August 28–December 18, 1972). We compare the performance of the ARMA(1, 1) and GARMA(1, 2; δ, 1) by using forecast values of the Daily Closing Value of Dow Jones Average (December 31, 2014–January 4, 2016) in Sect. 4. While, in Sect. 5 Daily Returns of the Dow Jones Utilities Average Index (January 1, 2015–May 5, 2016) data is fitted to ARMA and GARMA models. Finally, the conclusions are drawn in Sect. 6.

2 Estimation of Parameters

There are many estimation methods to estimate the parameters of ARMA models. The proposed preliminary estimation for GARMA models to obtain Whittle's Estimation (WE) and Maximum Likelihood Estimation (MLE) is Hannan-Rissanen Algorithm (HRA). We use the HRA technique to find a suitable set of starting up values for the WE and MLE.

2.1 Hannan-Rissanen Algorithm (HRA)

The estimation of the parameters of AR was done using Burg's algorithm [2]. Burg's algorithm usually gives higher likelihoods than the Yule-Walker (YW) equations for pure AR models [2]. The innovations algorithm gives slightly higher likelihoods than the HRA algorithm for MA models. However, the Hannan-Rissanen Algorithm (HRA) is usually successful for mixed models such as ARMA [2]. Hence, the initial start up values for the numerical minimization are obtained using the HRA method.

The HRA technique is one of the preliminary techniques used to estimate the parameters of ARMA models where $p > 0$ and $q > 0$ [2]. We have to do some modifications to the parameters of ARMA (p, q) that obtained from the HRA estimation method to suit the parameters of GARMA models. The HRA estimation values will be used as a start up values for WE and MLE estimations.

2.2 Whittle's Estimation (WE)

The Whittle's Estimator (WE) is considered as an accurate estimator [2]. In this section, we discuss the Whittle's Estimation of the parameters of the GARMA models. The Whittle's estimates were obtained by minimizing the function,

$$\ln\left(\frac{1}{T}\sum_{j}\frac{I_T(w_j)}{g(w_j)}\right)+\frac{1}{T}\sum_{j}\ln g(w_j) \tag{7}$$

where $I_T(w_j)$ is the periodogram of the series given by,

$$I_T(w_j = \frac{2\pi j}{T}) = \frac{1}{T}|\sum_{s=1}^{T} x_s \exp(-i\frac{2\pi sj}{T})|^2 \tag{8}$$

and $g(w_j = \frac{2\pi j}{T}) = \frac{|1-\beta_1\exp(-i\frac{2\pi j}{T})-\beta_2\exp(-i\frac{4\pi j}{T})|}{|1-\alpha\exp(-i\frac{2\pi j}{T})|^{2\delta}}, j = -[\frac{T-1}{2}], ..., [\frac{T}{2}]$ is the true spectrum of the process [2]. The corresponding estimate for σ^2 is given as,

$$\hat{\sigma}^2 = \frac{1}{T}\sum_{j}\frac{I_T(w_j)}{g(w_j)}. \tag{9}$$

2.3 Maximum Likelihood Estimation (MLE)

The Maximum Likelihood Estimator (MLE) is a popular method of parameter estimation and is an indispensable tool for many statistical techniques [15]. The Maximum Likelihood Estimates (MLE) for the parameters of the GARMA models are obtained by numerically minimizing the function,

$$-2\ln f(x) = T\ln(2\pi) + \ln|\Sigma| + x'\Sigma^{-1}x,$$

where T is the number of observations, x is the observed vector and Σ denotes the covariance matrix. The entries of Σ are the autocovariance of the model which consists of the parameters to be estimated [2].

In Sect. 3, we use these estimation methods in a real time series data set and illustrate the applicability and superiority of GARMA models over the traditional ARMA models.

3 Applications of ARMA and GARMA Modelling to Dow Jones Utilities Index Data Set

In this section, two examples of ARMA modelling and two examples of GARMA modelling are given. The time series data that have been taken into consideration is the Dow Jones Utilities Index data set (August 28–December 18, 1972) [2]. Brockwell and Davis differenced this data set at lag 1 and the mean was corrected to stationarize it [2]. Forecast point for ARMA and GARMA models are given as:

$$X_t = \Sigma_{k=0}^{\infty} \Psi_k Z_{t-k} \tag{10}$$

where Ψ_k depends on the models.

The Ψ_k values for ARMA and GARMA models are given. Apparently, the $\psi_0 = 1$ and $\psi_k = \alpha^k$ for AR(1) while the $\psi_0 = 1$ and $\psi_k = (\alpha - \beta)\alpha^{k-1}$ for $k \geq 1$ for ARMA (1, 1) [2]. It is obvious that the $\psi_0 = 1$ and $\psi_k = \alpha^k + \sum_{j=1}^{k} \tau_j \beta^j \alpha^{k-j}$ for $k \geq 1$ for GARMA (1, 1; 1, δ) and finally, the $\psi(0) = 1$, $\psi(1) = \pi_1\alpha - \beta_1\pi_0$ and $\psi(k) = \pi_k\alpha^k - \beta_1\pi_{k-1}\alpha^{k-1} - \beta_2\pi_{k-2}\alpha^{k-2}$ for $k \geq 2$ for GARMA(1, 2; δ, 1) [13]. The first three point forecasts and the corresponding 95% confidence interval will be obtained.

3.1 *First-Order Autoregression (AR(1))*

AR(1) model was fitted to the Dow Jones Utilities Index data set that has been differenced and mean corrected. Brockwell and Davis fitted an AR(1) model as follows:

$$(1 - 0.4219B)Y_t = Z_t, Z_t \sim WN(0, 0.1479) \tag{11}$$

where, $Y_t = (1 - B)(X_t - 0.1336)$ using the YW estimation method [2]. In other words, Y_t is the differenced and mean corrected data. They obtained

Table 1 First three forecasts and the 95% confidence interval of Dow Jones data using AR(1) model and ARMA(1, 1) model

	AR(1)	AR(1)	AR(1)	ARMA(1,1)	ARMA(1, 1)	ARMA(1, 1)
	HRA	WE	MLE	HRA	WE	MLE
1	51.75 ± 0.2383	53.62 ± 0.2256	54.85 ± 0.0340	62.55 ± 0.2380	59.00 ± 0.1800	81.40 ± 0.0668
2	51.78 ± 0.2807	53.65 ± 0.2687	54.87 ± 0.0408	62.52 ± 0.2608	58.08 ± 0.1912	81.59 ± 0.0951
3	51.83 ± 0.2882	53.70 ± 0.2770	54.93 ± 0.0422	62.57 ± 0.2719	59.01 ± 0.1969	81.65 ± 0.1080

$$(1 - 0.4371B)Y_t = Z_t, Z_t \sim WN(0, 0.1423) \tag{12}$$

model using the Burg's estimation method and

$$(1 - 0.4471B)Y_t = Z_t, Z_t \sim WN(0, 0.0217) \tag{13}$$

model using the MLE method [2]. Point forecast for the stationarized Dow Jones data set for the next three time periods ahead and the corresponding 95% forecast intervals were obtained and are listed in the Table 1. Point forecasts obtained from MLE are closer to true values than the other methods. MLE gives the best forecast values for AR(1) compared to YW and Burg's estimation methods.

3.2 ARMA(1, 1)

ARMA(1, 1) model was fitted to the Dow Jones Utilities Index data set that has been differenced and mean corrected. The HRA estimation is obtained for the ARMA(1, 1) model and the fitted model is

$$(1 - 0.6436B)Y_t = (1 - 0.2735B)Z_t \tag{14}$$

where $Z_t \sim WN(0, 0.1477)$. While, the ARMA(1, 1) fitted models are,

$$(1 - 0.6703B)Y_t = (1 - 0.3655B)Z_t \tag{15}$$

where $Z_t \sim WN(0, 0.1053)$, by the WE estimation method and

$$(1 - 0.5772B)Y_t = (1 + 0.2602B)Z_t \tag{16}$$

where $Z_t \sim WN(0, 0.0676)$, by the MLE method.

Point forecast for the stationarized Dow Jones data set for the next three time periods ahead and the forecast intervals were obtained and are listed in the Table 1. MLE estimation method gives a better forecast values compared to HRA and WE for ARMA(1, 1) model as in the case of AR(1).

3.3 GARMA(1, 1; 1, δ)

GARMA(1, 1; 1, δ) model was fitted to the Dow Jones Utilities Index data set that has been differenced and mean corrected. The HRA estimation is obtained for the GARMA(1, 1; 1, δ) model and the fitted model is,

$$(1 - 0.9895B)Y_t = (1 + 0.7798B)^{0.7798}Z_t \tag{17}$$

where $Z_t \sim WN(0, 0.3846)$. Whereas, the GARMA(1, 1; 1, δ) fitted models are,

$$(1 - 0.9982B)Y_t = (1 + 0.9999B)^{0.5066}Z_t \tag{18}$$

where $Z_t \sim WN(0, 18.2033)$, by the WE estimation method and

$$(1 - 0.9048B)Y_t = (1 + 0.6864B)^{0.9434}Z_t \tag{19}$$

where $Z_t \sim WN(0, 0.0713)$, by the MLE method. Point forecast for the stationarized Dow Jones data set for the next three time periods ahead and the forecast intervals were obtained and are listed in the Table 2. HRA provides better forecast values and followed by MLE and finally WE estimation method for GARMA(1, 1; 1, δ). GARMA(1, 1; 1, δ) results are closer to the true values than the traditional AR(1) and ARMA(1, 1).

Table 2 First three forecasts and the 95% confidence interval of Dow Jones data using GARMA(1, 1; 1, δ)(GARMA1) model and GARMA(1, 2; δ, 1)(GARMA2) model

	GARMA1 HRA	GARMA1 WE	GARMA1 MLE	GARMA2 HRA	GARMA2 WE	GARMA2 MLE
1	121.36 ± 0.7477	111.75 ± 34.1378	115.03 ± 0.1394	122.94 ± 0.2463	122.84 ± 0.0694	122.94 ± 0.1440
2	122.29 ± 0.7527	111.79 ± 35.6778	115.97 ± 0.1396	122.89 ± 0.2463	122.78 ± 0.0694	122.88 ± 0.1440
3	121.84 ± 0.7522	111.90 ± 35.4022	115.54 ± 0.1400	122.88 ± 0.2463	122.82 ± 0.0694	122.88 ± 0.1440

3.4 GARMA(1, 2; δ, 1)

GARMA(1, 2; δ, 1) model was fitted to the Dow Jones Utilities Index data set that has been differenced and mean corrected. The HRA estimation is obtained for the GARMA(1, 2; δ, 1) model and the fitted model is,

$$(1 - 0.4983B)^{0.4983} Y_t = (1 + 0.1949B + 0.3711B^2)Z_t \tag{20}$$

where $Z_t \sim WN(0, 0.1698)$. While, the GARMA(1, 2; δ, 1) fitted model is,

$$(1 - 0.9038B)^{0.6344} Y_t = (1 - 0.2832B - 0.1356B^2)Z_t \tag{21}$$

where $Z_t \sim WN(0, 0.0473)$, by the WE estimation. The GARMA(1, 2; δ, 1) fitted model is,

$$(1 - 0.5173B)^{0.0811} Y_t = (1 + 0.2217B + 0.3671B^2)Z_t, \tag{22}$$

where $Z_t \sim WN(0, 0.0811)$, by the MLE method.

Point forecasts for the Dow Jones data set for the next three time periods and the forecast intervals are shown in Table 2. It can be seen from Table 2 that all the point forecasted values through HRA, WE and MLE estimation give a very close reading to the actual values. In this case, HRA gives the best forecast values because the true values fall in the given confidence interval.

3.5 Comparison of Performance of ARMA and GARMA Models in Forecasting of Dow Jones Utilities Index Data Set

The performance of the ARMA and GARMA models are compared using the 95% confidence intervals of the first three forecasts, the mean estimated bias (EB) of the point forecasts and the mean absolute percent error (MAPE). The first three forecasts and the corresponding 95% confidence intervals are given in Tables 1 and 2. It can be seen from the confidence interval that the GARMA results are closer to true values than ARMA models. The point forecast using AR(1) gives a poor forecast values. The point forecast values using ARMA(1, 1) model is better than AR(1) model. GARMA(1, 1; 1, δ) model gives better forecast values compared to the AR(1) and ARMA(1, 1) models. GARMA(1, 2; δ, 1) gives the best forecast values compared to the other models. The EB and the MAPE values of the point forecasts of all the models and estimation methods are given in the Table 3. As for the MAPE values, the values in the GARMA models are much smaller when compared with ARMA models.

We have evaluated the performance of the three estimators based on HRA, WE and MLE. It appears from this study, the MLE estimation procedure is relatively good

Table 3 Estimated bias and mean absolute percent error of the point forecast values for Dow Jones data

	HRA EB	HRA MAPE	YW EB	YW MAPE	WE EB	WE MAPE	BU EB	BU MAPE	MLE EB	MLE MAPE
AR(1)			71	0.5773			69	0.5620	67	0.5520
ARMA (1, 1)	60	0.4893			64	0.5183			41	0.3345
GARMA (1, 1; 1, δ)	0.71	0.0058			10.52	0.0860			6.96	0.0570
GARMA (1, 2; δ, 1)	0.51	0.0042			0.38	0.0031			0.51	0.0041

for AR(1) and ARMA(1, 1). HRA estimation method performs better for GARMA(1, 1; 1, δ) and GARMA(1, 2; δ, 1) model because the true values fall in the given confidence interval.

We can conclude that the GARMA models performance better than ARMA models. Furthermore, higher order GARMA(1, 2; δ, 1) out performs the other models for all the estimation methods.

4 Applications of ARMA and GARMA Modelling to Daily Closing Value of the Dow Jones Average

In this section, an example of ARMA modelling and an example of GARMA modelling are given. The time series data that have been taken into consideration is the Daily Closing Value of the Dow Jones Average (DCVDJA) data set (December 31, 2014–January 4, 2016) [16]. This data set was differenced at lag 1 and the mean was corrected. The point forecast and the 95% confidence interval of the data set using ARMA(1, 1) and GARMA(1, 2; δ, 1) models are given in the Table 4. The EB and the MAPE values of the point forecasts of ARMA(1, 1) and GARMA(1, 2; δ, 1) are given in the Table 5.

Table 4 First three forecasts and the 95% confidence interval of the DCVDJA using ARMA(1, 1) model and GARMA(1, 2; δ, 1)

ARMA HRA	ARMA WE	ARMA MLE	GARMA HRA	GARMA WE	GARMA MLE
5634 ± 162	2976 ± 50103	5635 ± 162	17518 ± 475	17551 ± 8208	17320 ± 1742
5775 ± 174	2981 ± 50143	5775 ± 174	17520 ± 475	17552 ± 8208	17323 ± 1742
5627 ± 185	2983 ± 50164	5627 ± 185	17520 ± 475	17534 ± 8208	17328 ± 1742

Table 5 Estimated bias mean absolute percent error of the point forecast values for DCVDJA data

	HRA	HRA	WE	WE	MLE	MLE
	EB	MAPE	EB	MAPE	EB	MAPE
ARMA (1, 1)	11788	0.6737	14510	0.8293	11788	0.6737
GARMA (1, 2; δ, 1)	132	0.0076	134	0.0077	229	0.0131

4.1 ARMA(1, 1)

ARMA(1, 1) model was fitted to the Daily Closing Value of the Dow Jones Average data set that has been differenced and mean corrected. The HRA estimation is obtained for the ARMA(1, 1) model and the fitted model is

$$(1 - 0.9640B)Y_t = (1 + 0.0427B)Z_t \tag{23}$$

where $Z_t \sim WN(0, 1269.416)$. While, the ARMA(1, 1) fitted models are,

$$(1 - 0.7731B)Y_t = (1 - 0.8198B)Z_t \tag{24}$$

where $Z_t \sim WN(0, 25609.18)$, by the WE estimation method and

$$(1 - 0.9640B)Y_t = (1 + 0.0427B)Z_t \tag{25}$$

where $Z_t \sim WN(0, 1269.416)$, by the MLE method.

4.2 GARMA(1, 2; δ, 1)

GARMA(1, 2; δ, 1) model was fitted to the Daily Closing Value of the Dow Jones Average data set that has been differenced and mean corrected. The HRA estimation is obtained for the GARMA(1, 2; δ, 1) model and the fitted model is,

$$(1 - 0.2459B)^{0.2459}Y_t = (1 + 0.1670B + 0.1336B^2)Z_t \tag{26}$$

where $Z_t \sim WN(0, 257.6487)$. While, the GARMA(1, 2; δ, 1) fitted model is,

$$(1 - 0.3378B)^{2.2985}Y_t = (1 - 0.9136B + 0.0861B^2)Z_t \tag{27}$$

where $Z_t \sim WN(0, 4210.847)$, by the WE estimation. The GARMA(1, 2; δ, 1) fitted model is,

$$(1 - 0.5527B)^{0.6526} Y_t = (1 - 0.3806B - 0.4736B^2) Z_t, \tag{28}$$

where $Z_t \sim WN(0, 891.5095)$, by the MLE method.

The performance of the ARMA(1, 1) and GARMA(1, 2; δ, 1) models are compared using the 95% confidence intervals of the forecasts, EB and MAPE. The 95% confidence intervals of the first three forecasts are given in Table 4. The GARMA(1, 2; δ, 1) gives better forecast values compared to the ARMA(1, 1) forecast values. The true values fall in the given confidence interval of the GARMA(1, 2; δ, 1) model. HRA estimation method performs better for GARMA(1, 2; δ, 1) model because it provides shorter forecast intervals. It can be seen clearly from Table 5 that the MAPE values of GARMA(1, 2; δ, 1) model is much smaller than ARMA (1, 1) model. The GARMA(1, 2; δ, 1) is far better than ARMA (1, 1) when compared using the MAPE values and the confidence interval.

5 Applications of ARMA and GARMA Modelling to Daily Total Return of the Dow Jones Utility Average

In this section, an example of ARMA modelling and an example of GARMA modelling are given. The time series data that have been taken into consideration is the Daily Total Return of the Dow Jones Utility Average (DRDJ) data set (January 1, 2015–May 5, 2016) [17]. This data set was differenced at lag 2 and the mean was corrected. The point forecast and the 95% of the confidence interval of the data set using ARMA(1, 1) and GARMA(1, 2; δ, 1) models are given in the Table 6. The EB and MAPE values and the estimation methods are given in the Table 7.

Table 6 First three forecasts and the 95% confidence interval of daily return of the Dow Jones data using ARMA(1, 1) model and GARMA(1, 2; δ, 1)

ARMA	ARMA	ARMA	GARMA	GARMA	GARMA
HRA	WE	MLE	HRA	WE	MLE
2458.80 ± 162.46	969.45 ± 7891.63	2458.93 ± 162.46	2522.509 ± 47.2737	2523.308 ± 40.8944	2537.559 ± 25.9643
2448.10 ± 173.99	971.57 ± 7934.99	2448.13 ± 173.99	2528.545 ± 47.2737	2526.819 ± 40.8944	2536.402 ± 25.9643
2462.50 ± 185.02	974.92 ± 7957.04	2462.63 ± 185.02	2538.083 ± 47.2737	2533.185 ± 40.8944	2547.512 ± 25.9643

Table 7 Mean estimated bias and mean absolute percent error of the point forecast values for daily return of Dow Jones utilities index

	HRA EB	HRA MAPE	WE EB	WE MAPE	MLE EB	MLE MAPE
ARMA (1, 1)	95.19	0.0370	1592.01	0.6193	95.07	0.0370
GARMA (1, 2; δ, 1)	14.70	0.0057	19.61	0.0076	19.43	0.0076

5.1 ARMA(1, 1)

ARMA(1, 1) model was fitted to the Daily Return of the Dow Jones Utilities Average Index data set that has been differenced and mean corrected. The HRA estimation is obtained for the ARMA(1, 1) model and the fitted model is

$$(1 - 0.9640B)Y_t = (1 + 0.0427B)Z_t \tag{29}$$

where $Z_t \sim WN(0, 1269.416)$. While, the ARMA(1, 1) fitted models are,

$$(1 - 0.7675B)Y_t = (1 - 0.9127B)Z_t \tag{30}$$

where $Z_t \sim WN(0, 4067.554)$, by the WE estimation method and

$$(1 - 0.9640B)Y_t = (1 + 0.0427B)Z_t \tag{31}$$

where $Z_t \sim WN(0, 1269.416)$, by the MLE method.

5.2 GARMA(1, 2; δ, 1)

GARMA(1, 2; δ, 1) model was fitted to the Daily Return of the Dow Jones Utilities Average Index data set that has been differenced and mean corrected. The HRA estimation is obtained for the GARMA(1, 2; δ, 1) model and the fitted model is,

$$(1 - 0.9810B)^{0.9810}Y_t = (1 - 0.1396B + 0.9056B^2)Z_t \tag{32}$$

where $Z_t \sim WN(0, 76.3224)$. While, the GARMA(1, 2; δ, 1) fitted model is,

$$(1 - 0.9697B)^{0.5405}Y_t = (1 - 0.9999B + 0.0049B^2)Z_t \tag{33}$$

where $Z_t \sim WN(0, 494.9163)$, by the WE estimation. The GARMA(1, 2; δ, 1) fitted model is,

$$(1 - 0.5431B)^{0.9733} Y_t = (1 - 0.2282B + 0.99996B^2) Z_t, \quad (34)$$

where $Z_t \sim WN(0, 76.3279)$, by the MLE method.

The GARMA(1, 2; δ, 1) is better than ARMA(1, 1) when compared using the the confidence interval, EB and MAPE values. The true values fall in the given confidence interval of the ARMA(1, 1) and GARMA(1, 2; δ, 1) models. However, GARMA(1, 2; δ, 1) results provide shorter forecast intervals in all the estimation methods. MLE estimation method performs better for GARMA(1, 2; δ, 1) model because it provides shorter forecast intervals. The MAPE values of GARMA(1, 2; δ, 1) model is much smaller than ARMA(1, 1) model.

The above three examples illustrate the ARMA and GARMA modelling. In the first example HRA performs better than the other estimation methods for GARMA(1, 2; δ, 1) model because the true values fall in the given confidence interval. In the second example HRA performs better than the other estimation methods for GARMA(1, 2; δ, 1) model because it provides shorter forecast intervals. In the third example, MLE performs better for ARMA(1, 1) and GARMA(1, 2; δ, 1) model. GARMA(1, 2; δ, 1) model is better than ARMA(1, 1) because it provides shorter forecast intervals. It seems that there is no single estimation method that uniformly out performs the other for all the parameter values of the models.

6 Conclusion

The objective of our study is to evaluate the performance of ARMA and GARMA models in forecasting. GARMA(1, 2; δ, 1) model gives a closer reading to the actual values compared to the other models. We have successfully illustrated the usefulness, applicability and superiority of GARMA(1, 2; δ, 1) model using the Dow Jones data set, Daily Closing Value of the Dow Jones Average and Daily Total Return of the Dow Jones Utility Average. GARMA(1, 2; δ, 1) or generally GARMA should be used as an alternative to ARMA to get better performance. The authors are currently using GARMA models in the medical field to justify the importance and the advantages of these type of models.

Acknowledgements This research work was supported by the Fundamental Research Grant Scheme under Ministry of Education Malaysia (FRGS/1/2014/SG04/TAYLOR/02/1).

References

1. Prapanna, M., Labani, S., Saptarsi, G.: Study of effectiveness of time series modeling (ARIMA) in forecasting stock prices. Int. J. Comput. Sci. Eng. Appl. **4**(2), 13–29 (2014)
2. Brockwell, P.J., Davis, R.A.: Introduction to Time Series and Forecasting, 2nd edn. Springer, New York (2001)

3. Chen, S., Lan, X., Hu, Y., Liu, Q., Deng, Y.: The time series forecasting: from the aspect of network. arXiv preprint arXiv:1403.1713 (2014)
4. Peiris, S., Thavaneswaran, A.: An Introduction to volatility models with indices. Appl. Math. Lett. **20**, 177–182 (2006)
5. Peiris, M.S.: Improving the quality of forecasting using generalized AR models: an application to statistical quality control. Stat. Method **5**(2), 156–171 (2003)
6. Michael, A.B., Robert, A.R., Mikis, D.S.: Generalized autoregressive moving average models. J. Am. Stat. Assoc. **98**(461), 214–223 (2003)
7. Abraham, B., Ledolter, J.: Statistical Methods for Forecasting. John Wiley, New York (1983)
8. Peiris, S., Allen, D., Thavaneswaran, A.: An introduction to generalized moving average models and applications. J. Appl. Stat. Sci. **13**(3), 251–267 (2004)
9. Box, G.E.P., Jenkins, G.M.: Time Series: Forecasting and Control. Holden-Day, San Francisco (1976)
10. Pillai, T.R., Shitan, M.: Application of GARMA(1, 1; 1, δ) model to GDP in Malaysia: an illustrative example. J. Glob. Bus. Econ. **3**(1), 138–145 (2011)
11. Pillai, T.R., Shitan, M.: An illustration of generalized ARMA (GARMA) time series modeling of forest area in Malaysia. Int. J. Mod. Phys. Conf. Series **9**, 390–397 (2012)
12. Shitan, M., Peiris, S.: Time series properties of the class of generalized first-order autoregressive processes with moving average errors. Commun. Stat. Theory Method **40**, 2259–2275 (2011)
13. Pillai, T.R., Shitan, M., Peiris, S.: Some properties of the generalized autoregressive moving average (GARMA(1, 1; δ_1, δ_2)) model. Commun. Stat. Theory Method **4**(41), 699–716 (2012)
14. Pillai, T.R.: Generalized autoregressive moving average models: an application to GDP in Malaysia. Third Malaysia Statistics Conference—MYSTATS (2015)
15. Myung, J.: Tutorial on maximum likelihood estimation. J. Math. Psychol. **47**, 90–100 (2003)
16. Daily Closing Value of the Dow Jones Average in the United States. https://measuringworth.com/DJA/result.php
17. Daily Total Return of the Dow Jones Utility Average. http://www.djaverages.com/?go=utility-index-data

SARMA Time Series for Microscopic Electrical Load Modeling

Martin Hupez, Jean-François Toubeau, Zacharie De Grève and François Vallée

Abstract In the current context of profound changes in the planning and operations of electrical systems, many *Distribution System Operators (DSOs)* are deploying *Smart Meters* at a large scale. The latter should participate in the effort of making the grid smarter through active management strategies such as storage or demand response. These considerations involve to model electrical quantities as locally as possible and on a sequential basis. This paper explores the possibility to model *microscopic loads* (individual loads) using *Seasonal Auto-Regressive Moving Average (SARMA)* time series based solely on Smart Meters data. A systematic definition of models for 18 customers has been applied using their consumption data. The main novelty is the qualitative analysis of complete SARMA models on different types of customers and an evaluation of their general performance in an LV network application. We find that residential loads are easily captured using a single SARMA model whereas other profiles of clients require segmentation due to strong additional seasonalities.

Keywords SARMA · Smart metering · Low voltage distribution networks · Microscopic load modeling

1 Introduction

In the last ten years, electrical systems have been undergoing dramatic changes. This is in fact the whole electricity sector that faces a revolution. Issues such as the need to reduce greenhouse gases (Kyoto protocol, EU 20/20/20 objective etc.), the distrust in nuclear energy generation and the growing integration of renewable energies combined with the deregulation of electricity markets have led to profound changes in the structure and operations all along the electricity supply chain.

M. Hupez (✉) · J.-F. Toubeau · Z. De Grève · F. Vallée
Electrical Power Engineering Unit, University of Mons, Mons, Belgium
e-mail: martin.hupez@umons.ac.be

I. Rojas et al. (eds.), *Advances in Time Series Analysis and Forecasting*,
Contributions to Statistics, DOI 10.1007/978-3-319-55789-2_10

Henceforth, electricity systems, both at distribution and transmission levels must adapt and anticipate this transformation. While transmission has yet been going through this adaptation process for over a decade, it appears that DSOs must now face many new challenges as well. Indeed, the penetration rate of decentralized and stochastic energy sources is expected to rise, and a large proportion of it should be integrated on the distribution level. Those present a strong random behavior resulting in periods of high injection and others of nearly no production at all. The distribution network was not designed nor sized for such conditions and technical issues such as overvoltages and congestions may arise. Moreover, there is an increasing risk of electricity shortages due to the change of production mix that includes a growing proportion of green energies and less traditional and stable production units.

A response by DSOs to address those concerns is to operate the network more actively. To this point, DSOs usually have very little measurements in order to implement more dynamic strategies. Electricity market related companies depend mostly on *Synthetic Load Profiles (SLPs)* for residential, administration and small business users. Those can reflect quite well clients' behavior when they are aggregated but they have strong limitations when it comes to more local matters. Smart Meters aim to resolve these concerns. More than enhancing new monitoring and billing opportunities for utility companies, some technical issues could be overcome. In this respect, many DSOs consider the large-scale deployment of Smart Meters in the next few years. This is an important step in achieving smarter grids. In addition, it is a key point for power utilities to allow the reduction in peak load through economic incentives (adaptive pricing, demand response) and technical solutions (curtailment, storage). This reduction would help postponing or preventing large infrastructure investments.

In this context, modeling local consumption and generation will be decisive. An efficient power system planning should determine the critical nodes and penetration rate of decentralized units for which power quality can be assured. Indeed, DSOs are responsible to keep steady-state voltages within certain limits. A microscopic approach, i.e. client by client, is therefore better suited. Studies have shown that a stochastic load flow framework is more appropriate than analytical methods because it allows for the increasing statistical behavior of units connected to the network to be considered [1]. Analytical methods tend to tremendously oversize solutions as they rely on worst case scenarios.

Previous work of Low Voltage (LV) network analysis used pseudo-sequential *Monte Carlo simulation* based on real Smart Meters data. For each time step, each client is assigned one (consumer) or two (prosumer) Cumulative Distribution Functions (CDF) established on real measurements. Monte Carlo sampling is then performed in order to highlight the system state statistics [2, 3]. While allowing the assessment of voltage profiles at all nodes of the LV network it cannot reproduce the temporal dependence pattern as the sampling is performed independently for each interval.

Load control techniques such as demand response and storage management are made possible using sequential models. The sequential approach is necessary when considering processes for which the dynamic is explicitly dependent of the time. The dependence between times steps in the simulation must indeed be expressed.

Sequential models applied to electrical loads forecasting are highly popular as they are decisive for accurate power system planning and operation. The literature, however, focus mostly on forecasting aggregated load profiles at the level of a large substation or even a whole electrical system. With the ongoing effort of making grids smarter and with the increasing penetration of dispersed generation on the distribution level, there is a recent need in modeling consumers' loads individually and sequentially. While in the past, the possibilities were constrained by the lack of data, the large-scale deployment of Smart Meters opens a whole new perspective. A recent contribution explores a Markov-Chains model [4] for different classes of home load. More advanced methods combine demand profiles with electrical characteristics in order to obtain detailed time-varying models [5].

Those advanced approaches usually require a set of customer information which can be difficult and laborious to obtain (type of heating, devices, user habits …). Our purpose of probabilistic distribution system simulation requires simple models with representative transition patterns. To this end, we explored the use of time series, more specifically the Seasonal Auto-Regressive Moving Average (SARMA). Existing examples of SARMA models are applied on forecasting aggregated load values of 24 homes [6]. Another paper [7] benchmarks a SARMA model with other Machine Learning techniques in order to describe the load forecasting accuracy at different aggregation levels. Though the model in that example was applied to a single user as well, the study focuses on the improvement in terms of model performances when aggregating clients, rather than on the accuracy of single client models. In this work, complete SARMA models were defined individually and a qualitative and quantitative analysis of the different types of LV consumers was achieved.

This paper is structured as follows. In Sect. 2, we highlight considerations met when using time series and we introduce our choice of a SARMA model. Section 3 describes the methodology adopted for generating simulated load series for an individual customer. The generation of load series was systematically conducted for 18 customers of an existing feeder in Flobecq (Belgium). One month of Smart Meter data was collected on a 15 min basis in order to define the different models. The analysis of those models, the comparison between the different types of consumers and a simple benchmark with an LV network application are presented in Sect. 4 before we discuss future work in Sect. 5. Finally, Sect. 6 concludes the paper.

2 Seasonal ARMA Load Modeling

2.1 Seasonality Considerations

Electrical Loads being directly related to human activity, they tend to present strong seasonal patterns whose frequency and intensity vary according to the type of appliance.

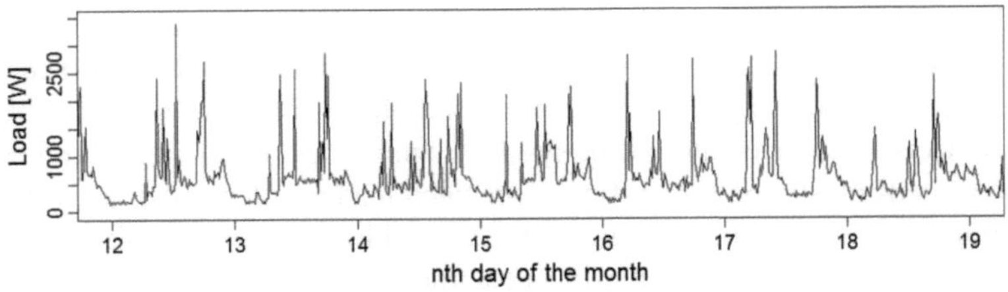

Fig. 1 Residential customer electrical load where a daily pattern due to routine activities can be noted though not purely deterministic

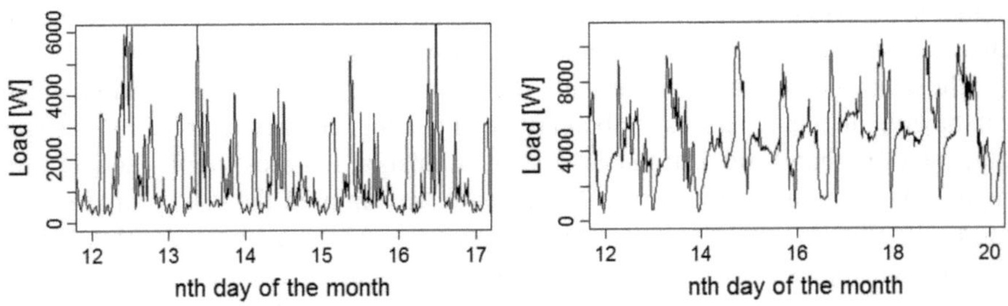

Fig. 2 Local businesses customers electrical load where other seasonalities are significant

Household consumers (Fig. 1) have a tendency to show significant daily and weekly patterns. However, those patterns are not entirely deterministic. While it can be relatively easy to identify periods of overall higher or lower consumption, the occasional activities and behavior as well as the daily variability in the time of activities introduces a more stochastic dimension. Indeed, the Smart Meters sampling frequency of 15 min implies that even routine activities do not occur always at the same quarter of the day.

The weekly pattern tend to be less obvious and more specific to each user because of the large variety of users' agenda (work schedules, vacation plans, activities …). The segmentation that would be needed in order to remove it requires much more information about consumers' habits.

Local businesses, administrations and services industry present daily and weekly pattern as well. However, the additional seasonality (weekly or other) is much more important to model as the activities depend strongly on the type of day. For instance, a school has zero activity during weekends and holidays, functioning half of Wednesday and has a full time activity the rest of the school year. Figure 2 shows on the left an example of local business where the activity is significantly higher on the weekend (12th and 13th day) as well as on Wednesday afternoon (16th day). The profile on the right presents an important base load typical of a small industrial process and requires the observation of a longer time period in order to retrieve the additional seasonality.

As the time window of the data considered in modeling the load is small (one month), the trend is not significant. Indeed, it is reasonable to think that a trend in the consumption cannot be identified unless a time span of several years is considered.

2.2 *Stationarity Considerations*

The seasonal pattern implies that the series are non-stationary. As the approach for modeling considered in this paper assumes that the time series are stationary, i.e. its statistical properties such as mean, variance, autocorrelation, etc. are constant over time, it is necessary to remove the seasonality. Two main options are possible.

The first option is to use *block averaging* technique. This is a very simple procedure that consists in subtracting the averaged observations at the same point of every season cycle. This technique suggests that the seasonality is strongly deterministic with both the pattern and some range of values repeating. This assumption is not valid in the present case because it fails to catch the variability of the seasonal component that the series present.

The second option consists in *differencing* over the period s of the seasonality (e.g. 96 lags of 15 min for a daily seasonality), i.e.

$$Y_t = \nabla_s X_t = X_t - X_{t-s} \tag{1}$$

2.3 *SARIMA Approach*

In this work, we explore the use of Auto-Regressive Moving Average (ARMA) related time series. If the seasonality is an exact repetition of the data, the series can be best modeled using an ARMA process after removing the seasonality as explained in the previous section. On the contrary, if the seasonality is not as deterministic and presents a stochastic aspect, the Seasonal ARIMA (SARIMA) approach will do better [8]. Taking into account the seasonality considerations developed in Sect. 2.1, we opted for the second option.

ARMA Process. A *stationary linear process* $\{X_t\}$ is called *ARMA* (p, q), $p \geq 0$, $q \geq 0$ if there are constants $a_1 \ldots a_p (a_k \neq 0)$ and $\theta_1 \ldots \theta_q (\theta_j \neq 0)$ and a process called innovations $\{\varepsilon_t\} \sim WN(0, \sigma_\varepsilon^2)$ so that:

$$X_t = \sum_{k=1}^{p} a_k X_{t-k} + \sum_{j=1}^{q} \theta_j \varepsilon_{t-j} + \varepsilon_t \tag{2}$$

The first term of the equation above is an Auto-Regressive (AR) process of order p and captures the deterministic part of the series, i.e. routine consumer activity.

The second term is a Moving Average (MA) process of order q and captures the stochastic part of the data.

Equation 2 can be rewritten using the backshift operator $B^i X_t = X_{t-i}$:

$$\left(1 - \sum_{k=1}^{p} a_k B^k\right) X_t = \left(1 + \sum_{j=1}^{q} \theta_j B^j\right) \varepsilon_t \tag{3}$$

SARIMA Process. Let

$$Y_t = \nabla^d \nabla_s^D X_t = (1-B)^d (1-B^s)^D X_t \tag{4}$$

where $\{X_t\}$ is called *SARIMA* $(p,d,q) \times (P,D,Q)_s$ of seasonality s if (4) is a stationary process that follows:

$$A(B)F(B^s)Y_t = \Theta(B)G(B^s)\varepsilon_t \tag{5}$$

where

$$A(B) = 1 - \sum_{k=1}^{p} a_k B^k, \quad \Theta(B) = 1 + \sum_{j=1}^{q} \theta_j B^j \tag{6}$$

and

$$F(B^s) = 1 - \sum_{k=1}^{P} \varphi_k B^{k\times s}, \ G(B^s) = 1 + \sum_{j=1}^{Q} \gamma_j B^{j\times s} \tag{7}$$

The two terms in (6) refer respectively to the AR and MA processes, the first term of (7) corresponds to the Seasonal AR process (SAR) of order P and the second term is the Seasonal MA process (SMA) of order Q.

3 Methodology

3.1 Normalization

This first step intends to transform the distribution in such a way that it follows a normal distribution. Indeed, the estimation of the model parameters used in this work assume a Gaussian distribution.[1] Considering that the load series present complicated distribution (very strong kurtosis and skewness), it is complex to find an analytical expression for the transformation (see Fig. 3).

[1]There exists more sophisticated techniques that can process non-Gaussian ARMA series, but they are significantly more complicated to implement.

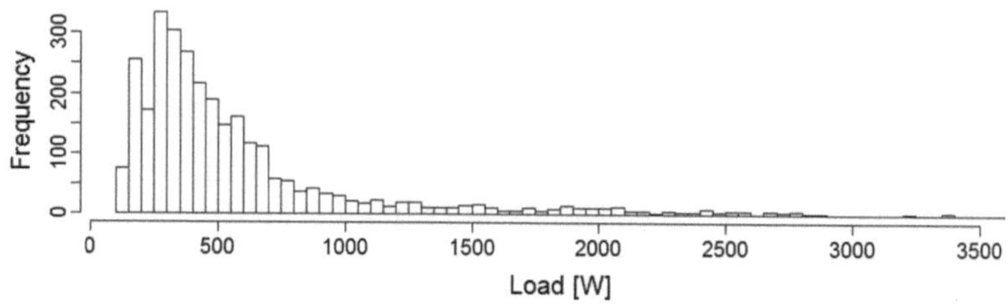

Fig. 3 Distribution example of a single residential customer

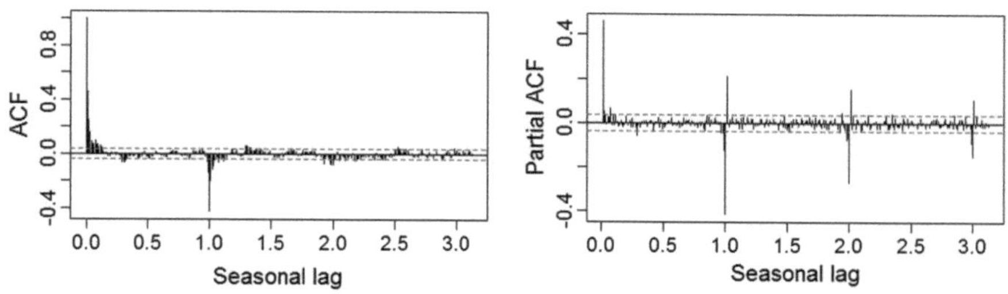

Fig. 4 ACF and PACF of the seasonally differenced series

Hence, the inversion of the cumulative distribution function has been considered [9]. The procedure consists in defining one CDF per customer based on historical data (one month). This allows to transform the original data into a uniform distribution $U(0, 1)$. Finally, we obtain a Gaussian using the inverse CDF of a $N(0, 1)$.

3.2 *Identification and Adjustment of SARIMA Models*

The Box-Jenkins analysis has been used to determine the 6 different parameters of the SARMA models:

1. d, D and s are chosen so that $\nabla^d \nabla_s^D X_t$ is stationary. Seasonality is logically $s = 96$ as the main seasonality has a time span of 96 quarters of an hour. A seasonal differencing of order $D = 1$ has shown to be sufficient while regular differencing is not needed $(d = 0)$ as there is no trend in the series (Kwiatkowski–Phillips–Schmidt–Shin test was conducted).
2. P and Q are determined based on the Auto-Correlation Function (ACF) and Partial Auto-Correlation Function (PACF) of the time series with the seasonality removed. The presence of a single significant spike in the ACF at lag 96 and the exponential decay of the spikes around the multiples of 96 lags in the PACF (Fig. 4) led us to introduce a SMA term of 1 $(Q = 1)$.
3. p and q are identified by determining the best combination that leads to a *SARIMA* $(p, 0, q) \times (0, 1, 1)_s$ model with the lowest corrected Akaike Information Criteria (AICc) [8].

4. The $2+p+q$ parameters are calculated using maximum likelihood estimation.
5. The quality of the model is verified by controlling the residuals (inspection of ACF and PACF and Box-Ljung test of white noise hypothesis).

It should be noted that an Auto-Arima function was applied as well for the sake of comparaison but it gave systematically worse results. Moreover, we find it to be very time-consuming as the algorithm fits many more models in order to find the most suitable one.

3.3 Simulation and De-normalization

Once a model for a customer is defined, it is possible to generate as many series as needed using different innovation series. Finally, de-normalization is necessary to reapply the original distribution of the load.

4 Analysis

4.1 Model Observation

For most of the residential customers (see Sect. 4.2), we managed to suit a satisfactory SARMA model to their electrical load. The next few figures emphasize this assertion. Figure 5 shows the general aspect of the initial series and one generated series from the model.

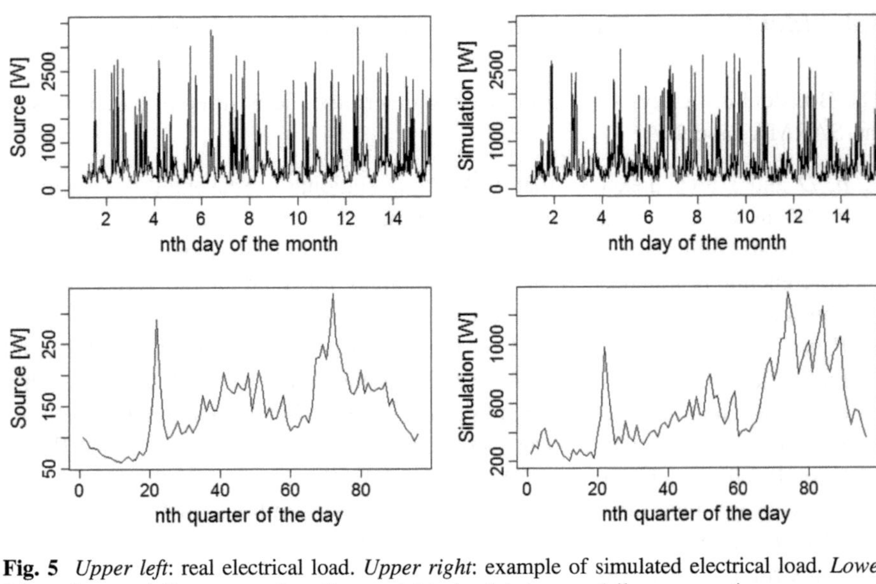

Fig. 5 *Upper left*: real electrical load. *Upper right*: example of simulated electrical load. *Lower left*: real mean daily consumption. *Lower right*: simulated mean daily consumption

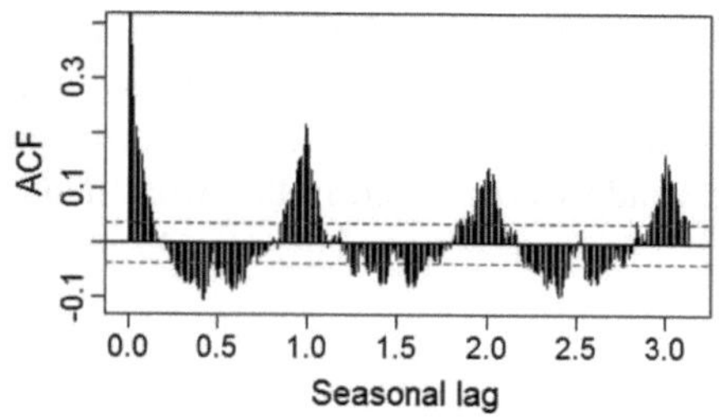

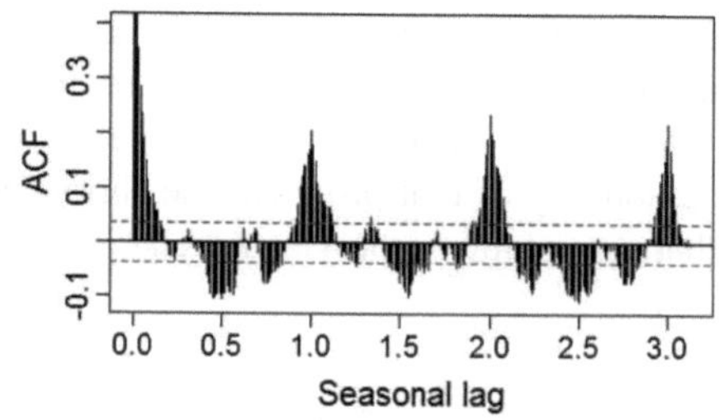

Fig. 6 ACF of the real electrical load (*left*) and an example of simulated electrical load (*right*)

Visual inspection alone does not allow to make any conclusions but it is clear and comforting to observe that the generated series bears a resemblance with the original one. Furthermore, their average daily load are very similar and can match even better with a longer generated period.

The observation of the ACF (see Fig. 6) and PACF of the two series show that the time correlation structure is well reproduced in most cases. This indicates that not only the distribution is retained but the statistics of transitions between load values is expressed.

The analysis of the residuals was systematically performed through the observation of their ACF and PACF as well as by executing the Box-Ljung test. Most of the models showed little or no remaining correlation. However, for some customers, the residuals indicate that there is some room for improvement (see Sect. 4.2).

4.2 *Qualitative Analysis*

Among the set of 18 customers, the following observations can be made:

Houses (14 customers). Most of the SARMA models are or could be satisfactory if further tuning is applied. Some models present, however, a remaining correlation in the residuals. This is expected as there can be another seasonality (weekly or other), although not as significant and clear as for commercial, administration or industrial buildings. Only one individual turned out to be impossible to model with our approach. The observation of the load showed that there were unusual steps in the pattern that are probably due to special circumstances (works during the considered month …).

Farms (2 customers). Both models gave good results with no remaining information in the residuals. Unlike other customers, the routine of a farm is indeed expected to be less subject to a weekly seasonality. Although there should probably be a strong yearly seasonality, the one month time span is too small to show it.

Others (2 customers). The remaining two customers are those of Fig. 2. One is a small commercial business and its activity is strongly dependent of the day of the week. It is clear that a segmentation with different models (e.g. peak, average and no activity days) is advised. This consideration requires, however, more historical measurements as the dataset is reduced by the segmentation. The other customer is a small industrial business where the process depicts another seasonality with a longer period of time. A longer time span would be required to identify the seasonality of that process. The short time window considered in this work makes that seasonality appear as a trend.

4.3 Application Example

A relevant mean to evaluate the performance of mathematical models is to study their implementation in the context of their intended application through a benchmark analysis. The application considered in this case, the assessment of electrical quantities on an LV network, is directly related to the motivations of this work. Two major concerns for LV networks are the voltage magnitudes and their evolution through time. Hence, for example, the proportion of a voltage limit violation and its duration are important considerations. Indeed, overvoltage and undervoltage can lead for most devices to malfunctioning or even to some damage depending on their duration. In addition, the latter information can tell on how good the models perform for capturing the transition pattern and therefore their ability to model sequential processes (e.g. storage strategies).

In this benchmark, we choose to study undervoltage indexes, but it is important to notice that the integration of distributed generation in the LV network (mainly photovoltaic panels) has led to significant overvoltage occurrences. Ongoing work focuses on modeling such generation using SARMA models as well.

In order to get a global performance, the 15 customers out of the 18 for which a reasonable model could be obtained are featured on a distribution network feeder. This means that among those 15 models, some perform better than others as discussed in the previous section.

The indexes (see Table 1) are calculated for the real Smart Meters data (RD), the individual random sampling of each quarter's distribution (DS, as used in pseudo-sequential frameworks [2, 3]) and the SARMA time series (SR). For the latter two, a Monte Carlo framework is used in order to have some significant statistics while the real indexes are calculated with the 30 days of Smart Meter data used for the models definition. The customers are spread out among the three phases and the three branches (with mutual influence), and the results are shown for a node at a terminal point (see Fig. 7) in order to have more substantial values. Figure 8 shows that the mean daily profile obtained by the combination of SARMA models can capture the real distribution satisfactorily. The Mean Absolute Error (MAE) of only 0.31 V and the very small difference in mean percentage of voltage under 227 V emphasize this assertion though it can obviously not perform as well as the sampling on the real distribution (DS).[2] More importantly, we can observe

[2]The value of 227 V is arbitrarily chosen in order to obtain significant indexes.

Table 1 Network study indexes (RD: real data, DS: distribution sampling, SR: SARMA)

Mean absolute error (MAE), [V]	NA (RD)	0.05 (DS)	0.31 (SR)
Mean percentage of voltage < 227 V, [%]	27.8 (RD)	27.2 (DS)	29.3 (SR)
Mean time < 227 V, [quarters of an hour]	3.37 (RD)	1.60 (DS)	2.81 (SR)

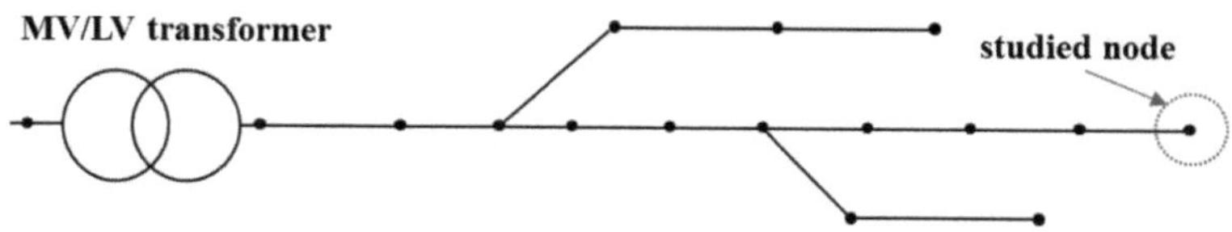

Fig. 7 LV network feeder with 15 different customers assigned on three branches

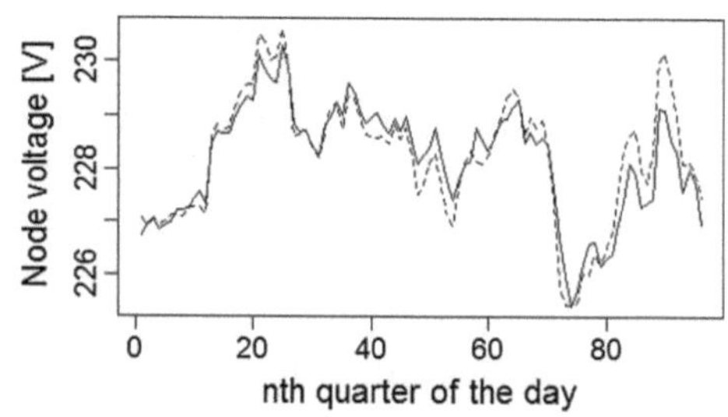

Fig. 8 Mean daily voltage profile at terminal node (*solid*: real data, *dashed*: sequential models)

that the mean duration of a voltage level under 227 V is much closer to the reality than a simple independent sampling (DS). Indeed, the real benefit of SARMA modeling resides in its ability to capture the time correlation structure.

5 Prospects

In order to study most of the possible scenarios on an LV network, the modeling of photovoltaic generation is in ongoing work. Along with the implementation of storage strategies and load management techniques, this is an entire sequential probabilistic tool using a Monte Carlo framework that is being developed.

Customers' models to include in this tool should be selected according to more systematical rules for either the segmentation or the more complex signal decomposition of the Smart Meters data. This should improve the general performances of the models. Future work shall focus on this issue by conducting advanced time-frequency analysis techniques.

In addition, the observations made in Sect. 4.2 bring to the notice that some similarities between clients are present. Yet, larger sets of consumers shall be analyzed in the probabilistic tool. It is therefore interesting to consider grouping techniques among clients so as to reduce the number of models. As it is delusive to obtain detailed characteristics of each customer (types of appliances, habits, number of persons …), a mathematical clustering should be most suited. The groups formed by such a clustering are consequently not expected to reflect any sort of reality.

6 Conclusion

The aim of this paper was to define effective individual sequential models solely based on Smart Meters data in order to introduce them in a probabilistic load flow tool. Unlike other approaches proposed in most of the literature, it requires no other information on customers. This work is the first step of a larger frame study that should open acknowledgement for more considerations developed in the previous paragraph. The main novelty resides in the application domain of the SARMA model. Electrical load patterns are indeed very particular and involve many considerations such as the complex seasonalities they retain. We find that the definition of a complete SARMA model on individual customers is possible. This time series approach appears to be effective and simple. We show that for residential users and farms, this approach is particularly well suited and can render the time correlation and the daily seasonality efficiently. The models could be improved by segmenting the database and define different models for a single customer. Defining groups of similar day patterns allows to take into account seasonalities of longer time periods (weekly, monthly, yearly …) and simplify the complexity of the time correlation pattern. However, this is for such users not critical and uneasy to achieve as individuals' behavior is not very obvious and quite changeable. Besides, multiplying models requires more computing effort and more data. With respect to local businesses and offices, this consideration of additional seasonalities and complexity is usually much more pronounced. Segmentation is advisable with this approach as the process or activity presents one or several strong additional seasonalities. Those being usually of longer time periods, more data should be collected.

Acknowledgements The authors would like to thank ORES, the operator in charge of managing the electricity and domestic gas distribution grids in 196 municipalities of Wallonia (Belgium), for its support in terms of financing and grid data supply both necessary for carrying out this research study.

References

1. Hernandez, J.C., Ruiz-Rodriguez, F.J., Jurado, F.: Technical impact of photovoltaic-distributed generation on radial distribution systems: stochastic simulations for a feeder in Spain. Int. J. Electr. Pow. Energy Syst. **50**(1), 25–32 (2013)
2. Klonari, V., Toubeau, J., De Grève, Z., Durieux, O., Lobry, J., Vallée, F.: Probabilistic simulation framework of the voltage profile in balanced and unbalanced low voltage networks, pp. 1–20

3. Vallée, F., Klonari, V., Lisiecki, T., Durieux, O., Moiny, F., Lobry, J.: Development of a probabilistic tool using Monte Carlo simulation and smart meters measurements for the long term analysis of low voltage distribution grids with photovoltaic generation. Int. J. Electr. Pow Energy Syst. **53**, 468–477 (2013)
4. Ardakanian, O., Keshav, S., Rosenberg, C.: Markovian models for home electricity consumption. In: *Proceedings of the 2nd ACM SIGCOMM Workshop on Green Networking '11*, p. 31 (2011)
5. Collin, A.J., Tsagarakis, G., Kiprakis, A.E., McLaughlin, S.: Development of low-voltage load models for the residential load sector. IEEE Trans. Pow. Syst. **29**(5), 2180–2188 (2014)
6. Singh, R.P., Gao, P.X., Lizotte, D.J.: On hourly home peak load prediction. In: *2012 IEEE 3rd International Conference on Smart Grid Communications (SmartGridComm)*, pp. 163–166 (2012)
7. Sevlian, R., Rajagopal, R.: Short term electricity load forecasting on varying levels of aggregation. pp. 1–8 (2014)
8. Von Sachs, R., Van Bellegem, S.: Séries Chronologiques, *notes de cours (Université Catholique de Louvain)*, p. 209 (2005)
9. Klöckl, B., Papaefthymiou, G.: Multivariate time series models for studies on stochastic generators in power systems. Electr. Pow. Syst. Res. **80**(3), 265–276 (2010)

Diagnostic Checks in Multiple Time Series Modelling

Huong Nguyen Thu

Abstract The multivariate relation between sample covariance matrices of errors and their residuals is an important tool in goodness-of-fit methods. This paper generalizes a widely used relation between sample covariance matrices of errors and their residuals proposed by Hosking (J Am Stat Assoc 75(371):602–608, 1980 [6]). Consequently, the asymptotic distribution of the residual correlation matrices is introduced. As an extension of Box and Pierce (J Am Stat Assoc 65(332):1509–1526, 1970 [11]), the asymptotic distribution recommends a graphical diagnostic method to select a proper VARMA(p, q) model. Several examples and simulations illustrate the findings.

Keywords Goodness-of-fit · Model selection · $VARMA(p, q)$ models

1 Introduction

A multivariate autoregressive moving average *VARMA(p, q)* model has been considered as one of the most influential and challenging models with a wide range of applications in economics. Diagnostic checking in modelling multiple time series is a crucial issue. However, it is still less developed than the univariate case. In the literature of goodness-of-fit, the ideas for multivariate time series originated from various work in univariate framework. For example, article [1] proposed a multivariate extension of [2]. Article [3] suggested a new portmanteau diagnostic test for *VARMA(p, q)* models based on the method of [4]. More diagnostic checking methods in multivariate framework have been studied in [5–9].

Properties of the residual autocorrelation matrices and their practical use play an important role in detecting model misspecification. The asymptotic distribution of the residual autocorrelation function from autoregressive models was first

H. Nguyen Thu (✉)
Department of Business Administration, Technology and Social Sciences,
Luleå University of Technology, 971 87 Luleå, Sweden
e-mail: huong.nguyen.thu@ltu.se

H. Nguyen Thu
Department of Mathematics, Foreign Trade University, Hanoi, Vietnam

I. Rojas et al. (eds.), *Advances in Time Series Analysis and Forecasting*,
Contributions to Statistics, DOI 10.1007/978-3-319-55789-2_11

documented in [10]. In a well-known paper, Box and Pierce in [11] derived a representation of residual autocorrelations as a linear transformation of their error version from *ARMA* models. Hosking in [6] extended distribution of the residual autocorrelation matrices for multiple time series models. More recently, Duchesne in [12] considered the case of *VARX* models where exogenous variables are included. A generalization from *VARMA(p, q)* models was proposed by [13]. Applying the idea of Box and Pierce in [11], this paper further suggests a practical implication of those results for examining the adequacy of fit. We provide graphical and numerical methods for diagnostic checking in multivariate time series.

The rest of the paper is organized as follows. In Sect. 2, some definitions and assumptions are introduced, where *VARMA(p, q)* models are defined, notations of covariance matrices and autocorrelation matrices are presented. Section 3 studies a generalized asymptotic distribution of residual autocovariance matrices. Section 4 proposes a graphical goodness-of-fit method to check model misspecification. It is a practical implication of the asymptotic behavior of residual autocorrelation matrices discussed in Sect. 3. Some simulation examples are presented in Sect. 5 for illustrations. Concluding remarks are given in the last section.

2 Definitions and Assumptions

A causal and invertible m-variate autoregressive moving average *VARMA(p, q)* process may be written as

$$\mathbf{\Phi}(B)(\mathbf{X}_t - \mu) = \mathbf{\Theta}(B)\varepsilon_t, \tag{1}$$

where B is backward shift operator $B\mathbf{X}_t = \mathbf{X}_{t-1}$, μ is the $m \times 1$ mean vector and $\varepsilon_t : t \in \mathbf{Z}$ is a zero mean white noise sequence $WN(\mathbf{0}, \mathbf{\Sigma})$. The $m \times m$ matrix $\mathbf{\Sigma}$ is positive definite. Additionally, $\mathbf{\Phi}(z) = \mathbf{I}_m - \mathbf{\Phi}_1 z - \cdots - \mathbf{\Phi}_p z^p$ and $\mathbf{\Theta}(z) = \mathbf{I}_m + \mathbf{\Theta}_1 z + \cdots + \mathbf{\Theta}_q z^q$ are matrix polynomials, where $\mathbf{I}_m$ is the $m \times m$ identity matrix, and $\mathbf{\Phi}_1, \dots, \mathbf{\Phi}_p, \mathbf{\Theta}_1, \dots, \mathbf{\Theta}_q$ are $m \times m$ real matrices such that the roots of the determinantal equations $|\mathbf{\Phi}(z)| = 0$ and $|\mathbf{\Theta}(z)| = 0$ all lie outside the unit circle. We assume also that both $\mathbf{\Phi}_p$ and $\mathbf{\Theta}_q$ are non-null matrices. The identifiability condition of [14], $rank(\mathbf{\Phi}_p, \mathbf{\Theta}_q) = m$, holds.

Let $P = \max(p, q)$ and define the $m \times mp$ matrix $\mathbf{\Phi} = (\mathbf{\Phi}_1, \dots, \mathbf{\Phi}_p)$, the $m \times mq$ matrix $\mathbf{\Theta} = (\mathbf{\Theta}_1, \dots, \mathbf{\Theta}_q)$, and the $m^2(p+q) \times 1$ vector of parameters $\mathbf{\Lambda} = \mathrm{vec}(\mathbf{\Phi}, \mathbf{\Theta})$. Given n observations $\mathbf{X}_1, \dots, \mathbf{X}_n$ from model (1), the mean vector μ can be estimated by the sample mean $\overline{\mathbf{X}}_n = n^{-1} \sum_{t=1}^{n} \mathbf{X}_t$. The remaining parameters $(\mathbf{\Phi}, \mathbf{\Theta}, \mathbf{\Sigma})$ can be derived by maximizing the Gaussian likelihood function following the procedure in [15, Sect. 12.2]. Once that $\widehat{\mathbf{\Lambda}} = \mathrm{vec}(\widehat{\mathbf{\Phi}}, \widehat{\mathbf{\Theta}})$ has been determined, the residual vectors $\widehat{\varepsilon}_t$, $t = 1, \dots, n$, are computed recursively in the form

$$\widehat{\varepsilon}_t = \varepsilon_t(\widehat{\mathbf{\Theta}}, \widehat{\mathbf{\Phi}}, \overline{\mathbf{X}}_n) = (\mathbf{X}_t - \overline{\mathbf{X}}_n) - \sum_{i=1}^{p} \widehat{\mathbf{\Phi}}_i(\mathbf{X}_{t-i} - \overline{\mathbf{X}}_n) - \sum_{j=1}^{q} \widehat{\mathbf{\Theta}}_j \widehat{\varepsilon}_{t-j}, \quad t = 1, \ldots, n, \tag{2}$$

with the usual conditions $\mathbf{X}_t - \overline{\mathbf{X}}_n \equiv \mathbf{0} \equiv \widehat{\varepsilon}_t$, for $t \leq 0$. In practice, only residual vectors for $t > P = \max(p, q)$ are considered. Define $m \times m$ sample error covariance matrix at lag k with the notation $\mathbf{C}_k = (1/n) \sum_{t=1}^{n-k} \varepsilon_t \varepsilon_{t+k}^{\top}$, $0 \leq k \leq n-1$. Similarly, the $m \times m$ kth residual covariance matrix is $\widehat{\mathbf{C}}_k = (1/n) \sum_{t>P}^{n-k} \widehat{\varepsilon}_t \widehat{\varepsilon}_{t+k}^{\top}$, $0 \leq k \leq n-(P+1)$. The relation between the residual and error covariance matrices derived in [6, p. 603] is given by

$$\widehat{\mathbf{C}}_k^{\top} = \mathbf{C}_k^{\top} - \sum_{i=1}^{p} \sum_{r=0}^{k-i} \mathbf{L}_{k-i-r}(\widehat{\mathbf{\Phi}}_i - \mathbf{\Phi}_i)\mathbf{\Omega}_r \mathbf{\Sigma} - \sum_{j=1}^{q} \mathbf{L}_{k-j}(\widehat{\mathbf{\Theta}}_j - \mathbf{\Theta}_j)\mathbf{\Sigma} + O_P(\frac{1}{n}). \tag{3}$$

Following [5], let $\mathbf{R}_k = \mathbf{C}_k^{\top} \mathbf{C}_0^{-1}$ be the kth sample correlation matrix of the errors ε_t. Its residual analogue is defined by $\widehat{\mathbf{R}}_k = \widehat{\mathbf{C}}_k^{\top} \widehat{\mathbf{C}}_0^{-1}$.

3 Preliminaries

This section presents some notation, definitions and some asymptotic results for *VARMA(p, q)* models. Define the sequences $\mathbf{\Omega}_j$ and $\mathbf{L}_j$ of the $m \times m$ coefficients of the series expansions $\mathbf{\Phi}^{-1}(z)\mathbf{\Theta}(z) = \sum_{j=0}^{\infty} \mathbf{\Omega}_j z^j$ and $\mathbf{\Theta}^{-1}(z) = \sum_{j=0}^{\infty} \mathbf{L}_j z^j$ where $\mathbf{\Omega}_0 = \mathbf{L}_0 = \mathbf{I}_m$. Consider the collection of matrices $\mathbf{G}_k = \sum_{j=0}^{k} (\mathbf{\Sigma}\mathbf{\Omega}_j^{\top} \otimes \mathbf{L}_{k-j})$ and $\mathbf{F}_k = \mathbf{\Sigma} \otimes \mathbf{L}_k$, $k \geq 0$, where $\otimes$ denotes the Kronecker product of matrices. By convention, $\mathbf{G}_k = \mathbf{F}_k = \mathbf{0}$ for $k < 0$. Define the sequence of $m^2 M \times m^2(p+q)$ matrices $\mathbf{Z}_M = (\mathbf{X}_M, \mathbf{Y}_M)$, $M \geq 1$, by

$$\mathbf{X}_M = \begin{pmatrix} \mathbf{G}_0 & \mathbf{0} & \mathbf{0} & \cdots & \mathbf{0} \\ \mathbf{G}_1 & \mathbf{G}_0 & \mathbf{0} & \cdots & \mathbf{0} \\ \mathbf{G}_2 & \mathbf{G}_1 & \mathbf{G}_0 & \cdots & \mathbf{0} \\ \vdots & \vdots & \vdots & \ddots & \vdots \\ \mathbf{G}_{M-1} & \mathbf{G}_{M-2} & \mathbf{G}_{M-3} & \cdots & \mathbf{G}_{M-p} \end{pmatrix},$$

and

$$\mathbf{Y}_M = \begin{pmatrix} \mathbf{F}_0 & \mathbf{0} & \mathbf{0} & \cdots & \mathbf{0} \\ \mathbf{F}_1 & \mathbf{F}_0 & \mathbf{0} & \cdots & \mathbf{0} \\ \mathbf{F}_2 & \mathbf{F}_1 & \mathbf{F}_0 & \cdots & \mathbf{0} \\ \vdots & \vdots & \vdots & \ddots & \vdots \\ \mathbf{F}_{M-1} & \mathbf{F}_{M-2} & \mathbf{F}_{M-3} & \cdots & \mathbf{F}_{M-q} \end{pmatrix}.$$

Define the $Mm^2 \times Mm^2$ block diagonal matrix $\mathbf{W} = \text{diag}(\mathbf{C}_0 \otimes \mathbf{C}_0, \overset{(M)}{\cdots}, \mathbf{C}_0 \otimes \mathbf{C}_0) = \mathbf{I}_M \otimes \mathbf{C}_0 \otimes \mathbf{C}_0$. The residual counterpart is $\widehat{\mathbf{W}} = \mathbf{I}_M \otimes \widehat{\mathbf{\Sigma}} \otimes \widehat{\mathbf{\Sigma}}$. Consider $Mm^2 \times 1$ random vectors $\widehat{\mathbf{H}}_M = [\text{vec}(\widehat{\mathbf{C}}_1^\top), \dots, \text{vec}(\widehat{\mathbf{C}}_M^\top)]^\top$ and $\mathbf{H}_M = [\text{vec}(\mathbf{C}_1^\top), \dots, \text{vec}(\mathbf{C}_M^\top)]^\top$ and $\mathcal{W} = \mathbf{I}_M \otimes \mathbf{\Sigma} \otimes \mathbf{\Sigma}$. Article [6] derived the asymptotic expansion

$$\widehat{\mathbf{W}}^{-1/2}\widehat{\mathbf{H}}_M = (\mathbf{I}_{Mm^2} - \mathbf{P}_H)\mathbf{W}^{-1/2}\mathbf{H}_M + O_P(\frac{1}{n}), \tag{4}$$

where $\mathbf{P}_H = \mathcal{W}^{-1/2}\mathbf{Z}_M(\mathbf{Z}_M^\top \mathcal{W}^{-1}\mathbf{Z}_M)^{-1}\mathbf{Z}_M^\top \mathcal{W}^{-1/2}$ is the $Mm^2 \times Mm^2$ orthogonal projection matrix onto the subspace spanned by the columns of $\mathcal{W}^{-1/2}\mathbf{Z}_M$.

As an alternative version of the relation (4), article [13] proposed a modification by considering a $Mm^2 \times Mm^2$ matrix $\mathbf{Q}_M = (\mathbf{I}_M \otimes \mathbf{a})(\mathbf{I}_M \otimes \mathbf{a}^\top)$, where $\mathbf{a} = \text{vec}(\mathbf{I}_m)/\sqrt{m}$. Furthermore, define $Mm^2 \times Mm^2$ matrix

$$\mathbf{P}_M = (\mathbf{I}_M \otimes \mathbf{a}^\top)\mathcal{W}^{-1/2}\mathbf{Z}_M(\mathbf{Z}_M^\top \mathcal{W}^{-1/2}\mathbf{Q}_M \mathcal{W}^{-1/2}\mathbf{Z}_M)^{-1}\mathbf{Z}_M^\top \mathcal{W}^{-1/2}(\mathbf{I}_M \otimes \mathbf{a}). \tag{5}$$

Consequently, the multivariate linear relation between the residual covariance matrices and their error versions is given by

$$\mathbf{Q}_M\widehat{\mathbf{W}}^{-1/2}\widehat{\mathbf{H}}_M = (\mathbf{I}_{Mm^2} - \mathbf{P}_M)\mathbf{W}^{-1/2}\mathbf{H}_M + O_P(\frac{1}{n}). \tag{6}$$

Finally, introduce $M \times 1$ random vectors $\widehat{\mathbf{T}}_M = [tr(\widehat{\mathbf{R}}_1), \dots, tr(\widehat{\mathbf{R}}_M)]^\top$ and $\mathbf{T}_M = [tr(\mathbf{R}_1), \dots, tr(\mathbf{R}_M)]^\top$. Since trace of a matrix is a singular number, the random vectors $\widehat{\mathbf{T}}_M$ partially deal with the curse of dimensionality relation (4). In other words, we use $M \times 1$ random vectors $\widehat{\mathbf{T}}_M$ instead of $Mm^2 \times 1$ random vectors $\widehat{\mathbf{H}}_M$ for dimension reduction purpose. As a result, the following section introduces a practical tool for modelling multivariate time series.

4 Application in Diagnostic Checking

This section begins with auxiliary asymptotic results imported from [13, 16, 17].

Lemma 1 *Suppose that the error vectors $\{\varepsilon_t\}$ are i.i.d. with $E[\varepsilon_t] = \mathbf{0}$; $Var[\varepsilon_t] = \mathbf{\Sigma} > 0$; and finite fourth order moments $E[\|\varepsilon_t\|^4] < +\infty$. Then, as $n \longrightarrow \infty$,*

$$\sqrt{n}\begin{pmatrix} vec(\mathbf{C}_1^\top) \\ vec(\mathbf{C}_2^\top) \\ \vdots \\ vec(\mathbf{C}_M^\top) \end{pmatrix} \xrightarrow{D} \mathcal{W}^{1/2}\begin{pmatrix} \mathbf{V}_1 \\ \mathbf{V}_2 \\ \vdots \\ \mathbf{V}_M \end{pmatrix}, \quad M \geq 1, \tag{7}$$

where the $\mathbf{V}_k$, $k = 1, \dots, M$, are i.i.d. $N_{m^2}(\mathbf{0}, \mathbf{I}_{m^2})$; and $\mathcal{W} = \mathbf{I}_M \otimes \mathbf{\Sigma} \otimes \mathbf{\Sigma}$.

Proof The proof is given in Appendix. ■

Theorem 1 *Under the same assumptions of Lemma 1, as $n \longrightarrow \infty$,*

$$\sqrt{n}\,[\frac{1}{\sqrt{m}}\begin{pmatrix} tr(\mathbf{R}_1) \\ tr(\mathbf{R}_2) \\ \vdots \\ tr(\mathbf{R}_M) \end{pmatrix}] \xrightarrow{D} N_M(\mathbf{0}, \mathbf{I}_M), \quad M \geq 1. \tag{8}$$

Proof Using the $Mm^2 \times Mm^2$ matrix $\mathbf{W} = \mathbf{I}_M \otimes \mathbf{C}_0 \otimes \mathbf{C}_0$, it can be written

$$\sqrt{n}\,[\frac{1}{\sqrt{m}}\begin{pmatrix} tr(\mathbf{R}_1) \\ tr(\mathbf{R}_2) \\ \vdots \\ tr(\mathbf{R}_M) \end{pmatrix}] = (\mathbf{I}_M \otimes \mathbf{a}_m^{\top})\mathbf{W}^{-1/2}[\sqrt{n}\begin{pmatrix} \text{vec}(\mathbf{C}_1^{\top}) \\ \text{vec}(\mathbf{C}_2^{\top}) \\ \vdots \\ \text{vec}(\mathbf{C}_M^{\top}) \end{pmatrix}].$$

By the law of the large numbers, $\mathbf{C}_0 \xrightarrow{P} \mathbf{\Sigma}$. Also, from a continuity argument similar to that in [18, Proposition 6.1.4], it can be checked that $\mathbf{W}^{-1/2} \xrightarrow{P} \mathcal{W}^{-1/2}$, where $\mathcal{W} = \mathbf{I}_M \otimes \mathbf{\Sigma} \otimes \mathbf{\Sigma} > \mathbf{0}$. Therefore, from Lemma 1 and Slutsky's theorem,

$$\sqrt{n}[\frac{1}{\sqrt{m}}\begin{pmatrix} tr(\mathbf{R}_1) \\ tr(\mathbf{R}_2) \\ \vdots \\ \text{tr}(\mathbf{R}_M) \end{pmatrix}] \xrightarrow{D} (\mathbf{I}_M \otimes \mathbf{a}_m^{\top})\mathcal{W}^{-1/2}\mathcal{W}^{1/2}\begin{pmatrix} \mathbf{V}_1 \\ \mathbf{V}_2 \\ \vdots \\ \mathbf{V}_M \end{pmatrix} = \begin{pmatrix} v_1 \\ v_2 \\ \vdots \\ v_M \end{pmatrix},$$

where the $v_k = \mathbf{a}_m^{\top}\mathbf{V}_k$, $k = 1, \ldots, M$. ■

We will also make use of the following result.

Theorem 2 *Under the same assumptions of Lemma 1, as $n \longrightarrow \infty$,*

$$\frac{1}{\sqrt{m}}\widehat{\mathbf{T}}_M = (\mathbf{I}_M - \mathbf{P}_M)\frac{1}{\sqrt{m}}\mathbf{T}_M + O_P(\frac{1}{n}). \tag{9}$$

Proof We recall the notion of a linear relationship for dimension-reduction purpose

$$\frac{1}{\sqrt{m}}\mathbf{T}_M = (\mathbf{I}_M \otimes \mathbf{a}^{\top})\mathbf{W}^{-1/2}\mathbf{H}_M\,, \quad M \geq 1\,. \tag{10}$$

Its residual version is given by

$$\frac{1}{\sqrt{m}}\widehat{\mathbf{T}}_M = (\mathbf{I}_M \otimes \mathbf{a}^{\top})\widehat{\mathbf{W}}^{-1/2}\widehat{\mathbf{H}}_M\,, \quad M \geq 1\,. \tag{11}$$

Combining (6), (10) and (11) finishes the proof of (9). ■

From Theorems 1 and 2, the aggregate measure of dependence structure in the observed family of statistics $\frac{1}{\sqrt{m}}\widehat{\mathbf{T}}_M$ can be characterized by

$$\sqrt{n}\frac{1}{\sqrt{m}}\widehat{\mathbf{T}}_M \cong N_M(\mathbf{0}, \mathbf{I}_M - \mathbf{P}_M)\,. \tag{12}$$

Put now

$$\boldsymbol{\Xi}_k^\top = (\boldsymbol{\Sigma}^{-1/2} \otimes \boldsymbol{\Sigma}^{-1/2})(\mathbf{G}_{k-1}, \ldots, \mathbf{G}_{k-p}\,; \mathbf{F}_{k-1}, \ldots, \mathbf{F}_{k-q}) \tag{13}$$

for the kth $m^2 \times m^2(p+q)$ row block of the matrix $\mathcal{W}^{-1/2}\mathbf{Z}_M$, $k = 1, \ldots, M$. Accordingly, the diagonal entries of the covariance matrix in (12) are of the form

$$1 - \mathbf{a}_m^\top \boldsymbol{\Xi}_k^\top (\mathbf{Z}_M^\top \mathcal{W}^{-1/2} \mathbf{Q}_M \mathcal{W}^{-1/2} \mathbf{Z}_M)^{-1} \boldsymbol{\Xi}_k \mathbf{a}_m\,. \tag{14}$$

It follows that

$$\sqrt{n}\,\mathrm{tr}(\widehat{\mathbf{R}}_k)/\sqrt{m} \overset{D}{\cong} N(0, 1 - \mathbf{a}_m^\top \boldsymbol{\Xi}_k^\top (\mathbf{Z}_M^\top \mathcal{W}^{-1/2} \mathbf{Q}_M \mathcal{W}^{-1/2} \mathbf{Z}_M)^{-1} \boldsymbol{\Xi}_k \mathbf{a}_m)\,.$$

Recall that the trace of a matrix is the sum of its (complex) eigenvalues, the expression (14) recommends a plot of the adjusted residual traces $\mathrm{tr}(\widehat{\mathbf{R}}_k)/\sqrt{m}$ with the residual version of bands

$$\pm\, z_{\alpha/2}\, n^{-1/2}\sqrt{1 - \mathbf{a}_m^\top \boldsymbol{\Xi}_k^\top (\mathbf{Z}_M^\top \mathcal{W}^{-1/2} \mathbf{Q}_M \mathcal{W}^{-1/2} \mathbf{Z}_M)^{-1} \boldsymbol{\Xi}_k \mathbf{a}_m}\,,\ 1 \le k \le M\,, \tag{15}$$

as a possible diagnostic checks in *VARMA(p, q)* processes, where $z_{\alpha/2}$ is a suitable quantile of a $N(0, 1)$ distribution. The value of M can be chosen as integer part of $\sqrt{n}$. This technique is an extension of a well-known result in univariate *ARMA(p, q)* models based on the residual correlations $\widehat{r}_k$. See [11] for more details. The proposed graphical approach is a practical method in multiple time series model selection.

5 Examples of *VARMA(p, q)* models

This section illustrates the critical bands (15) for five examples of *VARMA(p, q)* models. The sample size of the simulated series is $n = 250$. Consider a trivariate *VAR(1)* model for $m = 3$,

$$\mathbf{X}_t = \boldsymbol{\Phi}_1 \mathbf{X}_{t-1} + \varepsilon_t\,, \tag{16}$$

where

$$\boldsymbol{\Phi}_1 = \begin{pmatrix} 0.2673 & 0.1400 & -0.3275 \\ 0.0346 & 0.1646 & -0.1194 \\ 0.0693 & 0.0517 & -0.0413 \end{pmatrix}. \tag{17}$$

The matrix of (17) is obtained by taking eigenvalues $\delta_j = 0.1554 \pm 0.0354\,i, j = 1, 2$, $\delta_3 = 0.0797$ so that $|\delta_1| = |\delta_2| = 0.1594 < 1$. The covariance matrix of the errors ε_t in (16) will be given by

$$\mathbf{\Sigma} = \begin{pmatrix} 1.0 & 0.3 & 0.3 \\ 0.3 & 1.0 & 0.3 \\ 0.3 & 0.3 & 1.0 \end{pmatrix} . \tag{18}$$

For higher order vector autoregressive models, we construct autoregressive *VAR(p)* models as follows:

(1) For each $j = 1, \ldots, m$, select roots $\varsigma_{j,i}$ with $|\varsigma_{j,i}| > 1, i = 1, \ldots, p$.
(2) For each $j = 1, \ldots, m$, form the polynomial of degree p:

$$p_j(z) = 1 - d_{j,1}z - d_{j,2}z^2 - \cdots - d_{j,p}z^p , \tag{19}$$

so that its roots are $\varsigma_{j,i}$, $i = 1, \ldots, p$.
(3) Construct the $m \times m$ diagonal matrices

$$\mathbf{D}_i = \text{diag}(d_{1,i}, d_{2,i}, \ldots , d_{m,i}), \quad i = 1, \ldots , p . \tag{20}$$

Recall that $\mathbf{D}_i$ is associated to the coefficients of the power z^i in the polynomials $p_j(z)$ of (19), $j = 1, \ldots, m$.
(4) Consider an invertible matrix $\mathbf{A}$ of $m \times m$, and define

$$\boldsymbol{\Phi}_i = \mathbf{A}\mathbf{D}_i\mathbf{A}^{-1}, \quad i = 1, \ldots , p . \tag{21}$$

Under the construction (19)–(21), it follows that

$$|\boldsymbol{\Phi}(z)| = |\mathbf{I}_m - \boldsymbol{\Phi}_1 z - \cdots - \boldsymbol{\Phi}_p z^p| = |\mathbf{I}_m - \mathbf{D}_1 z - \cdots - \mathbf{D}_p z^p| = \prod_{j=1}^{m} p_j(z) . \tag{22}$$

Notice that the mp roots of the determinantal equation $|\boldsymbol{\Phi}(z)| = 0$ are $\varsigma_{j,i}, j = 1, \ldots, m$; $i = 1, \ldots, p$. These correspond to those of the polynomials $p_j(z)$ of (19). The systematic procedures provides a mechanism so that the assumptions of model (1) holds and a numerous number of *VAR(p)* models could be generated. For an example, we consider a trivariate *VAR(2)* model, $\mathbf{X}_t = \mathbf{\Phi}_1\mathbf{X}_{t-1} + \mathbf{\Phi}_2\mathbf{X}_{t-2} + \varepsilon_t$, where

$$\boldsymbol{\Phi}_1 = \begin{pmatrix} 0.1985 & 0.0180 & 0.0044 \\ -0.0113 & 0.2522 & -0.0029 \\ -0.0089 & 0.0082 & 0.2315 \end{pmatrix} \tag{23}$$

and

$$\boldsymbol{\Phi}_2 = \begin{pmatrix} -0.0218 & 0.0021 & 0.0019 \\ -0.0018 & -0.0147 & 0.0010 \\ -0.0021 & 0.0001 & -0.0127 \end{pmatrix} . \tag{24}$$

Table 1 Roots of the determinantal equation $|\boldsymbol{\Phi}(z)| = 0$ of the trivariate *VAR(2)* model

j	$\varsigma_{j,1}$	$\|\varsigma_{j,1}\|$	$\varsigma_{j,2}$	$\|\varsigma_{j,2}\|$
1	$4.8989 + 4.8989\,i$	6.9281	$4.8989 - 4.8989\,i$	6.9281
2	$7.4282 + 0.0000\,i$	7.4282	$8.9138 + 0.0000\,i$	8.9138
3	$7.9282 + 0.0000\,i$	7.9282	$9.5138 + 0.0000\,i$	9.5138

Table 2 Roots of the determinantal equation $|\boldsymbol{\Theta}(z)| = 0$ of the bivariate *VMA(2)* model

j	$\varsigma_{j,1}$	$\|\varsigma_{j,1}\|$	$\varsigma_{j,2}$	$\|\varsigma_{j,2}\|$
1	$2.0000 + 2.0000\,i$	2.8284	$2.0000 - 2.0000\,i$	2.8284
2	$3.3284 + 0.0000\,i$	3.3284	$3.9941 + 0.0000\,i$	3.9941

Table 1 provides six roots of the trivariate *VAR(2)* model.

Additionally, by taking eigenvalues $\delta_j = 0.0901 \pm 0.0433i$, $j = 1, 2$, with $|\delta_1| = |\delta_2| = 0.0999 < 1$, we generate a *VMA(1)* process of the form $\mathbf{X}_t = \boldsymbol{\varepsilon}_t + \boldsymbol{\Theta}_1\boldsymbol{\varepsilon}_{t-1}$, where

$$\boldsymbol{\Theta}_1 = \begin{pmatrix} 0.0589 \; 0.3047 \\ -0.0093 \; 0.1212 \end{pmatrix} . \tag{25}$$

The covariance matrix of the errors is selected as

$$\boldsymbol{\Sigma} = \begin{pmatrix} 1.0 & 0.2 \\ 0.2 & 1.0 \end{pmatrix} . \tag{26}$$

Using the idea to contruct *VAR(p)* models, we construct a bivariate *VMA(2)* model with roots given in Table 2, $\mathbf{X}_t = \boldsymbol{\varepsilon}_t + \boldsymbol{\Theta}_1\boldsymbol{\varepsilon}_{t-1} + \boldsymbol{\Theta}_2\boldsymbol{\varepsilon}_{t-2}$. The model is associated with the following 2×2 matrices:

$$\boldsymbol{\Theta}_1 = \begin{pmatrix} -0.4964 & -0.0218 \\ 0.0091 & -0.5544 \end{pmatrix} , \quad \boldsymbol{\Theta}_2 = \begin{pmatrix} 0.1286 & -0.0213 \\ 0.0089 & 0.0717 \end{pmatrix} . \tag{27}$$

The covariance matrix of the errors $\boldsymbol{\Sigma}$ is given in expression (26).

Finally, we simulate a bivariate *VARMA(1,1)* process

$$\mathbf{X}_t - \boldsymbol{\Phi}_1\mathbf{X}_{t-1} = \varepsilon_t + \boldsymbol{\Theta}_1\varepsilon_{t-1} , \tag{28}$$

where

$$\boldsymbol{\Phi}_1 = \begin{pmatrix} 0.2802 \; 0.2680 \\ -0.0183 \; 0.3152 \end{pmatrix} . \tag{29}$$

The matrix of (29) is derived by taking eigenvalues $\delta_j = 0.2977 \pm 0.0678\,i, j = 1, 2$, so that $|\delta_1| = |\delta_2| = 0.3053 < 1$. The model contains matrix $\boldsymbol{\Theta}_1$ of expression (25) and the covariance matrix $\boldsymbol{\Sigma}$ of expression (26).

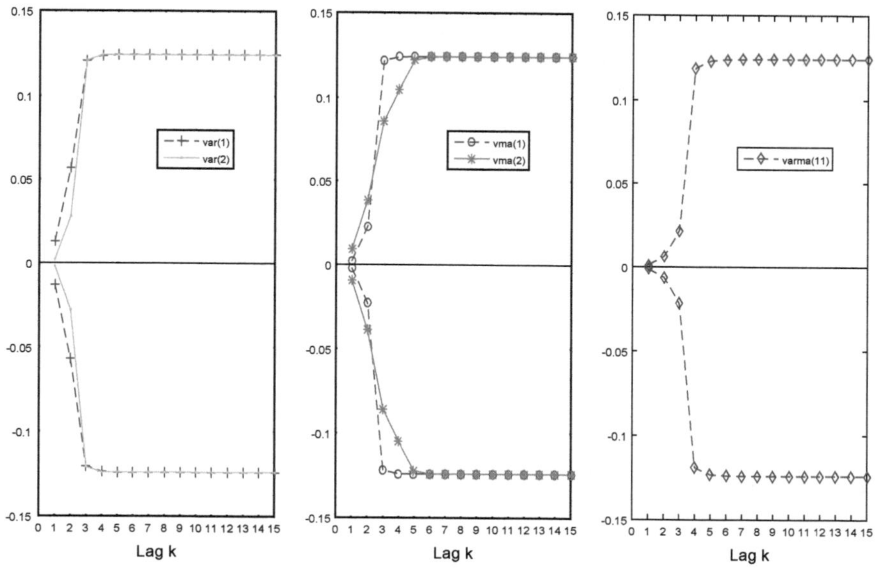

Fig. 1 Bands $\pm 1.96 n^{-1/2}(1 - \mathbf{a}_m^{\top} \Xi_k^{\top} (\mathbf{Z}_M^{\top} \mathcal{W}^{-1/2} \mathbf{Q}_M \mathcal{W}^{-1/2} \mathbf{Z}_M)^{-1} \Xi_k \mathbf{a}_m)^{1/2}$, $k = 1, \ldots, M$, with $n =$ 250 for the five models

Table 3 Asymptotic variances of the leading statistics $\sqrt{n}\,\mathrm{tr}(\widehat{\mathbf{R}}_k)/\sqrt{m}$, $k = 1, \ldots, M$, for the five models

Lag	*VAR(1)*	*VAR(2)*	*VMA(1)*	*VMA(2)*	*VARMA(1,1)*
1	0.0106	0.0003	0.0003	0.0055	0.0000
2	0.2112	0.0508	0.0340	0.0972	0.0532
3	0.9433	0.9504	0.9686	0.4817	0.0310
4	0.9968	1.0000	0.9995	0.7157	0.9200
5	0.9998	1.0000	1.0000	0.9706	0.9861
6	1.0000	–	–	0.9997	0.9978
7	–	–	–	1.0000	0.9997
8	–	–	–	–	1.0000

For the purpose of diagnostic checking, the empirical bands (15) are used to detect model misspecification. Figure 1 plots the bands

$$\pm 1.96\, n^{-1/2} (1 - \mathbf{a}_m^{\top} \Xi_k^{\top} (\mathbf{Z}_M^{\top} \mathcal{W}^{-1/2} \mathbf{Q}_M \mathcal{W}^{-1/2} \mathbf{Z}_M)^{-1} \Xi_k \mathbf{a}_m)^{1/2}, \quad 1 \leq k \leq M .$$

In the simulations, we generate the series with sample size of $n = 250$ and a nominal level $\alpha = .05$ for five time series models.

Figure 1 is confirmed numerically by the values of the asymptotic variances of the leading statistics $\sqrt{n}\,\mathrm{tr}(\widehat{\mathbf{R}}_k)/\sqrt{m}$ in Table 3. Figure 1 indicates that small lags are most useful in revealing model inadequacies. This remark is consistent with the one for *ARMA(p, q)* models in [11]. In practice, if plots of the adjusted traces $\mathrm{tr}(\widehat{\mathbf{R}}_k)/\sqrt{m}$ are outside of the confidence bands, they indicate a lack of fit. As a result, the plots of the adjusted traces lying inside the bands are in favour of identifying a proper multivariate time series model.

6 Conclusions

Graphical diagnostics is very practical for applications. However, very few of goodness-of-fit methods have considered the graphical methodology for multivariate time series model selection. This paper proposes the critical bands to detect a lack of fit in modelling multivariate time series. This tool is based on the properties of the generalized distribution of residual autocorrelation matrices in Sect. 4.

Acknowledgements I am grateful to Santiago Velilla for his sharing, encouragement and guidance as my Ph.D. advisor at Department of Statistics, Universidad Carlos III de Madrid. I also wish to thank Juan Romo, José Miguel Angulo Ibáñez, María Dolores Ugarte, Niels Richard Hansen and Thomas Strömberg for their helpful comments and suggestions. I thank the conference chairs of International Work-Conference on Time Series- ITISE 2016 and the Editors of the Springer series “Contributions to Statistics”. Finally, the Economics Unit, Luleå University of Technology is gratefully acknowledged. Any errors are mine.

Appendix: Proof of Lemma 1

Proof We define $\mathbf{C}_j = n^{-1}\sum_{t=1}^{n-j}\varepsilon_t\varepsilon_{t+j}^{\top}$, then

$$\sqrt{n}\begin{pmatrix}\mathrm{vec}(\mathbf{C}_1)\\ \mathrm{vec}(\mathbf{C}_2)\\ \vdots\\ \mathrm{vec}(\mathbf{C}_k)\end{pmatrix} = \frac{1}{\sqrt{n}}\sum_{t=1}^{n}\begin{pmatrix}\mathrm{vec}(\varepsilon_t\varepsilon_{t+1}^{\top})\\ \mathrm{vec}(\varepsilon_t\varepsilon_{t+2}^{\top})\\ \vdots\\ \mathrm{vec}(\varepsilon_t\varepsilon_{t+k}^{\top})\end{pmatrix}, \quad k \geq 1\,. \tag{30}$$

Consider the sequence of random variables $\{X_t : t \in \mathbb{Z}\}$ such that

$$X_t = \sum_{j=1}^{k}[\mathrm{vec}(\xi_j)]^{\top}\mathrm{vec}(\varepsilon_t\varepsilon_{t+j}^{\top}) = \sum_{j=1}^{k}\mathrm{tr}(\xi_j^{\top}\varepsilon_t\varepsilon_{t+j}^{\top}) = \sum_{j=1}^{k}\varepsilon_t^{\top}\xi_j\varepsilon_{t+j}\,, \tag{31}$$

where ξ_j is a constant $m \times m$ matrix, $j = 1, \ldots, k$. Under the i.i.d. assumption on the $\{\varepsilon_t\}$, the sequence $\{X_t : t \in \mathbb{Z}\}$ is strictly stationary. The sets $\{X_t : t \leq 0\}$ and $\{X_t : t \geq k+1\}$ are independent. Therefore, the sequence $\{X_t : t \in \mathbb{Z}\}$ is also k-dependent. Moreover,

$$E[X_t] = E[\varepsilon_t^{\top}\xi_j\varepsilon_{t+j}] = E[\mathrm{tr}(\xi_j^{\top}\varepsilon_t\varepsilon_{t+j}^{\top})] = \mathrm{tr}[\xi_j^{\top}E(\varepsilon_t\varepsilon_{t+j}^{\top})] = \mathrm{tr}[\xi_j^{\top}\mathrm{Cov}(\varepsilon_t\varepsilon_{t+j}^{\top})] = 0\,. \tag{32}$$

Additionally, the covariance function is given by

$$\gamma(h) = E[X_t X_{t+h}] =$$
$$= [\mathrm{vec}(\xi_1, \xi_2, \ldots, \xi_k)]^\top E\left[\begin{pmatrix} \mathrm{vec}(\varepsilon_t \varepsilon_{t+1}^\top) \\ \mathrm{vec}(\varepsilon_t \varepsilon_{t+2}^\top) \\ \vdots \\ \mathrm{vec}(\varepsilon_t \varepsilon_{t+k}^\top) \end{pmatrix} \begin{pmatrix} \mathrm{vec}(\varepsilon_{t+h} \varepsilon_{t+h+1}^\top) \\ \mathrm{vec}(\varepsilon_{t+h} \varepsilon_{t+h+2}^\top) \\ \vdots \\ \mathrm{vec}(\varepsilon_{t+h} \varepsilon_{t+h+k}^\top) \end{pmatrix}\right] \mathrm{vec}(\xi_1, \xi_2, \ldots, \xi_k) . \quad (33)$$

Recall that $\mathrm{vec}(\varepsilon_t \varepsilon_{t+j}^\top) = \varepsilon_{t+j} \otimes \varepsilon_t$, hence

$$E\{\mathrm{vec}(\varepsilon_t \varepsilon_{t+j}^\top)[\mathrm{vec}(\varepsilon_t \varepsilon_{t+h}^\top)]^\top\} = E[(\varepsilon_{t+j} \otimes \varepsilon_t)(\varepsilon_{t+h}^\top \otimes \varepsilon_t^\top)] = E[\varepsilon_{t+j}\varepsilon_{t+h}^\top \otimes \varepsilon_t \varepsilon_t^\top] .$$

By the law of iterated expectations,

$$\begin{aligned} E\{\mathrm{vec}(\varepsilon_t \varepsilon_{t+j}^\top)[\mathrm{vec}(\varepsilon_t \varepsilon_{t+h}^\top)]^\top\} \quad &= \quad E(E[\varepsilon_{t+j}\varepsilon_{t+h}^\top \otimes \varepsilon_t \varepsilon_t^\top \mid \varepsilon_t]) \quad = \\ = \quad E[\mathrm{Cov}(\varepsilon_{t+j}, \varepsilon_{t+h}) \otimes \varepsilon_t \varepsilon_t^\top] \quad &= \quad \mathrm{Cov}(\varepsilon_{t+j}, \varepsilon_{t+h}) \otimes E[\varepsilon_t \varepsilon_t^\top] . \end{aligned} \quad (34)$$

Note that the expectation given in (34) is $\mathbf{0}$, when $j \neq h$ and is $\boldsymbol{\Sigma} \otimes \boldsymbol{\Sigma}$, when $j = h$. From expression (33), it follows that $\gamma(h) = 0$ for $h \geq 1$, and

$$\gamma(0) = [\mathrm{vec}(\xi_1, \xi_2, \ldots, \xi_k)]^\top \mathcal{W} \mathrm{vec}(\xi_1, \xi_2, \ldots, \xi_k) .$$

By using theorem 6.4.2 in [18, p. 206], we obtain the below convergence

$$\left[\mathrm{vec}\left(\xi_1, \xi_2, \ldots, \xi_k\right)\right]^\top \left[\frac{1}{\sqrt{n}} \sum_{t=1}^{n} \begin{pmatrix} \mathrm{vec}\left(\varepsilon_t \varepsilon_{t+1}^\top\right) \\ \mathrm{vec}\left(\varepsilon_t \varepsilon_{t+2}^\top\right) \\ \vdots \\ \mathrm{vec}\left(\varepsilon_t \varepsilon_{t+k}^\top\right) \end{pmatrix}\right] = \sqrt{n}\left(\frac{1}{n}\sum_{t=1}^{n} X_t\right) \xrightarrow{D}$$
$$\xrightarrow{D} \left[\mathrm{vec}\left(\xi_1, \xi_2, \ldots, \xi_k\right)\right]^\top \mathcal{W}^{1/2} \begin{pmatrix} \mathbf{V}_1^* \\ \mathbf{V}_2^* \\ \vdots \\ \mathbf{V}_k^* \end{pmatrix} . \quad (35)$$

By the Cramér-Wold device, combining (30) and (35) leads to

$$\sqrt{n} \begin{pmatrix} \mathrm{vec}(\mathbf{C}_1) \\ \mathrm{vec}(\mathbf{C}_2) \\ \vdots \\ \mathrm{vec}(\mathbf{C}_k) \end{pmatrix} \xrightarrow{D} \mathcal{W}^{1/2} \begin{pmatrix} \mathbf{V}_1^* \\ \mathbf{V}_2^* \\ \vdots \\ \mathbf{V}_k^* \end{pmatrix}, \quad k \geq 1 . \quad (36)$$

Now consider a $m^2 \times m^2$ commutation matrix $\mathbf{K}_{mm}$ of order m, and the $km^2 \times km^2$ matrix $\mathbf{K} = \mathrm{diag}(\mathbf{K}_{mm}, \overset{(k)}{\cdots}, \mathbf{K}_{mm})$. Recall the identity $\mathrm{vec}(\mathbf{C}_k^\top) = \mathbf{K}_{mm}\mathrm{vec}(\mathbf{C}_k)$, it follows that

$$\sqrt{n}\begin{pmatrix} \mathrm{vec}(\mathbf{C}_1^\top) \\ \mathrm{vec}(\mathbf{C}_2^\top) \\ \vdots \\ \mathrm{vec}(\mathbf{C}_k^\top) \end{pmatrix} = \mathbf{K}\left[\sqrt{n}\begin{pmatrix} \mathrm{vec}(\mathbf{C}_1) \\ \mathrm{vec}(\mathbf{C}_2) \\ \vdots \\ \mathrm{vec}(\mathbf{C}_k) \end{pmatrix}\right] \xrightarrow{D} \mathbf{K}\mathcal{W}^{1/2}\begin{pmatrix} \mathbf{V}_1^* \\ \mathbf{V}_2^* \\ \vdots \\ \mathbf{V}_k^* \end{pmatrix} \overset{D}{\equiv} \mathcal{W}^{1/2}\begin{pmatrix} \mathbf{V}_1^* \\ \mathbf{V}_2^* \\ \vdots \\ \mathbf{V}_k^* \end{pmatrix}. \quad (37)$$

The equivalence in distribution at the right-hand side of (37) follows from the identity $\mathbf{K}_{mm}(\boldsymbol{\Sigma} \otimes \boldsymbol{\Sigma})\mathbf{K}_{mm} = \boldsymbol{\Sigma} \otimes \boldsymbol{\Sigma}$, that is a consequence of Eq. (24) in [15, p. 664]. Since $\mathbf{K}_{mm} = \mathbf{K}_{mm}^\top$, both $\mathbf{K}_{mm}(\boldsymbol{\Sigma}^{1/2} \otimes \boldsymbol{\Sigma}^{1/2})\mathbf{V}_j$ and $(\boldsymbol{\Sigma}^{1/2} \otimes \boldsymbol{\Sigma}^{1/2})\mathbf{V}_j$ have the same distribution $\mathbf{N}_{m^2}[\mathbf{0}, \boldsymbol{\Sigma} \otimes \boldsymbol{\Sigma}], j = 1, \ldots, k$.

References

1. Bouhaddioui, C., Roy, R.: A generalized portmanteau test for independence of two infinite-order vector autoregressive series. J. Time Ser. Anal. **27**(4), 505–544 (2006)
2. Hong, Y.: Consistent testing for serial correlation of unknown form. Econometrica **64**(4), 837–837 (1996)
3. Mahdi, E., McLeod, A.: Improved multivariate portmanteau test. J. Time Ser. Anal. **33**(2), 211–222 (2012)
4. Peña, D., Rodríguez, J.: A powerful portmanteau test of lack of fit for time series. J. Am. Stat. Assoc. **97**(458), 601–610 (2002)
5. Chitturi, R.V.: Distribution of residual autocorrelations in multiple autoregressive schemes. J. Am. Stat. Assoc. **69**(348), 928–934 (1974)
6. Hosking, J.R.M.: The multivariate portmanteau statistic. J. Am. Stat. Assoc. **75**(371), 602–608 (1980)
7. Li, W.K., McLeod, A.I.: Distribution of the residual autocorrelations in multivariate ARMA time series models. J. R. Stat. Soc. Ser. B (Methodological) **43**(2), 231–239 (1981)
8. Tiao, G.C., Box, G.E.P.: Modeling multiple times series with applications. J. Am. Stat. Assoc. **76**(376), 802–816 (1981)
9. Li, W.K., Hui, Y.V.: Robust multiple time series modelling. Biometrika **76**(2), 309–315 (1989)
10. Walker, A.: Some properties of the asymptotic power functions of goodness-of-fit tests for linear autoregressive schemes. J. R. Stat. Soc. Ser. B (Methodological) **14**(1), 117–134 (1952)
11. Box, G.E.P., Pierce, D.A.: Distribution of residual autocorrelations in autoregressive-integrated moving average time series models. J. Am. Stat. Assoc. **65**(332), 1509–1526 (1970)
12. Duchesne, P.: On the asymptotic distribution of residual autocovariances in VARX models with applications. TEST **14**(2), 449–473 (2005)
13. Nguyen Thu, H.: A note on the distribution of residual autocorrelations in VARMA(p, q) models. J. Stat. Econ. Methods **4**(3), 93–99 (2015)
14. Hannan, E.J.: The identification of vector mixed autoregressive-moving average systems. Biometrika **56**(1), 223–225 (1969)
15. Lütkepohl, H.: New Introduction to Multiple Time Series Analysis. Springer (2005)
16. Velilla, S., Nguyen, H.: A basic goodness-of-fit process for VARMA(p, q) models. Statistics and Econometrics Series 09, Carlos III University of Madrid (2011)
17. Nguyen, H.: Goodness-of-fit in Multivariate Time Series. Ph.D. dissertation, Carlos III University of Madrid (2014)
18. Brockwell, P.J., Davis, R.A.: Time Series: Theory and Methods, 2nd edn. Springer, New York (1991)

Mixed AR(1) Time Series Models with Marginals Having Approximated Beta Distribution

Tibor K. Pogány

Abstract Two different mixed first order AR(1) time series models are investigated when the marginal distribution is a two-parameter Beta $\mathrm{B}_2(p, q)$. The asymptotics of Laplace transform for marginal distribution for large values of the argument shows a way to define novel mixed time-series models which marginals we call asymptotic Beta. The new model's innovation sequences distributions are obtained using Laplace transform approximation techniques. Finally, the case of generalized functional Beta $\mathrm{B}_2(G)$ distribution's use is discussed as a new parent distribution. The chapter ends with an exhaustive references list.

Keywords Approximated beta distribution · First order mixed AR(1) model · Generalized beta distribution · Laplace transform integral · Erdélyi's theorem for Laplace's method · Watson's Lemma

MSC2010: 62M10 · 60E05 · 33C15 · 41A60 · 44A10 · 62F10

1 Introduction and Preliminaries

In standard time series analysis one assumes that its marginal distribution is normal (Gaussian in other words). However, in many cases the normal distribution is not always convenient. In earlier investigations stationary non-Gaussian time series models were developed for variables with positive and highly skewed distributions.

Dedicated to Professor Jovan Mališić to his 80th birthday anniversary.

T.K. Pogány (✉)
Applied Mathematics Institute, Óbuda University, Bécsi út 96/b,
Budapest 1034, Hungary
e-mail: poganj@pfri.hr
URL: http://www.pfri.uniri.hr/poganj

T.K. Pogány
Faculty of Maritime Studies, University of Rijeka, Studentska 2, 51000
Rijeka, Croatia

I. Rojas et al. (eds.), *Advances in Time Series Analysis and Forecasting*,
Contributions to Statistics, DOI 10.1007/978-3-319-55789-2_12

There still remain situations where Gaussian marginals are inappropriate, specially where the marginal time-series variable being modeled, although not skewed or inherently positive valued, has a large kurtosis and long-tailed distributions. There are plenty of real situations when the normal approach is not appropriate like in hydrology, meteorology, information theory and economics for instance.

The introductional studies began in late seventies and early eighties when simple models with exponential marginals or mixed exponential marginals by predecessors Lawrence, Lewis and coauthors [7, 9, 15–18]; also see [8]. Other marginals like Gamma and Weibull [7, 25, 36]; uniform [3, 35]; Laplace [25] are considered too. Finally, we point out autoregressive processes PBAR and NBAR constructed by McKenzie [21] for positively and negatively correlated pairs of Beta random variables, NEAR and NLAR time series by Karlsen and Tjøstheim [13] and others. Attention has to be drawn to the group of probabilist from Serbia was led by Mališić who initiated the investigations upon mixed AR and MA with exponential marginals (AREX models) either in his publications [19, 20] or in works of his contemporary PhD students Popović [27, 28, 35] and Jevremović [10, 11]; also the next 'Math Genealogy grandchild' generation Ph.D. students contribute to this research direction—Ristić [34, 35], Novković [25] and Popović [29–31].

These results presented here concern either the mixed AR(1) time series model [29]

$$X_t = \begin{cases} \alpha X_{t-1} & \text{w.p.} \quad \alpha^p \\ \beta X_{t-1} + \xi_t & \text{w.p.} \quad 1-\alpha^p \end{cases}, \qquad \alpha, \beta \in (0,1);\, p \in (0,1]\,, \qquad (\mathfrak{M}_1)$$

or the similar but a modestly developed AR(1) model investigated in [31]:

$$X_t = \begin{cases} \alpha X_{t-1} + \xi_t & \text{w.p.} \quad p_1 \\ \beta X_{t-1} + \xi_t & \text{w.p.} \quad 1-p_1 \end{cases}, \qquad \alpha, \beta, p_1 \in (0,1). \qquad (\mathfrak{M}_2)$$

In both models $\{X_t : t \in \mathbb{Z}\}$ possesses two-parameter Beta distribution $X_t \sim \mathrm{B}_2(p,q)$, while $\{\xi_t : t \in \mathbb{Z}\}$ stands for the innovation process of the models.

Let us recall that a rv $X \equiv X_t \sim \mathrm{B}_2(p,q)$ defined on a standard probability space $(\Omega, \mathscr{F}, \mathsf{P})$ when the related probability density function (PDF) equals

$$f(x) = \frac{\Gamma(p+q)}{\Gamma(p)\Gamma(q)}\, x^{p-1}(1-x)^{q-1} \cdot 1_{[0,1]}(x)\,, \qquad \min(p,q) > 0, \qquad (1)$$

where $1_S(x)$ signifies the indicator function of the set S.

The Laplace transform (LT) of the PDF (1) becomes[1]

[1]The LT of a PDF actually coincides with the moment generating function M_X of the input rv X with negative parameter $\mathsf{E}\,\mathrm{e}^{-sX} \equiv \mathsf{M}_X(-s)$.

$$\varphi_X(s) = \mathsf{E}e^{-sX} = \frac{\Gamma(p+q)}{\Gamma(p)\Gamma(q)} \int_0^1 e^{-sx} x^{p-1}(1-x)^{q-1} dx \qquad (2)$$
$$= {}_1F_1(p; q+p; -s);$$

here ${}_1F_1$ denotes the confluent hypergeometric function (or Kummer function of the first kind):

$${}_1F_1(a; c; z) = \sum_{n\geq 0} \frac{(a)_n}{(c)_n} \frac{z^n}{n!},$$

using the familiar Pochhamer symbol notation $(a)_n = a(a+1)\dots(a+n-1)$ taking conventionally $(0)_0 = 1$. The integral (2) is actually a special case of the Laplace type integral

$$\mathscr{I}_b(s) = \int_0^b e^{-sx}\, g(x)\, dx, \qquad 0 < b \leq \infty,$$

which asymptotics gives Watson's lemma [38, p. 133] when $s \to \infty$, provided $g(x) = x^{\alpha}\, h(x)$; $h(0) \neq 0$ is exponentially bounded and $h \in C^{\infty}$ in the neighborhood of origin. Then for any fixed b, the following asymptotic equivalence holds

$$\mathscr{I}_b(s) \sim \frac{\Gamma(\alpha+1)}{s^{\alpha+1}} \sum_{n\geq 0} \frac{h^{(n)}(0)\,(\alpha+1)_n}{n!\, s^n}.$$

Bearing in mind the asymptotics of Kummer's function [2, p. 508]

$${}_1F_1(a; c; z) \sim \left\{ \frac{\Gamma(c)(-z)^{-a}}{\Gamma(c-a)} + \frac{\Gamma(c)\, e^z z^{a-c}}{\Gamma(a)} \right\} \left(1 + \mathscr{O}(z^{-1})\right), \qquad |z| \to \infty$$

for $a = p, c = p+q, z = -s < 0$ we conclude that the second addend vanishes with the convergence rate $\mathscr{O}(s^{-q}\, e^{-s})$, which ensures that

$$\varphi_X(s) \sim \frac{\Gamma(p+q)}{\Gamma(q)\, s^p} \left(1 + \mathscr{O}(s^{-1})\right), \qquad s \to \infty. \qquad (3)$$

Our next task is to obtain the distribution of our innovation process ξ_t. However, this goal will be mainly relieved by assuming that in the sequel

(i) X_t is wide sense stationary, and

(ii) X_t and ξ_r are independent for all $t < r$,

consult [32] too. The related distribution specified by the approximant we call *approximated beta distribution* AB. This method of approximating of LT and the associated procedure in determining the approximate distribution of innovative series for the first time appear in literature in [29].

Since the wide sense stationarity of X_t and independence of X_t and ξ_t the LT of models $\mathfrak{M}_1, \mathfrak{M}_2$ become

$$\varphi_X(s) = \alpha^p\, \varphi_X(\alpha\, s) + (1-\alpha^p)\varphi_X(\beta\, s)\varphi_\xi(s),$$
$$\varphi_X(s) = p_1\, \varphi_X(\alpha\, s)\varphi_\xi(s) + (1-p_1)\varphi_X(\beta\, s)\varphi_\xi(s)$$

respectively, hence

$$\varphi_\xi(s) = \begin{cases} \dfrac{\varphi_X(s) - \alpha^p\, \varphi_X(\alpha\, s)}{(1-\alpha^p)\, \varphi_X(\beta\, s)} & (\mathfrak{M}_1) \\[2ex] \dfrac{\varphi_X(s)}{p_1\, \varphi_X(\alpha\, s) + (1-p_1)\, \varphi_X(\beta\, s)} & (\mathfrak{M}_2) \end{cases}. \tag{4}$$

After replacing the asymptotic formulæ for related $\varphi_X(\cdot)$ values the next step is inverting the derived expression (4) so, that the obtained formula either can be directly recognized in the inverse Laplace transforms table, or by certain further rewriting it into a linear combination of such expressions. Of course we should take care about the associated re-normalization constants. For the numerical inversion of LT we are referred to [1] and [33], for instance.

The parent distribution in both models is $B_2(p, q)$. By the approximation procedure and LT inversion results in expressions which should be non-negative and re-normalized to serve as PDF for a new generation time series. These will carry additional 'approximated Beta' $AB_{p,q}$ notation too. These results obtained will build the next two sections. In the closing section similar approximation procedure of the parent Generalized Beta distribution and subsequent mixed time series issues are discussed.

2 The $AB_{p,q}$AR(1) Model with Parent $\mathfrak{M}_1$

This chapter consists from result by Popović, Pogány and Nadarajah [29].

The mixed autoregressive first order time series model $\mathfrak{M}_1$ in which the time series X_t behaves according to the Beta distribution $B_2(p, q)$ with the parameter space $(p, q) \in (0, 1] \times (1, \infty)$. The related LT is approximated when the transformation argument is large. The resulting approximation (3) is re-defined by using the asymptotic relation [29, p. 1553, Eq. (3)]

$$\varphi_X(s) \sim \frac{\Gamma(p+q)}{\Gamma(q)\, s^p}\left(1 - e^{-p(q-1)s^p}\right). \tag{5}$$

Thus, for s large

$$\varphi_\xi(s) \sim \frac{B}{1-A}\,\frac{e^{-p(q-1)As^p} - e^{-p(q-1)s^p}}{e^{-p(q-1)Bs^p}},$$

where the shorthand $A = \alpha^p, B = \beta^p$ is used. This case of $\mathfrak{M}_1$ we call $\mathrm{AB}_{p,q}\mathrm{AR}(1)$ in the sequel.

Remark 1 The parent distribution $X \sim \mathrm{B}_2(1, q)$ is the power law distribution having PDF $f(x)q(1-x)^{q-1} \cdot 1_{[0,1]}(x)$, $q > 1$. The two parameters Kumaraswamy distribution $\mathrm{Kum}_2(p, q)$, $p \in (0, 1), q > 1$, see [14] and as an exhaustive account for the Kumaraswamy distribution the article [12], is determined by the PDF

$$f(x) = pqx^{p-1}(1-x)^{q-1} \cdot 1_{[0,1]}(x)\,.$$

The $\mathrm{B}_2(1, q)$ distribution generates the Kumaraswamy distribution in the way

$$\big[\mathrm{B}_2(1, q)\big]^{\frac{1}{p}} \stackrel{\mathscr{D}}{=} \mathrm{Kum}_2(p, q)\,.$$

The $\mathrm{Kum}_2(p, q)$, $p \in (0, 1), q > 1$ distribution is of importance in modeling e.g. the storage volume of a reservoir and another hydrological questions, and system design [6, 22, 23]. ■

Our first principal result concerns the distribution (or a related approximation) of the innovation sequence ξ_t in the case $p = 1$.

Theorem 1 [29, p. 1553, Theorem 1]
Consider the mixed $\mathrm{AB}_{p,q}\mathrm{AR}(1)$ *times series model* $\mathfrak{M}_1$ *having marginal distribution which* LT *behaves according to* (5) *for large values of argument. Let* $q > 1$, $\alpha, \beta \in (0, 1)$ *and* $\kappa(1) = (1-\alpha)\beta^{-1} \in \mathbb{N}$. *Then the i.i.d. sequence* $\{\xi_t : t \in \mathbb{Z}\}$ *possesses the uniform discrete distribution*

$$\mathsf{P}\big\{\xi_t = (q-1)(\alpha + j\beta)\big\} = \frac{1}{\kappa(1)}, \qquad j = \overline{0, \kappa(1) - 1}\,. \tag{6}$$

Remark 2 It is worth to mention that the same model $\mathfrak{M}_1$ but with the uniform $\mathscr{U}(0, 1)$ marginal distribution was considered in [35]. In this case it has been proven that the innovation sequence coincides with the discrete distribution (6) under $p = 1, q = 2$. ■

Theorem 2 [29, p. 1555, Theorem 3] *The conditional variance of* $\mathrm{AB}_{1,q}\mathrm{AR}(1)$ *model is*

$$\begin{aligned}\mathsf{D}(X_t|X_{t-1} = x) &= \alpha(1-\alpha)x[(\alpha-\beta)x - (q-1)(1+\alpha-\beta)] \\ &\quad + \frac{1}{12}(1-\alpha)(q-1)^2(1+\alpha+7\alpha^2+3\alpha^3 \\ &\quad - 6\alpha\beta + 3\alpha\beta^2 - -6\alpha^2\beta - 4\beta^2 + \beta^3)\,.\end{aligned}$$

The distribution of the innovation sequence turns out to be discrete uniform under above exposed constraints. However, as we will see the assumption $p \in (0,1)$ implies absolutely continuous $\mathrm{AB}_{p,q}$ distribution for ξ_t. To reach this result we need the definition of the Wright generalized hypergeometric series [37]

$$\Phi(a,c;z) = \sum_{n\geq 0} \frac{1}{\Gamma(a+cn)}\frac{z^n}{n!}\,.$$

The display

$$s^{-\tau}\,\mathrm{e}^{-Ts^p} = \int_0^\infty \mathrm{e}^{-sx}x^{\tau-1}\,\Phi(\tau,-p;-Tx^{-p})\,\mathrm{d}x, \qquad T>0$$

is equivalent to the Humbert–Pollard LT inversion formula [37]. By virtue of this relation we obtain the following

Theorem 3 [29, p. 1555, Theorem 4] *Consider the mixed* $\mathrm{AB}_{p,q}\mathrm{AR}(1)$ *times series mode. Let* $q>1,\ p,\alpha,\beta \in (0,1)$ *and* $\kappa(p) = (1-\alpha^p)\beta^{-p} \in \mathbb{N}$. *Then the i.i.d. sequence* $\{\xi_t : t \in \mathbb{Z}\}$ *possesses the PDF*

$$f(x) = \frac{1}{x\|\Phi_p\|}\sum_{j=0}^{\kappa(p)-1} \Phi\left(0,-p;-p(q-1)(A+Bj)\,x^{-p}\right)\,,$$

where by convention $A=\alpha^p$, $B=\beta^p$ *and*

$$\|\Phi_p\| = \sum_{j=0}^{\kappa(p)-1}\int_0^\infty \Phi\left(0,-p;-p(q-1)(A+Bj)\,x^{-p}\right)\frac{\mathrm{d}x}{x}\,.$$

Let us close this section with the important parameter estimation issue of $\mathrm{AB}_{p,q}\mathrm{AR}(1)$ model. Under the covariance and correlation functions of X_t the time with the lag $|\tau|<\infty$ we mean the functions

$$\gamma(\tau) = \mathsf{E}X_tX_{t-\tau} - \mathsf{E}X_t\mathsf{E}X_{t-\tau}; \qquad \rho(\tau) = \frac{\gamma(\tau)}{\gamma(0)}\,.$$

Theorem 4 [29, p. 1556, Theorem 5] *Assume that* $p\in(0,1]$ *is known. Then the correlation function of* $\mathrm{AB}_{p,q}\mathrm{AR}(1)$ *with respect to the integer lag* τ *reads:*

$$\rho(\tau) = \left[\alpha^{p+1}+\beta(1-\alpha^p)\right]^{|\tau|}, \qquad \tau\in\mathbb{Z}\,.$$

Theorem 5 [29, pp. 1556–7, Theorem 6; Eq. (14)] *For* $\beta\geq\alpha$ *the estimator*

$$\widehat{\alpha} = min_{2\leq t\leq n}\frac{X_t}{X_{t-1}}\,,$$

Table 1 Estimated parameters α and β with their standard deviations $\sigma(\widehat{\alpha})$, $\sigma(\widehat{\beta})$. Originally published in [29, p. 1557, Table 1]. Published with kind permission of (c) Elsevier 2017. All Rights Reserved

Sample size	$\widehat{\alpha}$	$\widehat{\beta}$	$\sigma(\widehat{\alpha})$	$\sigma(\widehat{\beta})$
500	0.6729	0.7981	0.0030	0.0540
1.000	0.6813	0.8001	0.0000	0.0024
5.000	0.6813	0.8214	0.0000	0.0015
10.000	0.6813	0.8176	0.0000	0.0010
50.000	0.6813	0.8179	0.0000	0.0006

is consistent for the parameter α. Moreover the parameter β has the estimator

$$\widehat{\beta} = \frac{\widehat{\rho(1)} - \widehat{\alpha^{p+1}}}{1 - \widehat{\alpha^p}},$$

where the correlation function

$$\widehat{\rho(1)} = \frac{\sum_{t=2}^{n}(X_t - \overline{X})(X_{t-1} - \overline{X})}{\sum_{t=1}^{n}(X_t - \overline{X})^2},$$

and $\overline{X}$ stands for the mean value of the generated sequence (X_t).

We close this section with results of a simulation study. Concentrate to parameters α, β in the $\mathrm{AB}_{p,q}\mathrm{AR}(1)$ model. In a numerical simulation of parameter estimation based on the Theorem 5 we draw 100 samples of sizes 100, 1000, 5000, 10.000 and 20.000 from $\mathfrak{M}_1$.

The true values of the parameters are $(\alpha, \beta) = (0.6813, 0.8182)$. Table 1 presents mean values estimated parameters $\widehat{\alpha}, \widehat{\beta}$ and their standard deviations $\sigma(\widehat{\alpha}), \sigma(\widehat{\beta})$.

3 The ABAR(1) Model with Parent $\mathfrak{M}_2$

This chapter mainly consists from results by Popović and Pogány [31].

The first order linear model $\mathfrak{M}_2$ with approximated distribution was considered in [31]. It was shown that distribution of innovation process coincides with uniform discrete distribution for $p = 1$. However, for $p \in (0, 1)$ the innovation process possesses continuous distribution.

It is shown that analytical inversion of Laplace transform is possible when $p = 1, q > 0, p = 3, q \geq (43\pi/6)^3, p = 4, q \geq (3\pi/4)^4$ and $p = 5, q \geq (9\pi/10)^5$. Therefore we obtain the associated PDF which approximates Beta distribution for large values of the transform's argument.

The technical part of the research also begins with approximating the derived LT of the parent model $\mathfrak{M}_2$, obtaining the inverse LT *mutatis mutandis* the distribution of the innovation sequence $\{\xi_t : t \in \mathbb{Z}\}$. Therefore, considering initially a time series model with AB, where the latter is in fact approximated $\mathrm{B}_2(p,q), (p,q > 0)$, we arrive at a new model called $ABAR(1)$. The presently considered approximation of the Watson's estimate (3) reads

$$\varphi_X(s) \sim \frac{\Gamma(p+q)}{\Gamma(q)\,(s^p+q)} =: C_X(s), \qquad s \to \infty\,. \tag{7}$$

For certain special cases for p, q, using the analytical inversion of the LT, it is possible to determine the exact PDF which approximates the $\mathrm{B}_2(p,q)$ distribution when the transformation argument $s \to \infty$.

Theorem 6 [31, pp. 585–6, Theorem 2] *For $s \to \infty$ the parent distribution $\mathrm{B}_2(p,q)$ generates the following* PDF *results for the approximated* LT (7) *of the model $\mathfrak{M}_2$:*

1. $\mathrm{B}_2(1,q), q > 0$. *The approximant's* PDF *is*

$$f_1(x) = \frac{q}{1-\mathrm{e}^{-q}}\,\mathrm{e}^{-qx} \cdot 1_{[0,1]}(x)\,.$$

3. $\mathrm{B}_2(1,q), q \geq (43\pi/6)^3 \approx 11413.04$. *Then the approximant's* PDF *becomes*

$$f_3(x) = \frac{2q\Big\{\mathrm{e}^{-\sqrt[3]{q}x} - \mathrm{e}^{\sqrt[3]{q}x/2}\cos\big(\pi/3 + \sqrt{3}\,\sqrt{3}qx/2\big)\Big\} \cdot 1_{\mathcal{I}_3}(x)}{\mathrm{e}^{-\pi/(3\sqrt{3})}\frac{1-\exp(-16\pi/\sqrt{3})}{1-\exp(-4\pi/\sqrt{3})} - \mathrm{e}^{-7\pi/6}\frac{1-\exp(-8\pi)}{1-\exp(-2\pi)} + C_1}, \tag{8}$$

where

$$C_1 = \mathrm{e}^{\pi/(6\sqrt{3})}\,C_3 - \mathrm{e}^{7\pi/2}\,C_2$$

$$C_2 = \cos\frac{7\sqrt{3}\pi}{12} + \mathrm{e}^{\pi}\cos\frac{19\sqrt{3}\pi}{12} + \mathrm{e}^{2\pi}\cos\frac{31\sqrt{3}\pi}{12} + \mathrm{e}^{3\pi}\cos\frac{43\sqrt{3}\pi}{12},$$

$$C_3 = \frac{\sqrt{3}}{2} + \mathrm{e}^{2\pi/\sqrt{3}}\cos\frac{13\pi}{6} + \mathrm{e}^{8\pi/\sqrt{3}}\cos\frac{25\pi}{6} + \mathrm{e}^{12\pi/\sqrt{3}}\cos\frac{37\pi}{6},$$

and

$$\mathcal{I}_3 = \bigcup_{k=0}^{3}\left[\frac{\pi(12k+1)}{3\sqrt{3}\,\sqrt[3]{q}}, \frac{\pi(12k+7)}{6\sqrt[3]{q}}\right].$$

4. $B_2(4, q)$, $q \geq (3\pi/4)^4 \approx 30.821$ *implies the approximant's* PDF

$$f_4(x) = 6\sqrt[4]{q}\,\frac{\mathrm{e}^{-\frac{\sqrt[4]{q}}{\sqrt{2}}x}\sin\left(\frac{\pi}{4}+\frac{\sqrt[4]{q}}{\sqrt{2}}x\right)-\mathrm{e}^{\frac{\sqrt[4]{q}}{\sqrt{2}}x}\cos\left(\frac{\pi}{4}+\frac{\sqrt[4]{q}}{\sqrt{2}}x\right)}{\mathrm{e}^{-\frac{\pi}{4}}+\mathrm{e}^{-\frac{3\pi}{4\sqrt{2}}}\sin\frac{3\pi}{4\sqrt{2}}-\mathrm{e}^{\frac{3\pi}{4\sqrt{2}}}\cos\frac{3\pi}{4\sqrt{2}}}\cdot 1_{\mathcal{I}_4}(x),$$

where

$$\mathcal{I}_4 = \left[\frac{\pi}{\sqrt[4]{4q}}, \frac{3\pi}{4\sqrt[4]{q}}\right].$$

Remark 3 The case $p = 2$ in [31, p. 586, Theorem 2] is unfortunately erroneous.

The case $p = 5, q \geq (9\pi/10)^5 \approx 180.70$ also belongs to the previous theorem. However the expressions obtained are to complicated to be presented here. The interested reader is referred to [31, p. 586, Theorem 2, Eq. (8) *et seq.*]. ■

The asymptotics (7) yields *via* (4) the following asymptotic expression of the LT for $\mathfrak{M}_2$:

$$\varphi_\xi(s) \sim \frac{(\alpha^p s^p + q)(\beta^p s^p + q)}{(s^p + q)(\mu^p s^p + q)} =: R_\xi(s). \tag{9}$$

Now, we are ready to represent the ABAR(1) model's probabilistic description.

Theorem 7 [31, p. 590, Theorem 3] *Consider the mixed times series model ABAR*(1) *related to* $\mathfrak{M}_2$. *Assume that* $\beta < \alpha$, *denote* $\mu^p = p_1\beta^p + (1-p_1)\alpha^p$ *for* $p > 0$ *and two scaling parameters* $\eta, \upsilon > 0$.

Then the i.i.d. sequence $\{\xi_t : t \in \mathbb{Z}\}$ *possesses the mixture of discrete component* 0 *and two continuous distributions:*

$$\xi_t = \begin{cases} 0 & w.p. \quad (\alpha\beta/\mu)^p \\ K_t & w.p. \quad \eta\,(1-\alpha^p)(1-\beta^p)(1-\mu^p)^{-1} \\ \mu K_t & w.p. \quad \upsilon\,(\beta^p-\mu^p)(\mu^p-\alpha^p)\mu^{-p}(1-\mu^p)^{-1}, \end{cases} \tag{10}$$

where K_t *is i.i.d. sequence of random variables such that have* PDF $f_p(x), p \in \{1, 3, 4\}$ *from the* Theorem 6, *for* $p = 5$ *the* PDF *is given as* [31, p. 586, Eq. (8)] *respectively with parameter* $q = q(\eta, \upsilon)$ *which satisfies the constraint*

$$\frac{\Gamma(p+q)}{\Gamma(q)} = \frac{\eta\mu^p(1-\alpha^p)(1-\beta^p)+\upsilon(\mu^p-\beta^p)(\alpha^p-\mu^p)}{(1-\mu^p)\big(\mu^p-(\alpha\beta)^p\big)}, \quad p = 1, 3, 4, 5 \tag{11}$$

so, that $q > 0$ *for* $p = 1$; $q \geq (43\pi/6)^3$ *for* $p = 3$, $q \geq (3\pi/4)^4$ *when* $p = 4$ *and finally* $q \geq (9\pi/10)^5$ *if* $p = 5$.

On the other hand the counterpart result of Theorem 7 reads as follows.

Table 2 Mean values of parameter α and the related deviations. Originally published in [31, p. 596, Table 1]. Published with kind permission of (c) Elsevier 2017. All Rights Reserved

Sample size	$\overline{\widehat{\alpha}}$	$\sigma(\overline{\widehat{\alpha}})$
1.000	0.1049	0.1125
5.000	0.1010	0.0462
10.000	0.0968	0.0334
50.000	0.1011	0.0149
100.000	0.1007	0.0107

Theorem 8 [31, p. 591, Theorem 4] *Let $\{\xi_t : t \in \mathbb{Z}\}$ be an i.i.d. sequence of rv having mixed distribution like* (10)*. If $0 < \beta < \alpha < 1$ and $\mu^p = p_1\beta^p + (1 - p_1)\alpha^p$, where $p = 1, 3, 4, 5$ and q satisfies* (11)*, then the mixed* ABAR(1) *model defines a time series $\{X_t : t \in \mathbb{Z}\}$ whose marginal distribution is specified by* LT (9).

In the sequel we are faced with some properties of the ABAR(1) process X_t.

Theorem 9 [31, pp. 591–2, Theorem 5] *The correlation function with integer lag $\tau > 0$ and the spectral density of the model $\mathfrak{M}_2$ are respectively given as:*

$$\rho(\tau) = (p_1\alpha + \beta(1 - p_1))^{|\tau|}$$

$$f(\nu) = \frac{\mathsf{D}X}{2\pi\left(1 + (p_1\alpha + (1 - p_1)\beta)(1 - 2\cos\nu)\right)}, \qquad \nu \in [-\pi, \pi]\,.$$

Here $\mathsf{D}X$ denotes the variance of the mixed ABAR(1) *process X_t.*

Finally, let us present a numerical simulation of the parameter estimation. Bearing in mind all earlier considerations, 100 samples of sizes 1.000, 5.000, 10.000, 50.000 and 100.000 were drawn using $\mathfrak{M}_2$. We will assume that parameters β, p_1 and μ^p are known. Since we had 100 samples for each size, mean value of all 100 estimates per each sample size is reported in Table 2 below. Mean value of estimates of parameter α is denoted by $\overline{\widehat{\alpha}}$ and its standard deviation by $\sigma(\overline{\widehat{\alpha}})$. True value of parameter α is 0.1; consult Table 2.

Estimator $\widehat{\alpha}$ converges very slowly to the true value of α, therefore it is necessary to generate huge samples for better accuracy of this kind estimators.

4 Generalized Beta as the Parent Distribution

A new type functional generalized Beta distribution has been considered by Cordeiro and de Castro in [4, p. 884]. Starting from a parent absolutely continuous CDF $G(x)$, having PDF $g'(x) = g(x)$ consider the rv X on a standard probability space $(\Omega, \mathscr{F}, \mathsf{P})$ defined by the associated PDF [4, p. 884, Eq. (3)]

$$f_G(x) = \frac{g(x)}{\mathrm{B}(p,q)}\,[G(x)]^{p-1}\,[1-G(x)]^{q-1}\cdot 1_{\mathrm{supp}(g)}(x), \qquad \min(p,q)>0.$$

Under the support supp(h) of a real function h we mean as usual, the subset of the domain containing those elements which are not mapped to nil.

Obviously G replaces the argument in $\mathrm{B}_2(p,q)$, so our notation $\mathrm{B}_2(G)$ for the functional generalized two-parameter Beta distribution; when G reduces to the identity, we arrived at $X \sim \mathrm{B}_2(p,q)$.

Theorem 10 *The asymptotic behavior of the* LT *of a rv* $X \sim \mathrm{B}_2(G)$

$$\varphi_G(s) \sim \frac{\Gamma\left(\frac{p}{\mu}\right) \mathrm{e}^{-sG^{-1}(0)}}{\mu\,\mathrm{B}(p,q)\,\mathfrak{g}_0^{\frac{p}{\mu}}\, s^{\frac{p}{\mu}}}\left(1+\mathscr{O}(s^{-\frac{1}{\mu}})\right), \qquad s\to\infty, \tag{12}$$

provided finite $G^{-1}(0)$ *and*

$$G^{-1}(x) \sim G^{-1}(0) + \sum_{k\ge 0} \mathfrak{g}_k x^{\mu+k}; \qquad \mu, \mathfrak{g}_0 > 0, \tag{13}$$

when $x \to 0^+$. *Here* G^{-1} *denotes the inverse of* G.

Proof Firstly, we are looking for the LT expression of the rv $X \sim \mathrm{B}_2(G)$:

$$\begin{aligned}\varphi_G(s) = \mathsf{E}\,\mathrm{e}^{-sX} &= \frac{1}{\mathrm{B}(p,q)}\int_{\mathrm{supp}(g)} \mathrm{e}^{-sx}\, g(x)\,[G(x)]^{p-1}\,[1-G(x)]^{q-1}\,\mathrm{d}x \\ &= \frac{1}{\mathrm{B}(p,q)}\int_0^1 \mathrm{e}^{-s\,G^{-1}(x)}\, x^{p-1}\,(1-x)^{q-1}\,\mathrm{d}x\,.\end{aligned}$$

Being G a CDF, it is monotone non–decreasing, hence the inverse there exists, at least the generalized $G^{-}(x) := \inf\{y\in\mathbb{R}\colon G(y)\ge x\}$, $x\in\mathbb{R}$. So, according to assumptions upon the parent CDF G with finite $G^{-1}(0)$ and (13), to fix the asymptotics of $\varphi_G(s)$ for growing s we apply the Erdélyi's expansion [5, 26] of Watson's lemma which details we skip here.[2] By Erdélyi's theorem we deduce

$$\varphi_G(s) \sim \frac{\mathrm{e}^{-sG^{-1}(0)}}{\mathrm{B}(p,q)}\sum_{n\ge 0}\Gamma\left(\frac{p+n}{\mu}\right) b_n\, s^{-\frac{p+n}{\mu}}, \qquad s\to\infty,$$

where $b_0 = (\mu\,\mathfrak{g}_0^{\frac{p}{\mu}})^{-1}$, which is equivalent to the statement. □

[2] Updated extensions of Erdélyi's theorem are obtained also in [24] and [39].

The leading term in (13) possesses inverse LT equal to

$$\frac{1}{\mu\, \mathrm{B}(p,q)\, \mathfrak{g}_0^{\frac{p}{\mu}}}\, (x - G^{-1}(0))^{\frac{p}{\mu}-1}\, \theta(x - G^{-1}(0))\,,$$

where $\theta(\cdot)$ denotes the Heaviside function. To become a PDF it should be re–normalized. So, the support set is a finite interval $\mathcal{I}_g := [G^{-1}(0), G^{-1}(0) + T]$, $T > 0$, as $p, \mu > 0$. This results in approximated $\mathrm{B}_2(G)$ distribution which belongs to the cutoff delayed power law (or Pareto) family. The related PDF is

$$f(x) = \frac{p}{\mu\, T^{\frac{p}{\mu}}}\, (x - G^{-1}(0))^{\frac{p}{\mu}-1} \cdot 1_{\mathcal{I}_G}(x)\,,$$

which does not contain q. This is understandable since q appears in higher order terms in (12) or take place in a suggested asymptotic forms for φ_G like the exponential (5) for $\mathrm{AB}_{p,q}\mathrm{AR}(1)$ or the rational (7) for ABAR(1) models. Now, it remains to apply this approach to the initial models $\mathfrak{M}_1, \mathfrak{M}_2$ with the derived approximated marginal. However, these problems deserve another separate study.

References

1. Abate, J., Whitt, W.: Numerical inversion of Laplace transforms of probability distributions. ORSA J. Comput. **7**(1), 36–43 (1995)
2. Abramowitz, M., Stegun, I.A. (eds.): Handbook of Mathematical Functions with Formulas, Graphs, and Mathematical Tables. Applied Mathematics Series, vol. 55. Tenth Printing, National Bureau of Standards (1972)
3. Chernick, M.: A limit theorem for the maximum of autoregressive processes with uniform marginal distribution. Ann. Probab. **9**, 145–149 (1981)
4. Cordeiro, G.M., de Castro, M.: A new family of generalized distributions. J. Stat. Comput. Simul. **81**(7), 883–898 (2011)
5. Erdélyi, A.: Asymptotic Expansions. Dover, New York (1956)
6. Fletcher, S., Ponnambalam, K.: Estimation of reservoir yield and storage distribution using moments analysis. J. Hydrol. **182**, 259–275 (1996)
7. Gaver, D., Lewis, P.: First order autoregressive Gamma sequences and point processes. Adv. Appl. Probab. **12**, 727–745 (1980)
8. Hamilton, J.: Time Series Analysis. Princeton University Press, Princeton (1994)
9. Jacobs, P.A., Lewis, P.A.W.: A mixed autoregressive moving average exponential sequence and point process EARMA (1, 1). Adv. Appl. Probab. **9**, 87–104 (1977)
10. Jevremović, V.: Two examples of nonlinear process with mixed exponential marginal distribution. Stat. Probab. Lett. **10**, 221–224 (1990)
11. Jevremović, V.: Statistical properties of mixed time series with exponentially distributed marginals. PhD Thesis. University of Belgrade, Faculty of Science [Serbian] (1991)
12. Jones, M.: Kumaraswamy's distribution: a beta-type distribution with some tractability advantages. Stat. Methodol. **6**, 70–81 (2009)
13. Karlsen, H., Tjøstheim, D.: Consistent estimates for the NEAR (2) and NLAR (2) time series models. J. Roy. Stat. Soc. B **50**(2), 120–313 (1988)

14. Kumaraswamy, P.: A generalized probability density function for double-bounded random processes. J. Hydrol. **46**, 79–88 (1980)
15. Lawrence, A.J.: Some autoregressive models for point processes. In: Bártfai, P., Tomkó, J. (eds.) Point Processes and Queuing Problems, Colloquia Mathematica Societatis János Bolyai 24. North Holland, Amsterdam (1980)
16. Lawrence, A.J.: The mixed exponential solution to the first order autoregressive model. J. Appl. Probab. **17**, 546–552 (1980)
17. Lawrence, A.J., Lewis, P.A.W.: A new autoregressive time series model in exponential variables (near(1)). Adv. Appl. Probab. **13**, 826–845 (1980)
18. Lawrence, A.J., Lewis, P.A.W.: A mixed exponential time-series model. Manage. Sci. **28**(9), 1045–1053 (1982)
19. Mališić, J.: On exponential autoregressive time series models. In: Bauer, P., et al. (eds.) Proceedings of Mathematical Statistics and Probability Theory (Bad Tatzmannsdorf, 1986), vol. B, pp. 147–153. Reidel, Dordrecht (1987)
20. Mališić, J.: Some properties of the variances of the sample means in autoregressive time series models. Zb. Rad. (Kragujevac) **8**, 73–79 (1987)
21. McKenzie, E.: An autoregressive process for beta random variables. Manage. Sci. **31**, 988–997 (1985)
22. Nadarajah, S.: Probability models for unit hydrograph derivation. J. Hydrol. **344**, 185–189 (2007)
23. Nadarajah, S.: On the distribution of Kumaraswamy. J. Hydrol. **348**, 568–569 (2008)
24. Nemes, G.: An explicit formula for the coefficients in Laplace's method. Constr. Approx. **38**(3), 471–487 (2013)
25. Novković, M.: Autoregressive time series models with Gamma and Laplace distribution. MSc Thesis. University of Belgrade, Faculty of Mathematics [Serbian] (1997)
26. Olver, F.W.J., Olde Daalhuis, A.B., Lozier, D.W., Schneider, B.I., Boisvert, R.F., Clark, C.W., Miller, B.R., Saunders, B.V. (eds.): NIST Digital Library of Mathematical Functions. §2.3. (iii) Laplace's Method. Release 1.0.13 of 2016-09-16. http://dlmf.nist.gov/
27. Popović, B.Č.: Prediction and estimates of parameters of exponentially distributed *ARMA* series. PhD Thesis. University of Belgrade, Faculty of Science [Serbian] (1990)
28. Popović, B.Č.: Estimation of parameters of RCA with exponential marginals. Publ. Inst. Math. (Belgrade) (N.S.) **54**, 135–143 (1993)
29. Popović, B.V., Pogány, T.K., Nadarajah, S.: On mixed *AR*(1) time series model with approximated Beta marginal. Stat. Probab. Lett. **80**, 1551–1558 (2010)
30. Popović, B.V.: Some time series models with approximated beta marginals. PhD Thesis. University of Niš, Faculty of Science [Serbian] (2011)
31. Popović, B.V., Pogány, T.K.: New mixed *AR*(1) time series models having approximated beta marginals. Math. Comput. Model. **54**, 584–597 (2011)
32. Pourahmadi, M.: Stationarity of the solution of $x_t = a_t x_{t-1} + \xi_t$ and analysis of non-gaussian dependent random variables. J. Time Ser. Anal. **9**, 225–239 (1988)
33. Ridout, M.: Generating random numbers from a distribution specified by its Laplace transform. Stat. Comput. **19**, 439–450 (2009)
34. Ristić, M.M.: Stationary autoregressive uniformly distributed time series. PhD Thesis. University of Niš, Faculty of Science (2002)
35. Ristić, M.M., Popović, B.Č.: The uniform autoregressive process of the second order. Stat. Probab. Lett. **57**, 113–119 (2002)
36. Sim, C.H.: Simulation of Weibull and gamma autoregressive stationary process. Comm. Stat. B-Simul. Comput. **15**(4), 1141–1146 (1986)
37. Stanković, B.: On the function of E.M. Wright. Publ. Inst. Math. (Belgrade) (N.S.) **10**, 113–124 (1970)
38. Watson, G.N.: The harmonic functions associated with the parabolic cylinder. Proc. London Math. Soc. **2**(17), 116–148 (1918)
39. Wojdylo, J.: On the coefficients that arise from Laplace's method. J. Comput. Appl. Math. **196**(1), 241–266 (2006)

Prediction of Noisy ARIMA Time Series via Butterworth Digital Filter

Livio Fenga

Abstract The problem of predicting noisy time series, realization of processes of the type ARIMA (Auto Regressive Integrated Moving Average), is addressed in the framework of digital signal processing in conjunction with an iterative forecast procedure. Other than Gaussian random noise, deterministic shocks either superimposed to the signal at hand or embedded in the ARIMA excitation sequence, are considered. Standard ARIMA forecasting performances are enhanced by pre-filtering the observed time series according to a digital filter of the type Butterworth, whose cut-off frequency, iteratively determined, is the minimizer of a suitable loss function. An empirical study, involving computer generated time series with different noise levels, as well as real-life ones (macroeconomic and tourism data), will also be presented.

Keywords ARIMA models · Butterworth filter · Noisy time series · Time series forecast

1 Introduction

Virtually all the domains subjected to empirical investigation are affected, to different extents, by some source of noise, regardless the accuracy of the measurement device adopted. Noise-free signals are practically an unachievable goal, belonging to the realm of abstraction or of lab-controlled experiments. Noise, in fact, is simply ubiquitous, an all-permeating entity affecting physical and non-physical world at all scales and dimensions, whose uncountable expressions can be only partially controlled and never fully removed nor exactly pinpointed. A common yet important source of noise attains to the measurement processes, e.g. related to telemetry systems, data recover algorithms and storage devices. Non-electronic world is no exception: a simple set of data in a paper questionnaire can embody a number of noisy components, for example in the form of different types of mistakes—such as

L. Fenga (✉)
UCSD, University of California San Diego, San Diego, CA, USA
e-mail: lfenga@math.ucsd.edu; fenga@istat.it

I. Rojas et al. (eds.), *Advances in Time Series Analysis and Forecasting*,
Contributions to Statistics, DOI 10.1007/978-3-319-55789-2_13

recording errors, lies or failure in detecting and correcting outlying observations—sampling bias, attrition and changes in data collecting procedures. However, even in an ideal world—with infinitely precise devices, 0-noise transmission lines, error proof storage devices and so forth—going down to scale, interference with the theoretically pure information can still be found at thermodynamic and quantum levels (the so called quantum noise) [1].

Many are the areas where ad hoc noise reduction techniques are routinely employed, such as data mining, satellite data, radio communications, radar, sonar and automatic speech recognition, just to cite a few. Often, the treatment of noisy data is critical, as in the case of bio-medical signals, where crucial details might be lost as a result of over-smoothing, aerospace—as tracking data are of little or no use if not properly filtered—or economics, where noise components can mask important features of the investigated system.

In time series analysis, noise is one common element of obstruction for accurate predictions and its ability to impair even simple procedures—such as visual extrapolation of meaningful patterns—is unfortunately well known. Far from ideal for being noisy and always of finite—and in many case limited—length, the portion of information exploitable in real time series might not adequately support model selection and inference procedures, so that the task of detecting, extracting and projecting into the future only the relevant dynamic structures and discarding the rest is not trivial. This is especially true if one considers that gauging the level of noise present in a given system is usually difficult, so that a precise discrimination between weak signals—which can justify modelling efforts—and absence of any structure, might not be possible. The proposed method deals with this problem from a signal processing perspective, assuming the information being carried by a time series realization of an ARIMA (Auto Regressive Integrated Moving Average) [2] data generating process (DGP). Such an assumption is mainly motivated by the widespread use of this class of models for the representation of many linear (or linear-approximable) phenomena and the strong theoretical foundation it is grounded upon.

2 Noise Reduction Techniques

In general, statistical models performances are heavily dependent on the level of the signal compared with the system noise. This relation is commonly expressed in terms of signal-to-noise-ratio (SNR), which essentially measures the signal strength relative to the background noise. By construction, this indicator indirectly provides an estimate of the uncertainty present in a given system and therefore its degree of predictability. In order to maximize the SNR, a number of de-noising signal processing methods and noise-robust statistical procedures have been devised. A popular approach, which has gained widespread acceptance among theoretical statisticians and practitioners in recent years, is based on Wavelet theory [3, 4]. Successfully applied for noise removal from a variety of signals, its main focus is the extraction and the treatment of noise components at each wavelet scale. In the same spirit, thresholds wavelet approaches are grounded upon the fact that information tend to be captured by those few coefficients showing larger absolute values, so that a properly set threshold can discriminate their relative magnitude and thus it is likely to retain only the useful information. However, the choice of the threshold is critical and might not fully account for the noise distribution across the different scales. Noise reduction computer intensive methods, such as artificial neural networks [5–7], dynamic genetic programming [8], kernel support vector machines [9] and self-organizing maps [10], are also widely employed. More traditional, model-based procedures encompass a broad range of approaches: from the type autoregressive [11], to those based on Bayesian hierarchical hidden Markov chain [12] and Markov chain Monte Carlo techniques [13].

Linear filter theory can be considered a fundamental source of many signal extraction and de-noising methods of massive employment in industry—e.g. aerospace, terrestrial and satellite communications, audio and video—as well as in a wide range of scientific areas, e.g. physics, engineering and econometrics. In such a large diffusion, their computational and algebraical tractability played a significant role along with their remarkable capabilities in the attenuation of additive noise of the type Gaussian. Hodrick–Prescott, Baxter–King, Christiano–Fitzgerald as well as random walk band pass filter or simple moving averages, are all examples of linear filters of common use in econometric and finance, mainly to the end of extracting the different frequency components of a time series (e.g. trend-cycle, expansion–recession phases) . In other domains of application, e.g. electric engineering, geophysics, astronomy or neuroscience, linear filters are usually of different types: a popular class is the Elliptic filters, which encompasses as special case filters of common use such as Cauer, Chebyshev, reverse Chebyshev and the one employed in the present paper: Butterworth. These filters are commonly employed for the separation of audio signals, to enhance the radio signal by rejecting simultaneously unwanted sources, and more in general for information reconstruction and linear time invariant systems de-noising.

2.1 The Proposed ARIMA-ω Procedure

The proposed method is an automatic, auto-adaptive, partially self-adjusting data driven procedure, conceived to improve the forecast performances of a linear prediction model of the type ARIMA—detailed in Eqs. (2–3)—by removing noisy components embedded in the high frequency spectral portion of the signal under investigation. The attenuation of those components, responsible for higher forecast errors, is performed through an Infinite Impulse Response, time invariant, low-pass digital filter of the type Butterworth (BW), which is characterized by a monotonic frequency response and is entirely specified by two parameters: cut-off frequency and order. Also referred to as maximally flat magnitude filter, it belongs to the class of Wiener–Kolmogorov solution of the signal extraction Mean Squared Error minimization problem. Its performances in extracting trend and cycle components in economic time series[1] has been studied in [14]. The design of the type BW has been chosen here mainly for being ripple-free in both the stop-band and pass-band and for possessing good attenuation capabilities in the former and for showing more linear phase response in the latter, in comparison with popular filters such as *Tchebychev* Type 1 and Type 2. On the other hand, it has a significant drawback involving the roll-off rate, which is slow and therefore implies higher filter orders to achieve an acceptable sharpness at cut-off frequencies. BW n-order squared amplitude response function can be expressed in terms of the transfer function $H(s)$ as $G^2(\omega) = H|(j\omega)|^2 = \frac{G_0^2}{1+(\omega/\tilde{\omega})^{2\nu}}$, where with $\tilde{\omega}$, ν and G_0 the cut-off frequency, the filter order and gain at 0-frequency are respectively denoted. The pass-band/stop-band width of the transition region (filter's sharpness) at $\tilde{\omega}$, is controlled by the parameter ν, so that for $\nu \to \infty$ the gain becomes a rectangle determining for all the frequencies below (above) $\tilde{\omega}$ to be passed (suppressed). By setting G_0 to 1, the squared amplitude response function becomes:

$$G^2(\omega) = \frac{1}{1 + (\omega/\tilde{\omega})^{2\nu}}. \tag{1}$$

The proposed procedure employs the digital version of BW filter, for it ensures a consistently better flatness in the pass-stop band than its analogue counterpart and superior attenuation capabilities in the stop-band. The digital design has been obtained by redefining the analogue transfer function from the complex s-plane $H(s)$ to the z-plane $H(z)$, by means of the bilinear transform [15]. This approximation is performed by replacing the variable s in the analog transfer function by an expression in z, i.e. $2f_s\frac{z-1}{z+1}$, f_s being the sampling frequency, so that the filter is now expressed as $\frac{1}{1+\lambda}(f_s\frac{1-z}{1+z})^{2\nu}$, with $\lambda = [1/\tan(\tilde{\omega})]^{2\nu}$. Being based on a statistical model of the type ARIMA, which is fed with a smoothed version of the time series according to a BW-type filter tuned at a particular frequency $\tilde{\omega}$, the method is called ARIMA-ω.

To adequately perform, it requires the optimal, system-specific calibration of the cut-off frequency parameter. This is a crucial step, as its incorrect estimate might severely impact the expected filtering performances and introduce biasing elements in the method's outcomes. In such circumstances, manual adjusting strategies, i.e. conducted on a trial and error basis, might be a tedious and in many cases not a workable solution, considering the high number of operations required, such as visual inspection and comparison of the original and filtered signals, spectral and residual analysis, evaluation of the trade-off between signal distortion and noise passed, and so forth.

[1]From now on, the term signal and time series will be used interchangeably.

2.2 The Underlying Stochastic Process and the Noise Model

Throughout the paper, the signal of interest is assumed to be a finite realization of a process of the type ARIMA, which envisions the current observation of a time series to be a linear combination of the previous realizations, an error term related to the present realization and a weighted average of past error terms.

Let $\{X_t\}_{t\in\mathbb{Z}^+}$ be a real 2nd order stationary process, with mean μ. It is said [2] to admit a Autoregressive Moving Average representation of order p and q—i.e. $x_t \sim ARMA(p,q)$, with $(p,q) \in \mathbb{Z}^+$—if for some constant $\phi_1 \dots \phi_p$, $\theta_1 \dots \theta_q$, it is: $\sum_{j=0}^{p} \phi_j(X_{t-j} - \mu) = \sum_{j=0}^{q} \theta_j \alpha_{t-j}$, assuming: (a) $\phi_0 = \theta_0 = 1$; (b) $E\left\{\alpha(t)|S_{t-1}\right\} = 0$; (c) $E\left\{\alpha^2(t)|S_{t-1}\right\} = \sigma^2$; (d) $E\alpha^4(t) < \infty$; (e) $\sum_{j=0}^{p} \phi_j z^j \neq 0$, $\sum_{j=0}^{p} \theta_j z^j \neq 0, |z| \leq 1$. Here S_t denotes the sigma algebra induced by $\{\alpha(j), j \leq t\}$, whereas $\sum_{j=0}^{p} \phi_j z^j$ and $\sum_{j=0}^{p} \theta_j Z^j$ are assumed not to have common zeros. In what follows, X_t is assumed to be 0-mean and either realization of a stationary *ARMA* process, as above defined, or integrated of order d, so that stationarity is achieved by differencing d times the original time series. This differencing factor is embodied in the *ARMA* scheme by adding the integration term, denoted as $I(d)$, with d being a positive integer, so that we have $x_t \sim ARIMA(p,d,q)$. By using the back-shift operator L, i.e. $LX_t = X_{t-1}$ (therefore $L^n X = X_{t-n}$) and the difference operator $\nabla^d X_t = (1-L)^d X_t \quad d = 0, 1, \dots, D$, the ARIMA model is synthetically expressed as follows:

$$\nabla^d x_t = \frac{\theta(L)}{\phi(L)} \alpha_t, \tag{2}$$

with $\phi_p(L) = 1 - \phi_1 L - \phi_2 L^2 - \dots - \phi_p L^p$; $\theta_q(L) = 1 - \theta_1 L - \theta_2 L^2 - \dots - \theta_q L^q$, and difference operator applied d times until stationarity is reached. Here ϕ, θ and α_t are, respectively, the autoregressive and moving average parameters and the 0-mean, finite variance white noise. The model can be estimated when the stationary and invertibility conditions are met for both the autoregressive and moving average polynomials respectively, that is when $\phi_p(L)\theta_q(L) = 0$ has roots lying outside the unit circle.

In order to mimic actual time series, data are always supposed to be observed with error. Therefore, no direct access to the theoretical, uncorrupted realizations $x_t \sim$ ARIMA(p, d, q) (2) is possible, but only to the signal $y_t \in \{Y_t\}_{t\in\mathbb{Z}^+}^T$, which is measured with additive, independent noise δ_t, i.e.

$$y_t = \nabla^d x_t + \delta_t. \tag{3}$$

In presence of noisy data, virtually the whole model building procedure can be affected. For example, assessment of the probabilistic structure of the time series via the Empirical Autocorrelation Functions (global and partial), Maximum Likelihood (ML)-based inference and model order selecting procedures or validation and diagnostic checks can be, to a different extent, all biased. The consequences are in general not negligible and range from instability of the parameter estimates (able to introduce significant amount of model uncertainty), to the selection of a wrong model, whose outcomes can be totally unreliable or, in the best case scenario, require additional validation efforts.

ARIMA-ω requires no particular assumptions on the noise structure nor efforts to pinpoint its sources. Noise is simply treated in an agnostic way, as a nuisance element to get rid of *if* and inasmuch as it is detrimental to the predictions generated by (3). The proposed method has been tested considering two theoretical forms of noise: one directly affecting the data generating mechanism, which enters the system through the excitation sequence α_t (2) in the form of outliers of the type innovation (*IO*) and the other superimposed to the data in the form of both Gaussian noise and additive outlier (*AO*), accounted for by the term δ_t in (3). Affecting the *DGP* fundamental equation, the first type of noise is system-dependent whereas the latter is assumed superimposed to the theoretical clean signal. As such, it arises in the measurement process (e.g. sensor or telemetry related) in the form of Gaussian noise or as a result of exogenous shocks or mistakes made in the observation or data recording stages (*AO*).

2.3 The ARIMA-BW Filter Unified Framework

In order to keep the explanation of the method as simple as possible, in this paragraph the parameter $\nu \in \mathbb{Z}^+$ (1), which controls the filter order, is assumed to be known. Such an assumption is reasonable also by an operative point of view, as the number of the candidates is in general limited to few integers. Therefore, one can ground its choice on experience and/or best practices—possibly in conjunction with a trial and error approach. However, a more structured approach, envisioning its automatic estimation as integrated in the optimal cut-off frequency searching algorithm, will be pursued here (Sect. 2.4). On the other hand, it has to be said that, in general, such an approach unfortunately comes at much greater computational expense.

In essence, ARIMA-ω procedure is based on the idea of choosing the "best" cut-off frequency—conditional to an optimal filter order ν_0, say $(\omega_0|\nu_0)$—as the minimizer of the vector of outcomes $\bar{\mathfrak{L}}$ of a suitable loss function, $\mathfrak{L}(\cdot)$. This vector is generated by iteratively computing $\mathfrak{L}(\cdot)$ on the out-of-sample predictions yielded by a set of "best" ARIMA(p, d, q) models, fitted to a set of filtered versions of a given time series, according to different cut-off frequencies. Optimality of the ARIMA structure is granted by an Information Criterion (IC)-based model order selector, as explained below. Formally, ω_0 is the cut-off frequency minimizing the loss function $\mathfrak{L}(\cdot)$ computed on the *best* forecasting function $f(Y_t) : y \rightarrow \mathbb{R}$, estimated on the J filtered values $y_t(\omega_j)$ $j = 1, 2, \ldots, J$. However, it will be taken as the winner one—in the sense that the final predictions will be generated by the original series transformed accordingly—if the correspondent $\mathfrak{L}(\omega_0)$-value is smaller than the one obtained on the unfiltered data, i.e. $\mathfrak{L}(y_t, f_{\omega_0}(y_t)) < \mathfrak{L}(y_t, f_{\omega_j}(y_t))$, $j = 1, 2, \ldots, J - 1$, *iff.* $\mathfrak{L}(y_t, f_{\omega_0}(y_t)) < \mathfrak{L}(\hat{y}_t, f(y_t))$. Such a design, guarantees the low-pass filter to operate only to the extent needed and *when* needed, according to the inherent structure of the series under investigation: should $\mathfrak{L}(\cdot)$ show no improvements after the BW filter intervention, the procedure automatically cuts the filter off and provides at its output the predictions generated by the ARIMA model fitted on the original data. The adopted loss function is the *RMSFE* (Root Mean Square Forecast Error) which requires a validation set of appropriate length. Based on the L_2-norm and widely employed in the construction of many de-noising algorithms [16], the *RMSFE* in general takes the following form:

$$\mathfrak{L}(y_i, \hat{y}_i) = [R^{-1} \sum_{i=1}^{R} |e_i|^2]^{\frac{1}{2}}, \tag{4}$$

with y_i and $\hat{y}_i$ denoting the observed values and the predictions respectively, e their difference and R the sample size.

The estimation of the ARIMA order $(\hat{p}, \hat{d}, \hat{q})$, is driven by the Akaike Information Criterion (*AIC*) [17], which is defined as $-2 \max \log(L(\hat{\theta}|y)) + 2K$, with K the model dimension and $(L(\hat{\theta}|y))$ the log-likelihood function. The related selection strategy, called *MAICE* (short for Minimum *AIC* Expectation) [18], is a procedure aimed at extracting, among the candidate models, the order $(\hat{p}, \hat{d}, \hat{q})$ satisfying:

$$(\hat{p}, \hat{d}, \hat{q}) = \arg \min_{p \le P_0, d \le D, q \le Q_0} AIC(p, d, q). \tag{5}$$

MAICE procedure requires the definition of an upper bound for p, d and q, i.e. (P, D, Q), as a maximum order a given process can reach. This choice, unfortunately, is a priori and arbitrary.

Under *MAICE*-driven optimal *ARIMA* searching strategy, the selected cut-off frequency ω_0 yields the desired response in terms of attenuation of those signal frequencies affecting the quality of the predictions. ARIMA-ω promising performances can be justified in terms of parameter estimation inference. Let $\boldsymbol{\Theta} \equiv (\boldsymbol{\phi}, \boldsymbol{\theta})$ be the vector of the *ARMA* parameters as defined in (2), with y_t having spectral density $f_\Theta(\omega)$ and $T \times T$ variance covariance matrix $\Sigma_{T,\Theta}$. The *MLE* estimate for $\boldsymbol{\Theta}$, i.e. $\hat{\boldsymbol{\Theta}}$, is the minimizer of the log-likelihood function, defined as usual as $-2 \log lik(\Theta) = T \log(2\pi) + \log |\Sigma_{T,\Theta}| + y' \ \Sigma_{T,\Theta}^{-1} \ y$. By virtue of the assumptions made in formulating the model (3), the BW-filtered time series $y_t(\omega_0)$ shows a less noisy spectral density $f_\Theta(\omega_0)$ and thus enters the Likelihood function with a smaller variance, say $\sigma_y^2(\omega)$, which by construction is in general closer to the one of the original signal, i.e. σ_y^2. Therefore, the matrix $\Sigma_{T,\Theta}$ exhibits less disperse values and as a result the parameters estimation will be more precise and adherent to the "pure", uncorrupted *DGP*. On the other hand, once the optimal cut-off frequency is reached, trying to further reduce σ_y^2 would imply necessarily progressively suboptimal results, proportionally to the relevance of the portion of relevant information filtered-out. In this framework, the good results achieved by the method when the signal is corrupted by impulsive noise are also explained. In fact, its potentially catastrophic impact—e.g. on the *ML* function in terms of departure from the assumed data distribution and suboptimal maximization procedure—is mitigated by the smoothing properties of the BW filter.

2.4 The Algorithm

As already pointed out, ARIMA-ω procedure is aimed at finding the optimal cut-off frequency ω_0 of a digital version of a filter of the type Butterworth and its optimal order ν_0. In practice, $(\omega_0; \nu_0)$ is the minimizer of a quadratic loss function $\mathfrak{L}(\cdot)$ (4) computed recursively on the predicted values generated by a set of best ARMA models (in the MAICE sense) on a long "enough" validation set. The predictions obtained using the ω_0-filtered version of the time series under investigation, instead of the original one, are in general more accurate. In what follows, ARIMA$-\omega$ procedure is detailed in a step-by-step fashion.

Let y_t be the time series of interest (2),

1. y_t is split in three disjoint segments: training set $\{y^A\}_1^{T-(S+V)}$, validation set, $\{y^U\}_{T-(S+V+1)}^{(T-V)}$ and test set, $\{y^E\}_{T-(V+1)}^{T}$, where with V and S respectively, the length of validation and test set are denoted;
2. the two 1-dimensional grids of:
   (a) tentative cut-off frequencies
   $$\{(\omega_j;\ j = 1, 2, \dots, J) \subset \Omega\}$$
   (b) tentative filter orders
   $$\{(\nu_w;\ w = 1, 2, \dots, W) \subset M\}$$
   are built;
3. a maximum ARIMA order (P, D, Q), likely to encompass the true model order, is arbitrarily chosen;
4. y_t is BW-filtered J times according to a given filter order ν_w, and a set of cut-off frequencies $j \in J$, so that the matrix $\boldsymbol{\Psi}$ with dimensions $(T - (S + V) \times J)$ whose column-vectors are the filtered time series $y_t(\omega_j)j = 1, 2, \dots, J$, is generated;
5. an exhaustive set of tentative ARIMA(p, d, q)—of size $(D + 1)[(Q) + 1)]^2$ (assuming, as it will be done in the empirical section, $P = Q$)—is fitted recursively, up to the order (P, D, Q), to the original, unfiltered time series y_t;
6. *AIC* is computed for all the "candidate" triple (p, d, q) and the winner one, called (p^*, d^*, q^*), is extracted according to *MAICE* procedure (Eq. 5);

7. steps (5, 6) are performed for each column vector of $\boldsymbol{\Psi}$, i.e. $y_t(\omega_j)$ $j = 1, 2, \ldots, J$, so that the optimal ARIMA, *MAICE*-based, is determined for each filtered time series conditional to ν_w, i.e. $(p^*, d^*, q^*)_{\omega_j} \equiv \omega_j^* | \nu_w$;
8. horizon h_γ (h_γ; $\gamma = 1, 2, \cdots, \Gamma$) predictions for the validation set Y^U are generated according to the ARIMA models selected respectively in steps (5, 6 and 7). Without loss of generality, it is supposed that only one horizon, say h_0, is considered;
9. the loss function $\mathfrak{L}(\cdot)$ is computed on the $J+1$ vectors containing the predictions, i.e. $\mathfrak{L}(y_E, \hat{y}_E) = [E^{-1} \sum_{i=T-(S+V+1)}^{T-V} |e_i|^2]^{\frac{1}{2}}$ so that the vector, called $\bar{\mathfrak{L}}(y_E, \hat{y}_E)$, containing the *RMSFE* of all the filtered series and the unfiltered one is generated, i.e. $\bar{\mathfrak{L}}(y_U, \hat{y}_U) \equiv \Big[\mathfrak{L}(y_U, \hat{y}_U^*), \mathfrak{L}(y_U, \hat{y}_U(\omega_1^* | \nu_w)), \ldots, \mathfrak{L}(y_U, \hat{y}_U(\omega_j^* | \nu_w)), \ldots, \mathfrak{L}(y_U, \hat{y}_U(\omega_J^* | \nu_w))\Big]'$. Here, with $\hat{y}_U^*$ and $\hat{y}_U(\omega_j^*)$ the predictions of the best ARIMA models, respectively for the original and j-filtered time series, are denoted;
10. the cut-off frequency ω_0 satisfying

$$^{w}\omega_0 = \omega_0 | \nu_w = \underset{(\omega \subset \Omega) | \nu_w}{\arg\min}\ \bar{\mathfrak{L}}(y_U, \hat{y}_U^*) \iff \mathfrak{L}(y_U, \hat{y}_U^*) > \mathfrak{L}(y_U, \hat{y_U}(\omega_0^*)) \tag{6}$$

    is the winner one conditional to ν_w;
11. steps 4 to 10 are repeated $(M-1)$ times, i.e. for all the remaining grid values in M. The value ω_0 minimizing (6) is the final cut-off frequency, i.e.

$$\omega_0 = (\omega_0; \nu_0) = \underset{(\omega \subset \Omega) | (\nu \subset M)}{\arg\min}\ \bar{\mathfrak{L}}(y_U, \hat{y}_U) \iff \mathfrak{L}(y_U, \hat{y}_U^*) > \mathfrak{L}(y_U, \hat{y_U}(\omega_0^*));$$

12. final performances evaluations are made on the predictions obtained by the best ARIMA structure—fitted on $\{y^A(\omega_0; \nu_0)\}$—for the test set y_t^E.

3 Empirical Experiment

This section focuses on the illustration of the design and the outcomes of an empirical study implemented[2] to evaluate the performances delivered by ARIMA-ω procedure. Based on a Monte Carlo experiment and on the analysis of real-life time series, it envisions two different sets of data: in particular, the Monte Carlo experiment uses an artificial one, consisting of four subsets of time series generated according to four pre-specified *DGPs* (detailed in Table 1) under a variety of noise conditions, whereas the set of actual data consists of eight time series, related to macroeconomic and tourism related variables (summarized in Table 2). The quality of the proposed

[2]Part of the elaborations has been performed using the computing resource "Pythagoras", maintained by the Mathematical Department of the University of California San Diego.

Table 1 Parametrization of the simulated *DGP*s

dgp number	ARIMA order	ϕ	θ
1	$(0, 1, 1)$	–	−0.6
2	$(1, 1, 2)$	−0.65	0.6; −0.45
3	$(2, 0, 1)$	0.7; −0.5	−0.5
4	$(1, 0, 2)$	−0.6;	0.5; −0.4

method has been assessed comparatively using the classical ARIMA-MAICE procedure as a benchmark.

The artificial time series employ the same random sequence for each of the parameter combinations ($\boldsymbol{\phi}$, $\boldsymbol{\theta}$) considered, identical background noise structure (functional form and degrees of intensity) and impulsive shocks characteristics (magnitude and location). All the algorithms employed in the Monte Carlo experiment, i.e. for (*i*) time series generation, (*ii*) parameters' estimation, (*iii*) model order selection, share the same design and settings for both the competing methods. Conditions (*ii*) and (*iii*) hold also for the part of the experiment involving real time series. Such a framework can reasonably guarantee the impartial judgment of the performances recorded and to connect them to the suppression of the perturbing components affecting the signal at hand.

The size of both the validation and test sets are equal and kept fixed throughout the whole experiment, i.e. $\mathbf{card}(Y_t^U) \equiv \mathbf{card}(Y_t^E) = 24$. The former is used to select the optimal cut-off frequencies and filter orders for three predefined time horizons $\{h_i;\ i = 1, 2, 3\}$, according to the objective function (4), i.e. $RMSFE^{(h)} = \sqrt{\frac{1}{V} \sum |y^U - \hat{y}^U|^2}$, whereas the overall performances obtained by the method are quantitatively evaluated on the test set Y_t^E, in terms of out-of-sample forecasting accuracy $\forall\ h$. The employed metrics are the $RMSFE^{(h)}(Y; Y^E(\omega_j))$ and the Mean Absolute Error ($MAE^{(h)} = \frac{1}{S} \sum |y^E - \hat{y}^E|$). Finally, the maximum ARMA order searched has been set to 5 for both the AR and MA parts ($P = Q = 5$) for the artificial time series, whereas for the actual ones, the maximum ARIMA order considered is ($P = Q = 5, D = 1$).

The algorithm (Sect. 2.4) requires a computationally critical pre-setting stage, i.e. the construction of the sequence of the cut-off frequencies $\{\omega_j \subset \Omega\}$. This has been performed taking as a starting point the cut-off frequency minimizer of the in-sample RMSE—computed on the set $\{y^A\}$—say ω^A, and then by going bidirectionally by increments of $\frac{1}{1000}\omega^A$ for each direction. Regarding the choice of the parameter ν, even though critical and computationally significant, in the present context it seemed not to involve a particularly large set of candidates. On the contrary, in all the performed simulations, the selection of a limited number of tentative ν parameters, chosen as a result of a visual inspection approach, has proven to be a fruitful strategy for the selection of a "good" filter order. In particular, the related grid set M has

Table 2 Macroeconomic time series employed in the empirical section: sources and main details

Code	Variable	Source	SA	Frequency	Units	Data range (Number of obs.)
X1	Gross Domestic Product	US. Bureau of Economic Analysis	Yes	Quarterly	Billions of Dollars	Jan. 2000 to Feb. 2015 (142 obs.)
X2	ISM Manufacturing: New Orders Index©	US. Bureau of Labor Statistics	Yes	Monthly	Index	Jan. 2000 to Jul. 2015 (187 obs.)
X3	S&P/Case-Shiller 20-City Composite Home Price Index©	S&P-Dow Jones Indices LLC	Yes	Monthly	Index Jan 2000 = 100	Jan. 2000 to May 2015 (185 obs.)
X4	Manufacture of oils and fats	Italian National Institute of Statistics	No	Monthly	Index 2010 = 100	Jan. 2010 to Apr. 2015 (64 obs.)
X5	Overseas visits to UK	U.K. Office for National Statistics	No	Monthly	Thousand of visitors	2000-01-01 to 2015-04-01 (192 obs.)
X6	UK visits abroad	U.K. Office for National Statistics	No	Monthly	Thousand of visitors	2000-01-01 to 2015-04-01 (192 obs.)
X7	UK visits abroad: Expenditure	U.K. Office for National Statistics	No	Quarterly	£ Millions	1980-01-01 to 2015-04-01 (144 obs.)
X8	Overseas visits to Italy	Italian National Institute of Statistics	No	Monthly	Thousand of visitors	2000-01-01 to 2015-04-01 (168 obs.)

been limited to only three integers: $M \equiv \{2, 3, 4\}$. Through the whole experiment, the integer $\nu = 2$ has been always selected by the algorithm for the filter order.

3.1 Simulated Time Series

As already mentioned, four different DGPs—whose parametrization is given in Table 1 along with the codification used for brevity and reported in the column labeled "dgp number"—have been employed to generate 4000 realizations (1000 realizations for each model), with sample sizes equal to 150 and 300, in the sequel referred to as T_A and T_B respectively for brevity. Two reasons are behind the choice of series showing such limited sample sizes: to study the behavior of ARIMA-ω procedure in the very potentially dangerous situation of short time series which are also noisy and to keep the computational time of the whole experiment at a reasonable level.

In order to mimic reality, realizations of dgp1-4 are corrupted with both a iid Gaussian, time independent continuous observation noise and a non-Gaussian short bursts of noise (iid shocks). This framework is formalized as follows: let $\mathcal{N}[\cdot\,,\cdot]$ be the normal distribution and i_t and j_t 0/1 binary switching variables between background Gaussian noise, with variance σ_δ^2 and σ_α^2, and impulsive non-Gaussian noise, with variances respectively $g_t^2\sigma_\delta^2$ and $h_t^2\sigma_\alpha^2$, the error terms in (2) and (3) are of the form:

$$\alpha_t \sim \mathcal{N}\Big[0, (1 - j_t)\sigma_\alpha^2 + j_t h_t^2 \sigma_\alpha^2\Big], \tag{7}$$

$$\delta_t \sim \mathcal{N}\Big[0, (1 - i_t)\sigma_\delta^2 + i_t g_t^2 \sigma_\delta^2\Big], \tag{8}$$

being g_t and h_t time dependent unknown mixing parameters. Although compact, this formalization covers a wide range of disturbances one might encounter in practice, i.e.: (*i*) a mixture of iid Gaussian noise and heavy tailed scale mixture of Gaussian distributions [19], (*ii*) heavy-tailed scale mixture of normal distributions (i_t or $j_t = 1$, $\forall t$) and (*iii*) pure Gaussian noise (i_t or $j_t = 0$, $\forall t$). Noise intensity is quantified in terms of *SNR* expressed in decibel (*DB*), i.e. $SNR = 10\log_{10} A^2 = 20\log_{10} A$, with A denoting the amplitude of the signal. In practice, each simulated realization has been injected with: (*i*) three different levels of background Gaussian noise ($SNR_1 = 20$ *DB*, $SNR_2 = 2.5$ *DB*, $SNR_3 = 1.0$ *DB*); (*ii*) Impulsive noise in the excitation sequence ($SNR = 20$ *DB*, localized at $t = \frac{T}{2}$); (*iii*) two addictive outliers ($SNR \approx 23.5$ *DB*, localized at $t = \frac{T}{4}$ and $t = \frac{T}{3}$). Tables 3 and 4 provide results for the case (*i*), whereas Tables 5 and 6 consider the case where all the three types of disturbances are simultaneously present. Finally, as an example, in Fig. 1 the same realization of dgp3 is reported as a pure signal (left graph) and corrupted with both a Gaussian noise (SNR = 3 DB) and impulsive disturbances, as above described under (*ii*) and (*iii*) (right graph).

Table 3 Outcomes of the empirical experiment with artificially generated time series of sample sizes T_A, corrupted with continuous Gaussian noise with different SNRs

	Model	h	SNR_1			SNR_2			SNR_3		
			RMSFE	MAE	ω_0	RMSFE	MAPE	ω_0	RMSFE	MAPE	ω_0
ARIMA	1	h = 1	1.69	1.41	–	1.86	1.55	–	2.03	1.69	–
		h = 2	2.30	1.90	–	3.10	2.56	–	3.10	2.66	–
		h = 3	3.18	2.62	–	3.70	3.27	–	4.60	4.11	–
	2	h = 1	1.75	1.41	–	1.93	1.55	–	2.10	1.69	–
		h = 2	2.36	1.90	–	3.19	2.56	–	2.10	1.69	–
		h = 3	3.26	2.62	–	4.08	3.27	–	4.15	4.19	–
	3	h = 1	1.57	1.92	–	1.73	2.11	–	1.88	2.31	–
		h = 2	2.12	2.59	–	3.10	3.50	–	2.87	3.63	–
		h = 3	2.92	3.58	–	3.65	4.02	–	4.24	5.72	–
	4	h = 1	1.88	1.77	–	2.70	1.95	–	2.26	2.13	–
		h = 2	2.54	2.39	–	3.43	3.22	–	3.43	3.35	–
		h = 3	3.50	3.29	–	4.38	4.11	–	4.97	4.40	–
ARIMA-ω	1	h = 1	1.10	0.92	0.3722	1.49	1.24	0.3749	1.90	1.59	0.3732
		h = 2	1.97	1.62	0.3722	2.93	2.39	0.3749	2.88	2.47	0.3732
		h = 3	3.02	2.49	0.3722	3.39	3.18	0.3749	4.35	4.01	0.3732
	2	h = 1	1.14	0.92	0.3785	1.54	1.24	0.3755	1.97	1.59	0.3745
		h = 2	2.02	1.61	0.3785	2.95	2.32	0.3755	3.07	2.41	0.3745
		h = 3	3.09	2.50	0.3785	3.40	3.02	0.3732	4.06	4.11	0.3748
	3	h = 1	1.02	1.25	0.4025	1.38	1.69	0.3905	1.77	2.16	0.3915
		h = 2	1.82	2.20	0.4025	2.58	3.20	0.3905	2.78	3.48	0.3915
		h = 3	2.77	3.40	0.4025	3.55	3.20	0.3905	4.16	5.60	0.3915
	4	h = 1	1.22	1.15	0.3905	1.65	1.56	0.3925	2.12	2.00	0.3843
		h = 2	2.18	2.03	0.3905	3.19	2.96	0.3925	3.36	3.28	0.3843
		h = 3	3.36	3.20	0.3905	4.33	4.04	0.3925	4.43	4.30	0.3843

Table 4 Outcomes of the empirical experiment with artificially generated time series of sample sizes T_B, corrupted with continuous Gaussian noise with different SNR

	Model	h	SNR_1			SNR_2			SNR_3		
			RMSFE	MAE	ω_0	RMSFE	MAPE	ω_0	RMSFE	MAPE	ω_0
ARIMA	1	h = 1	1.88	1.50	–	1.97	1.58	–	2.05	1.64	–
		h = 2	2.26	1.83	–	3.00	2.47	–	3.25	2.56	–
		h = 3	2.96	2.52	–	3.69	3.41	–	4.73	4.39	–
	2	h = 1	1.98	1.41	–	2.20	1.58	–	2.04	1.48	–
		h = 2	2.57	1.92	–	3.14	2.32	–	3.11	2.47	–
		h = 3	3.37	2.67	–	3.78	3.33	–	4.55	3.97	–
	3	h = 1	1.79	2.14	–	1.79	2.06	–	1.86	2.20	–
		h = 2	2.42	2.89	–	2.83	3.58	–	3.27	4.04	–
		h = 3	3.02	3.61	–	3.66	3.47	–	4.05	4.10	–
	4	h = 1	2.07	1.95	–	2.04	1.96	–	2.48	2.34	–
		h = 2	2.80	2.63	–	3.46	3.23	–	3.36	3.15	–
		h = 3	3.86	3.62	–	4.34	4.13	–	5.02	5.03	–
ARIMA-ω	1	h = 1	1.23	0.98	0.3755	1.88	1.50	0.3745	1.87	1.49	0.3765
		h = 2	1.94	1.56	0.3755	2.83	2.31	0.3745	3.02	2.39	0.3765
		h = 3	2.81	2.40	0.3755	3.59	3.31	0.3745	4.16	3.86	0.3765
	2	h = 1	1.29	0.92	0.3725	1.72	1.26	0.3755	1.92	1.39	0.3785
		h = 2	2.20	1.63	0.3745	2.91	2.10	0.3757	2.99	2.25	0.3786
		h = 3	3.20	2.54	0.3745	3.68	3.27	0.3750	4.02	3.51	0.3788
	3	h = 1	1.16	1.39	0.3975	1.55	1.87	0.3965	1.51	1.79	0.3925
		h = 2	2.08	2.46	0.3975	2.82	3.38	0.3965	3.17	3.96	0.3925
		h = 3	2.87	3.43	0.3975	3.48	3.40	0.3965	4.02	4.00	0.3926
	4	h = 1	1.35	1.27	0.397	2.02	1.90	0.3902	2.27	2.13	0.3895
		h = 2	2.41	2.24	0.397	3.45	3.12	0.3902	3.23	3.10	0.3895
		h = 3	3.71	3.53	0.397	4.09	4.00	0.3902	4.74	4.24	0.3895

Table 5 Outcomes of the empirical experiment with artificially generated time series of sample sizes T_A, corrupted with continuous Gaussian and impulsive noise with different SNR

	Model	h	SNR_1			SNR_2			SNR_3		
			RMSFE	MAE	ω_0	RMSFE	MAPE	ω_0	RMSFE	MAPE	ω_0
ARIMA	1	h = 1	2.09	1.97	–	2.59	2.55	–	2.82	2.89	–
		h = 2	2.64	2.19	–	3.10	2.56	–	3.93	3.49	–
		h = 3	3.43	2.93	–	3.97	3.27	–	4.44	4.05	–
	2	h = 1	2.10	1.68	–	2.29	2.13	–	2.92	2.52	–
		h = 2	2.62	2.31	–	3.19	2.56	–	3.71	3.47	–
		h = 3	3.55	2.96	–	3.71	3.27	–	3.95	3.97	–
	3	h = 1	1.86	2.22	–	1.87	2.11	–	2.71	3.18	–
		h = 2	2.39	2.94	–	3.09	3.50	–	3.71	4.44	–
		h = 3	3.29	3.90	–	4.17	3.47	–	3.95	4.51	–
	4	h = 1	2.21	2.10	0.3695	2.81	1.95	0.3695	2.99	2.93	0.3705
		h = 2	2.90	2.70	0.3695	3.94	3.22	0.3695	4.25	4.09	0.3705
		h = 3	3.67	3.65	0.3695	4.80	4.11	0.3695	4.78	4.89	0.3705
ARIMA-ω	1	h = 1	1.63	1.43	0.3725	2.15	2.00	0.3705	2.33	2.36	0.3750
		h = 2	2.35	2.03	0.3725	3.00	2.50	0.3705	3.54	3.23	0.3750
		h = 3	3.30	2.56	0.3725	3.54	3.13	0.3705	4.41	4.03	0.3752
	2	h = 1	1.55	1.20	0.3939	1.81	1.68	0.3893	2.37	2.02	0.3835
		h = 2	2.43	2.30	0.3939	3.00	2.46	0.3893	3.12	3.10	0.3835
		h = 3	3.33	2.63	0.3939	3.45	3.00	0.3894	3.91	3.90	0.3835
	3	h = 1	1.33	1.75	0.3985	1.68	2.02	0.38257	2.40	3.00	0.3808
		h = 2	2.01	2.48	0.3985	2.77	3.24	0.38257	3.20	4.43	0.3808
		h = 3	3.21	3.88	0.3985	4.14	3.33	0.38257	3.95	4.51	0.3808
	4	h = 1	1.71	1.53	0.3985	2.12	1.73	0.3825	2.60	2.85	0.3805
		h = 2	2.47	2.70	0.3985	3.43	3.06	0.3825	4.14	4.00	0.3805
		h = 3	3.67	3.63	0.3985	4.56	4.05	0.3824	4.70	4.75	0.3807

Table 6 Outcomes of the empirical experiment with artificially generated time series of sample sizes T_B, corrupted with continuous Gaussian and impulsive noise with different SNR

	Model	h	SNR_1			SNR_2			SNR_3		
			RMSFE	MAE	ω_0	RMSFE	MAPE	ω_0	RMSFE	MAPE	ω_0
ARIMA	1	h = 1	2.08	1.90	–	2.58	2.61	–	2.82	2.85	–
		h = 2	2.69	2.22	–	3.11	2.61	–	3.85	3.47	–
		h = 3	3.39	3.19	–	3.93	4.10	–	4.54	4.03	–
	2	h = 1	2.08	1.64	–	2.31	2.12	–	2.90	2.54	–
		h = 2	2.70	2.30	–	3.16	2.51	–	3.70	3.53	–
		h = 3	3.60	3.10	–	3.80	3.18	–	3.92	3.85	–
	3	h = 1	1.85	2.21	–	1.93	2.29	–	2.59	3.23	–
		h = 2	2.36	2.91	–	3.27	3.90	–	3.76	4.46	–
		h = 3	3.27	3.83	–	3.69	4.34	–	4.04	4.54	–
	4	h = 1	2.17	2.04	–	2.28	2.15	–	3.02	2.83	–
		h = 2	2.89	2.73	–	3.78	3.54	–	4.11	4.03	–
		h = 3	3.59	3.61	–	4.63	4.31	–	4.93	4.39	–
ARIMA-ω	1	h = 1	1.59	1.30	0.3735	2.06	2.00	0.3750	2.21	2.31	0.3782
		h = 2	2.60	2.16	0.3735	3.04	2.46	0.3750	3.81	3.38	0.3785
		h = 3	3.38	3.02	0.3735	3.81	3.92	0.3759	4.34	3.92	0.3792
	2	h = 1	1.42	1.19	0.3735	1.79	1.58	0.3755	2.25	1.91	0.3727
		h = 2	2.53	2.16	0.3735	3.03	2.42	0.3755	3.59	3.38	0.3725
		h = 3	3.50	2.99	0.3735	3.71	3.00	0.3755	3.92	3.82	0.3724
	3	h = 1	1.24	1.72	0.3955	1.84	2.18	0.3945	2.39	2.84	0.3906
		h = 2	2.25	2.90	0.3955	3.04	3.59	.3945	3.60	4.38	0.3906
		h = 3	3.14	3.74	0.3955	3.59	4.22	0.3945	4.03	4.41	0.3902
	4	h = 1	1.95	1.42	0.3895	2.18	2.06	0.3921	2.58	2.76	0.3885
		h = 2	2.87	2.54	0.3894	3.66	3.37	0.3921	3.99	3.93	0.3885
		h = 3	3.59	3.56	0.3895	4.58	4.23	0.3921	4.93	4.39	0.3889

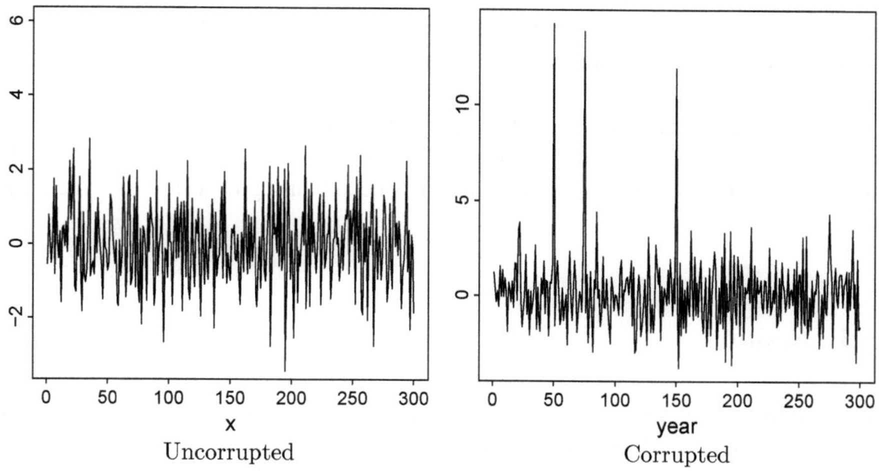

Fig. 1 DGP 3: Uncorrupted and corrupted artificially generated realizations. The y-axes are differently scaled for a better visual inspection

3.2 *Real Time Series*

In Table 2, the eight time series employed in the empirical study are detailed along with their conventional name, in the sequel adopted for brevity, stored in the column labeled "Code". Series X1 through X4 are of the type macroeconomic, whereas the remaining refer to tourism-related variables. In addition, two different sampling frequencies are considered, that is quarterly for X1 and X7 and monthly for the remaining series. All the time series are characterized by a limited sample sizes (not too far from T_A), the presence of outliers—e.g. of the type level shift, as clearly noticeable in the series X3 (June 2006) and X4 (August 2011, July 2013)—and, to a different extent, non stationary behaviors. Different degrees of roughness can also be noticed, e.g. X1 and X8 show smoother shapes than those of X2 and X5. Also, the macroeconomic series exhibit different trend patterns: a linear one for X1 and X2, a polynomial trend for X3, whereas X4 exhibits a multiple regime-type structure. Regarding the tourism time series, different overall patterns can be easily noticed, e.g. by comparing X8 and X7, with the former exhibiting a consistent behavior over time, in terms of both trend evolution and regularity of the oscillations, and the latter, which shows non-stationary dynamics at both seasonal and and non-seasonal frequencies. The empirical analysis has been carried out using both seasonally adjusted time series—as in the case of X1, X2 and X3—whereas the remaining ones have been considered in their raw format. The series denominated X4 has been included in the study not only because of its small sample size—in fact, being publicly available only from January 2010, is the shortest one—but also in that it is characterized by two significant level shifts—clearly noticeable in Fig. 2—with the most recent one,[3] being located toward the end of the observation period (July 2013), is particularly dangerous.

Finally, for this data set, two additional tests have been applied: specifically, the choice of the integration order has been driven by a test of the type *KPSS* [20, 21] whereas the presence of units roots has been checked through a test of the type *ADF* [22].

3.3 Results

The quality of the predictions provided by the proposed method can be noticed by inspecting Tables 3, 4, 5, 6 and 7, where the main results of both the simulated and real time series experiments are summarized. Regarding the former, substantial improvements are noticeable for the shortest prediction horizon considered (h = 1), whereas they tend to degrade as the forecasting horizon increases, especially when the time series are injected with the highest Gaussian noise level SNR3 (with or without impulsive noise added).

For example, considering the case of DGP4, very little—possibly not significant—improvements were recorded for h = 3, where the percentage reduction of MAE with respect to standard ARIMA procedure is less than 3% across all the background noise intensities for T_A (figures for T_B are only slightly better). As expected, such a pattern holds when the DGPs are injected with additional impulsive noise, in this case the performances of the method worsen considerably. Averaging over all the models at noise level SNR3 (impulsive plus background) and horizon h = 3, both RMSFE and MAE show very little differences between the methods: considering T_A the values of 4.3 and 4.4 for ARIMA and 4.2 and 4.3 for ARIMA-ω have been respectively recorded. Not significant benefits are recorded considering the larger sample size T_B. The increasing performance pattern noticeable for the intermediate horizon h = 2, unlike what recorded for h = 3, is affected by the type of noise injected other than by the SNR level: considering again DGP4 under simple Gaussian noise and sample size T_A, the percentage difference in the MAE between standard and ARIMA-ω procedure is of approximately 8.5% and 4%, respectively for SNR2-3 in favor of the latter, whereas such figures reduces to 7.4% and 0.3% when impulsive noise is added. Averaging over all the models and the noises levels, the RMSFE values recorded at the same horizon h = 2, range from 2.9 (ARIMA) and 2.6 (ARIMA-ω), when only Gaussian noise and length T_A are considered, whereas they increase respectively to 3.3 and 3.2 (considering the case of impulsive noise and T_B).

As already pointed out, ARIMA-ω procedure delivers remarkably more precise predictions than its competitor for h = 1, especially in the case of pure Gaussian background noise with signal to noise ratios equal to SNR1-2. At these noise lev-

[3]With all probabilities, this outlier is due to a setback in the production of olive oil as a result of a serious disease affecting olive trees in certain areas of Southern Italy.

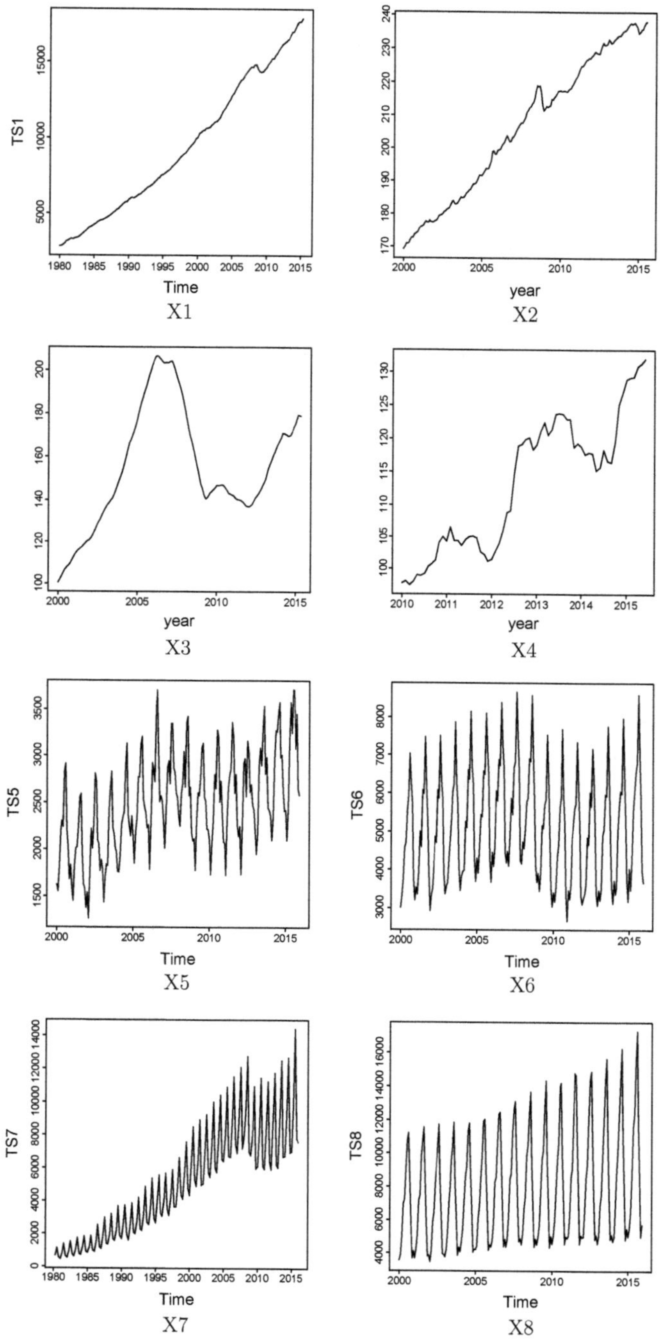

Fig. 2 Real time series analyzed (details provided in Table 2)

Table 7 Outcomes of the empirical experiment conducted on real time series

Variable	ARIMA order	ARIMA−ω order	h	*ARIMA*		ARIMA-ω		ω_0
				RMSFE	MAE	RMSFE	MAE	
X1	(4, 1, 1)	(5, 1, 1)	h = 1	101.49	90.28	71.12	62.21	0.3425
			h = 2	174.98	123.23	150.35	118.34	0.3212
			h = 3	219.23	200.24	0.217.75	196.35	0.3037
X2	(2, 0, 1)	(5, 0, 0)	h = 1	0.688	0.486	0.473	0.3252	0.4280
			h = 2	1.22	1.56	1.01	1.342	0.3525
			h = 3	2.55	2.57	2.51	2.47	0.3327
X3	(1, 2, 0)	(5, 2, 0)	h = 1	0.659	0.449	0.537	0.444	0.4300
			h = 2	1.16	0.713	1.07	0.646	0.4547
			h = 3	2.65	2.06	2.38	1.95	0.4583
X4	$(1, 1, 0)(0, 0, 0)_{12}$	$(1, 1, 0)(0, 0, 0)_{12}$	h = 1	2.09	1.49	1.602	1.269	0.4275
			h = 2	3.75	2.90	3.62	2.84	0.4275
			h = 3	5.14	4.25	5.04	4.24	0.4308
X5	$(1, 1, 0)(1, 0, 0)_{12}$	$(2, 1, 0)(2, 1, 1)_{12}$	h = 1	138.1	121.4	71.6	62.5	0.4001
			h = 2	180.6	145.3	99.5	101.3	0.4006
			h = 3	252.9	219.3	230.5	207.6	0.3989
X6	$(1, 1, 2)(1, 1, 2)_{12}$	$(2, 1, 1)(2, 1, 1)_{12}$	h = 1	395.7	322.7	307.6	267.7	0.3545
			h = 2	458.2	400.6	420.7	356.3	0.3667
			h = 3	750.4	625.6	649.6	608.7	0.3555
X7	$(1, 0, 1)(0, 1, 2)_3$	$P(2, 1, 1)(0, 1, 1)_4$	h = 1	722.2	520.9	648.5	430.2	0.3122
			h = 2	916.1	882.7	897.4	718.2	0.3085
			h = 3	1503.8	1107.3	1480.7	1003.1	0.3083
X8	$(1, 0, 2)(0, 1, 0)_{12}$	$(1, 1, 1)(0, 1, 1)_{12}$	h = 1	1057.3	771.7	1001.7	671.4	0.2645
			h = 2	1381.5	982.9	1288.3	1121.5	0.2646
			h = 3	1572.5	1125.4	1500.8	1116.5	0.2640

els, considering the sample size T_A and averaging over all the DGPs, the RMSFE drops from 1.7, in case of SNR1 and 1.9 in case of SNR2 to 1.2 and 1.5 respectively. For the larger sample size T_B, the performances appear to worsen slightly but still seem to be good, being the RMSFE now 1.4 and 1.8—for simple background noise with SNR1-2 respectively, where the standard ARIMA procedure delivers 1.9 and 2.0 always for SNR1-2. Considering only DGP2 with sample size T_A affected with non impulsive noise of intensity SNR1, the reduction versus the standard procedure reaches the remarkable values of around 35% in both RMSFE and MAE (the approx values amount respectively to 1.14 and 0.92). Focusing on the DGPs corrupted by both impulsive and background noise, the results—reported in Tables 5 and 6—show, as expected, less impressive performances for both the procedures. However, the predictions generated by ARIMA-ω can be still considered acceptable under the following experimental conditions: short prediction horizon (h = 1), high SNR (SNR1 and possibly SNR2) and sample size equal to T_B. In this case, averaging over all the models, the RMSFE is equal to 1.6 versus 2.0 of the standard procedure. Departure from such conditions determines a quick deterioration of the performances until, as already highlighted, ARIMA-ω tends to break down for h = 3 and SNR3. Regarding the cut off frequencies, it can be said that the variability of its mean values, within each of the considered DGPs and considering all the experimental parameters (prediction horizon, SNR, sample size and type of noise) appear to be very small and always insensitive to the prediction horizon when the noise level SNR1 is considered. Slight variations are noticeable with decreasing levels of the SNR and sample size equal to T_B. The ω parameter, on the other hand, shows more variability when the real time series are considered. This is consistent with the dynamics involved, which in this case are much more complicated and naturally far from the simple ones artificially generated.

Pointing our attention on the real time series of the type macroeconomic, it appears clearly, by inspecting Table 7 and Fig. 3, how those who benefit the most from the application of the proposed procedure at lag 1—and to less extent at lag 2—are the series X1 and X2, where the percentage variation in terms of RMSFE between the two methods reaches approximately the values of 29.9% and 31.3% respectively. At lag 2, only the first two series seem to show noticeable gains from ARIMA-ω whereas for h = 3 they might be considered negligible, being basically in line with the standard ARIMA procedure's results. In term of prediction horizon, the recorded overall behavior therefore is consistent with what found in the case of the artificial time series. With regard to the tourism time series, the best results have been recorded in the case of X5, where a reduction of 48.1% and 48.5% have been achieved—for $h = 1$—in the values of the RMSFE and the MAPE respectively. For $h = 2$, the proposed procedure still seems to deliver noticeable improvements: in fact, the percentage reduction for the RMSFE and the MAPE respectively is approximately equal to 45% and 30%. Such performances can be attributable to the inherent level of roughness of the original time series, which makes the ARIMA-ω procedure particularly effective. On the other hand, the least impressive results have been obtained in the case of the series X8, which, as already pointed out, shows a regular and smooth pattern. In more details, the RMSFE computed on the original time

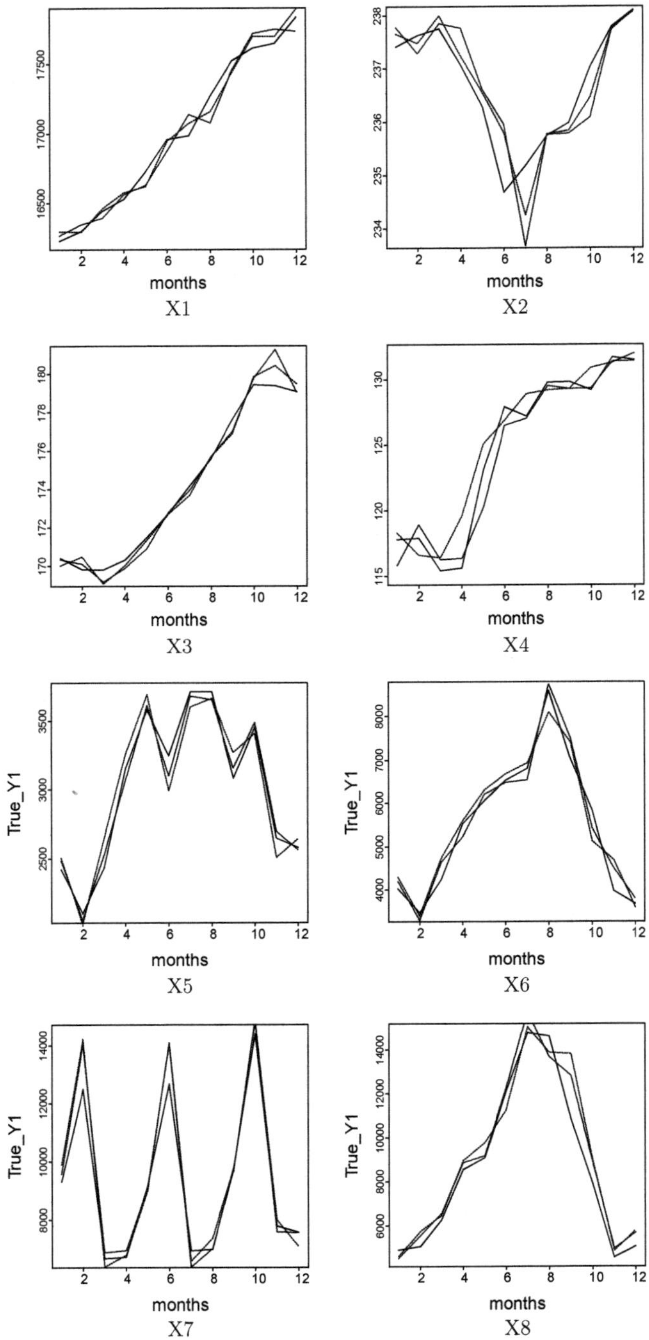

Fig. 3 1-step ahead predictions delivered by standard ARIMA models (*dashed lines*) and by ARIMA-ω procedure (*dotted lines*)

series, equal to 1057.3 for $h = 1$, becomes 998.1 on the filtered data, for a reduction of approximately 5%. Slightly better results have been achieved considering the MAPE: here, for the same horizon, the recorded values are 771.7 (raw data) and 671.4 (filtered data), for an improvement approximately equal to 13%.

3.4 Concluding Remarks and Future Work

The results of the empirical study presented in the previous section, show the better prediction performances, under specific conditions, achieved by the proposed method using both artificial and real data. However, it is important to stress that while ARIMA-ω enjoys the same automatic MAICE-based ARIMA selection framework, this plus comes at the expense of a program execution time, which, using standard hardware resources, can become unreasonable. In fact, considering for example the maximum orders (P, D, Q) chosen in the empirical study, each step of the iterative searching procedure envisions a model space of cardinality $2 \cdot (6)^2$. Such a situation can be mitigated by reducing P and/or Q and/or by considering a smaller Ω set. Another viable alternative is the non homogeneous reduction of the model space, as a result of the suppression of certain lags—e.g. on the basis of prior knowledge of the phenomenon at hand or previous studies. This strategy is especially recommended when DGPs with sparse matrix of coefficients are suspected e.g. in presence of a big sample size and if one or more seasonal components are present. Unfortunately, actions aimed at minimizing the exploitation of the computational resources can induce less accurate predictions but, on the other hand, make feasible the application of the method.

In order to cope with this computational issue, future work—aimed at studying the performances of the method under heuristic searching procedures in conjunction with different filter designs—has already been planned. In particular, a possible focus can be the employment of a genetic algorithm-based approach possibly associated with a different, hopefully computationally faster, statistical model, e.g. of the type exponential smoothing.

References

1. Gardiner, C., Zoller, P.: Quantum Noise: A Handbook of Markovian and Non-markovian Quantum Stochastic Methods with Applications to Quantum Optics, vol. 56. Springer Science & Business Media (2004)
2. Box, G.E., Jenkins, G.M., Reinsel, G.C.: Time Series Analysis: Forecasting and Control, vol. 734. Wiley (2011)
3. Donoho, D.L.: De-noising by soft-thresholding. IEEE Trans. Inf. Theory **41**, 613–627 (1995)
4. Motwani, M.C., Gadiya, M.C., Motwani, R.C., Harris, F.C.: Survey of image denoising techniques. In: Proceedings of GSPX, pp. 27–30 (2004)
5. Azoff, E.M.: Reducing error in neural network time series forecasting. Neural Comput. Appl. **1**, 240–247 (1993)

6. Tamura, S.: An analysis of a noise reduction neural network. In: 1989 International Conference on Acoustics, Speech, and Signal Processing, 1989. ICASSP-89, pp. 2001–2004. IEEE (1989)
7. Matsuura, T., Hiei, T., Itoh, H., Torikoshi, K.: Active noise control by using prediction of time series data with a neural network. In: IEEE International Conference on Systems, Man and Cybernetics, 1995. Intelligent Systems for the 21st Century, vol. 3, pp. 2070–2075. IEEE (1995)
8. Wagner, N., Michalewicz, Z., Khouja, M., McGregor, R.R.: Time series forecasting for dynamic environments: the dyfor genetic program model. IEEE Trans. Evol. Comput. **11**, 433–452 (2007)
9. Wachman, G.: Kernel methods and their application to structured data. Ph.D. thesis, Tufts University (2009)
10. Toivanen, P.J., Laukkanen, M., Kaarna, A., Mielikainen, J.S.: Noise reduction in multispectral images using the self-organizing map. In: AeroSense 2002, International Society for Optics and Photonics, pp. 195–201 (2002)
11. Takalo, R., Hytti, H., Ihalainen, H.: Adaptive autoregressive model for reduction of poisson noise in scintigraphic images. J. Nucl. Med. Technol. **39**, 19–26 (2011)
12. Pesaran, M.H., Pettenuzzo, D., Timmermann, A.: Forecasting time series subject to multiple structural breaks. Rev. Econ. Stud. **73**, 1057–1084 (2006)
13. Godsill, S.J.: Robust modelling of noisy arma signals. In: 1997 IEEE International Conference on Acoustics, Speech, and Signal Processing, 1997. ICASSP-97, vol. 5, pp. 3797–3800. IEEE (1997)
14. Gomez, V.: The use of butterworth filters for trend and cycle estimation in economic time series. J. Bus. Econ. Stat. **19**, 365–373 (2001)
15. Denbigh, P.: System Analysis and Signal Processing: With Emphasis on the Use of MATLAB. Addison-Wesley Longman Publishing Co., Inc. (1998)
16. Kohler, T., Lorenz, D.: A Comparison of Denoising Methods for One Dimensional Time Series, vol. 131. Bremen, Germany, University of Bremen (2005)
17. Akaike, H.: A new look at the statistical model identification. IEEE Trans. Autom. Control **19**, 716–723 (1974)
18. Ozaki, T.: On the order determination of arima models. Appl. Stat. 290–301 (1977)
19. Gneiting, T.: Normal scale mixtures and dual probability densities. J. Stat. Comput. Simul. **59**, 375–384 (1997)
20. Kwiatkowski, D., Phillips, P.C., Schmidt, P., Shin, Y.: Testing the null hypothesis of stationarity against the alternative of a unit root: how sure are we that economic time series have a unit root? J. Econom. **54**, 159–178 (1992)
21. Peter, K.: A Guide to Econometrics (1998)
22. Dickey, D.A., Hasza, D.P., Fuller, W.A.: Testing for unit roots in seasonal time series. J. Am. Stat. Assoc. **79**, 355–367 (1984)

Mandelbrot's 1/*f* Fractional Renewal Models of 1963–67: The Non-ergodic Missing Link Between Change Points and Long Range Dependence

Nicholas Wynn Watkins

Abstract The problem of 1/*f* noise was identified by physicists about a century ago, while the puzzle posed by Hurst's eponymous effect, originally identified by statisticians, hydrologists and time series analysts, is over 60 years old. Because these communities so often frame the problems in Fourier spectral language, the most famous solutions have tended to be the stationary ergodic long range dependent (LRD) models such as Mandelbrot's fractional Gaussian noise. In view of the increasing importance to physics of non-ergodic fractional renewal processes (FRP), I present the first results of my research into the history of Mandelbrot's very little known work on the FRP in 1963–67. I discuss the differences between the Hurst effect, 1/*f* noise and LRD, concepts which are often treated as equivalent, and finally speculate about how the lack of awareness of his FRP papers in the physics and statistics communities may have affected the development of complexity science.

Keywords Long range dependence · Mandelbrot · Change points · Fractional renewal models · Weak ergodicity breaking

---

N. Wynn Watkins (✉)
Centre for Fusion, Space and Astrophysics, University of Warwick, Coventry, UK
e-mail: N.Watkins2@lse.ac.uk

N. Wynn Watkins
Universität Potsdam, Institut für Physik und Astronomie, Campus Golm, Potsdam-golm, Germany

N. Wynn Watkins
Max Planck Institute for the Physics of Complex Systems, Dresden, Germany

N. Wynn Watkins
Centre for the Analysis of Time Series, London School of Economics, London, UK

N. Wynn Watkins
Faculty of Science, Technology, Engineering and Mathematics, Open University, Milton Keynes, UK

I. Rojas et al. (eds.), *Advances in Time Series Analysis and Forecasting*, Contributions to Statistics, DOI 10.1007/978-3-319-55789-2_14

1 Ergodic and Non-ergodic Solutions to the Paradoxes of 1/*f* Noise and Long Range Dependence

This paper is about historical and conceptual aspects of a topic—"the infrared catastrophe" in sample periodograms, variously studied as long range dependence and 1/*f* noise—which has long been seen as a theoretical puzzle by time series analysts and physicists. Its purpose is twofold, to report new historical research into Mandelbrot's little-known work on fractional renewal processes in the mid 1960s, and to then use these findings to better classify the approaches to 1/*f* noise and LRD.

The physicist's "problem of 1/*f* noise" has been with us since the pioneering work of Schottky and Johnson in the early 20th century on fluctuating currents in vacuum tubes [1–3]. It is usually framed as a spectral paradox, i.e. "how can the Fourier spectral density $S'(f)$ of a stationary process take the form $S'(f) \sim 1/f$ and thus be singular at the lowest frequency", or equivalently "how can the autocorrelation function "blow up" at large lags and thus not be summable?". The framing of the problem in spectral terms has, as we will see, conditioned the type of solutions sought.

In the 1950s an analogous time domain effect (the Hurst phenomenon) was seen in the statistical growth of Hurst's rescaled range in the minima of levels of the Nile river [1]. Rather than a dependence of $\tau^{1/2}$ on the observation time scale τ, many time series were seen to show a dependance τ^J where J, the "Hurst exponent" was typically greater than 0.5. This soon presented a conceptual problem because Feller quickly proved that an iid sequence must asymptotically have $J = 1/2$. Although many of the observed Hurst effects may indeed arise from pre-asymptotic effects, nonstationarity, or other possibilities, the desire for a stationary solution to the problem with a satisfying level of generality remained.

It was thus a key step when in 1965–67 Mandelbrot presented a stationary process, fractional Gaussian noise, which could exhibit both the Hurst effect and 1/*f* noise. The fGn process is effectively the derivative of fractional Brownian motion (fBm), and was subsequently developed by him with Van Ness and Wallis, particularly in a hydrological context [1, 4]. fGn is a stationary ergodic process, for which a power spectral density is a natural, well-defined concept, the paradox here residing in the singular behaviour of $S'(f)$ at zero frequency (in the $H < 1/2$ case). Skepticism about fGn and another related LRD process, Granger and Hosking's autoregressive fractionally integrated moving average (ARFIMA) as a universal explanation for observed Hurst effects remained (and remains) considerable, though, because of their highly non-Markovian properties. Many authors, particularly in statistics and econometrics, have found models based on change points to be better motivated, not least because many datasets are known to have clear change points that need to be handled.

In the last two decades, however, it has increasingly been realised in physics [5] that another class of models, the fractional renewal processes (FRPs), give rise to 1/*f* spectra in a very different way. They have discretised amplitude values, with sometimes as few as 2 or 3 levels, and random intervals between the changes of level. In that sense they can be seen as part of the broader class of random change point models. Unlike most random change point models, however, they also have heavy tailed distributions for the times beween changes in amplitude. They are non-ergodic, and nonstationary but bounded, and require us to interpret Fourier periodograms differently from the familiar Wiener-Khinchine power spectrum. Physical interest in the FRP has come from phenomena such as weak ergodicity breaking (in e.g. blinking quantum dots [6–9]) and the related question of how many different classes of model can share the common property of the $1/f$ spectral shape (e.g. [10, 11]). In Sect. 2 I briefly recap the key properties of the FRP and fGn and the differences between them.

In view of this new interest, my first main aim in this paper is to report in Sect. 3 the results of historical research which has found, to my great surprise, that this dichotomy between ergodic and non-ergodic origins for $1/f$ periodograms was not only recognised but also published by Mandelbrot about 50 years ago; work that seems remarkably little known [12–15]. He developed his FRPs in parallel with his seminal and much more visible work on ergodic, stationary fGn which is today very much better known to physicists, geoscientists and many other time series analysts [1, 16]. In the FRP papers, and the bridging essays he wrote when he revisited and lightly edited them late in life for republication in his collected Selecta volumes, particularly [4, 17], he developed several models. For copyright reasons, quotations in this chapter are taken from the Selecta, and readers are urged to consult the originals when available. In his FRP models the periodogram, the empirical autocorrelation function (acf), and the observed waiting time distributions, were all found to grow in extent with the length of time over which they are measured. Mandelbrot explicitly [15] drew attention to this non-ergodicity and its origins in what he called "conditional" stationarity. He explicitly contrasted the fractional renewal models with fGn. Mandelbrot's work at IBM was not the only contemporary paper on point processes with heavy tailed waiting times, at least one other example being the work of Pierre Mertz [18, 19] at RAND on modelling telephone errors, so this article will not attempt to assign priority. I plan to return to the history of this period in more detail in future articles.

The other main purpose of this contribution, in Sect. 4, is to clarify the subtle differences between 3 phenomena: the empirical Hurst effect, the appearance of $1/f$ noise in periodograms, and the concept of LRD as embodied by the stationary ergodic fGn model, and to set out their hierarchy with respect to each other, aided in part by this historical perspective. This relatively short paper does not deal with multiplicative models (e.g. [10, 17]), although these remain a very important alternative source of $1/f$ spectra, and particularly those which arise from turbulent cascades. I also do not consider $1/f$-type periodograms arising from nonstationary self-similar walks such as fBm. Such walks are intrinsically unbounded and so the periodogram must a priori be different from a stationary power spectrum.

I will (Sect. 5) conclude by arguing that the relative neglect of [12–15] at the time of their publication must have had long-term effects, particularly on the nascent field of complexity science as it developed in the 70s and 80s.

2 fGn and the Fractional Renewal Process Compared

fGn [16] is effectively a derivative of fractional Brownian motion $Y_{H,2}(t)$:

$$Y_{H,2}(t) = \frac{1}{C_{H,2}} \int_R dL_2(s)\, K_{H,2}(t-s) \tag{1}$$

which in turn extends the Wiener process to include a self-similar memory kernel $K_{H,2}(t-s)$, such that

$$K_{H,2}(t-s) = [(t-s)_+^{H-1/2} - (-s)_+^{H-1/2}] \tag{2}$$

thus giving a decaying, non-zero weight to all of the values in the time integral over dL. In consequence fGn shows long range dependence by construction, and it became the original paradigmatic model for LRD. The attention paid to its $1/f$ spectrum and long-tailed acf as diagnostics of LRD, has often led to it being forgotten that its stationarity is the other essential ingredient for LRD in this sense. Intuitively one can see that without stationarity there can be no LRD because there is no infinitely long past history over which sample values of the process can be dependent. Models like fGn, and also fractionally integrated noise (FIN) and the ARFIMA process, which have been widely studied in the statistics community (e.g. [16, 20]) exhibit LRD *by construction*, i.e. stationarity is assumed at the outset in defining them. More subtly, this notion of LRD also appears to require the stronger property of ergodicity, in order that their conventional meanings can be ascribed to the power spectrum and autocorrelation function.

While undeniably important to time series analysis and the development of complexity science, we can already see from the restriction to stationary processes, however, that the LRD concept as embodied by fGn might be insufficient to describe the whole range of either $1/f$ or Hurst behaviour that observations may present us with. Full awareness of this fundamental limitation has been slow, however. I think this has probably been due to three widespread, deeply-ingrained, but unfortunately erroneous "folk beliefs": (i) that an observed Fourier periodogram can *always* be taken to estimate a power spectrum, (ii) that the Fourier transform of an empirically obtained periodogram is *always* a meaningful estimator of an autocorrelation function, and (iii) that the observation of a $1/f$ Fourier periodogram in a time series *must* imply the kind of long range dependence that is embodied in the ergodic fractional Gaussian noise model. The first two beliefs are of course routinely cautioned against in any good course or book on time series analysis, including classics like Bendat's [21]. The third belief remains highly topical, however, because it is only relatively recently being appreciated in the theoretical physics literature just how distinct two of the paradigmatic classes of $1/f$ noise model are, and how these differences relate not only to LRD but also to the fundamental physical question of weak ergodicity breaking (e.g. [5, 7, 22]).

The second paradigm for 1/f noise mentioned above is the fractional renewal class, which is a descendent of the classic random telegraph model [21]. It looks at first sight to be stationary and Markovian, but has switching times at power law distributed intervals. A particularly well studied variant is the alternating fractal renewal process (AFRP, e.g. [23, 24]), which is also closely connected to the renewal reward process in mathematics. When studied in the telecommunications context, however, the AFRP has often had a cutoff applied to its switching time distribution for large times to allow analytical tractability. The use of an upper cutoff unfortunately masks some of its most physically interesting behaviour, because when the cutoffs are not used the periodogram, the empirical acf, and *observed* waiting time distributions, all *grow* with the length of time over which they are measured, rendering the process both non-ergodic and non-stationary in an important sense (Mandelbrot preferred his own term "conditionally stationary"). In particular, Mandelbrot stressed that the process no longer obeys the necessary conditions on the Wiener-Khinchine theorem for its empirical periodogram to be interpreted as an estimate of the power spectrum. This property of weak ergodicity breaking (named by Bouchaud in the early 1990s [22]) is now attracting much interest in physics, see e.g. [7], on the resolution of the low frequency cutoff paradox, and subsequent developments [10, 11, 25, 26].

The existence of this alternative, nonstationary, nonergodic fractional renewal model makes it clear that there is a difference between the observation of an empirical 1/f noise alone, and the presence of the type of LRD that is embodied in the stationary ergodic fGn model. We will develop this point further in Sect. 4, but will first go back to the 1960s to survey Mandelbrot's twin tracks to 1/f.

3 Mandelbrot's Fractional Renewal Route to 1/f

Mandelbrot was not only aware of the distinction between fGn and fractional renewal models [4, 17], but also published a nonstationary model of the AFRP type in 1965 [13, 14] and had explicitly discussed the time dependence of its power spectrum as a symptom of non-ergodicity by 1967 [15].

There are 4 key papers in Mandelbrot's consideration of fractional renewal models. The first, cowritten with physicist Jay Berger [12], appeared in IBM Journal of Research and Development. Concerned with with errors in telephone circuits, its main point was the power law distribution of times between errors, which were themselves assumed to have discrete states. Switching models, particularly the state dependent switching models, were already being looked at in order to study clustering of errors. Berger and Mandelbrot acknowledged that Pierre Mertz of RAND had already studied a power law switching model [18], but Mandelbrot's early exposure to the extended central limit theorem, and the fact that he was studying heavy tailed models in economics and neuroscience among other applications, seem to have enabled him to see a broader significance for the FRP class.

The second, a sole author paper [13], was in the IEEE Transactions on Communication Technology, and essentially also used the model published with Berger. The abstract notes that it describes:

> ... a model of certain random perturbations that appear to come in clusters, or bursts. This is achieved by introducing the concept of a "self-similar stochastic point process in continuous time." From the mathematical viewpoint, the resulting mechanism presents fascinating peculiarities. In order to make them more palatable, as well as to help in the search for further developments, the basic concept of "conditional stationarity" is discussed in greater detail than would be strictly necessary from the viewpoint of engineering.

It is clear that by 1965 Mandelbrot had come to appreciate that the application of the Fourier periodogram to the FRP would give ambiguous results, saying in [13] that:

> The now classical technique of spectral analysis is inapplicable to the processes examined in this paper but it is sometimes unavoidable. [Ref 18 in [13]] will examine what happens when the scientist applies the algorithms of spectral analysis without testing whether or not they have the usual meaning. This investigation will lead to fresh concepts that appear most promising indeed in the context of a statistical study of turbulence, excess noise, and other phenomena where interesting events are intermittent and bunched together. [See also Ref 19 in [13]]

The "other publication ... Ref 18", became the third key paper in the sequence, and resulted from an IEEE conference talk in 1965. It [14] is now available in the *post hoc* edited form that papers take in his Selecta volumes [4, 17]. "Reference 19" seems from the description of its subject matter to have been intended to be a paper in the physics literature. I have not yet been able to determine that paper's fate but its role was effectively taken over by the fourth key paper [15] which rather than a physics journal, appeared in the electrical engineering literature. With the proviso that the Selecta version of [14] may not fully reflect the original content, it is clear that by mid-1965 Mandelbrot was already focusing on the implications for ergodicity of the conditional stationarity idea. He remarked that:

> In other words, the existence of f^{D-2} noises challenges the mathematician to reinterpret spectral measurements otherwise than in "Wiener-Khinchin" terms. [...] operations meant to measure the Wiener-Khinchin spectrum may unintentionally measure something else, to be called the "conditional spectrum" of a "conditionally covariance stationary" random function. [15]

Taking the two papers [14, 15] together we can see that Mandelbrot expanded on his initial vision by discussing several FRP models, including in [14] a three state, explicitly nonstationary model with waiting times whose probability density function decayed as a power law $p(t) \sim t^{-(1+D)}$. This stochastic process was intended as a "cartoon" to model intermittency, in which "off" periods of no activity were interrupted by jumps to a negative (or positive) "on" active state. His key finding, confirmed in [15] for a model with an arbitrary number of discrete levels, was that the traditional Wiener-Khinchine spectral diagnostics would return a $1/f$ periodogram and thus a spectral "infrared catastrophe" when viewed with traditional methods, but, building on the notion of conditional stationarity proposed in [13], a conditional power spectrum $S(f, T)$ could be defined that was decomposable into a stationary part in which no catastrophe was seen, and one that depended on the time series' length T, multiplying a slowly varying function $L(f)$. He found

$$S(f, T) \sim f^{D-1} L(f) Q(T) \tag{3}$$

where $Q(T)T^{1-D}$ was slowly varying, so that the conditional spectral density $S'(f, T)$ obeyed

$$S'(f, T) = \frac{d}{df} S(f, T) \sim f^{D-2} T^{D-1} L(f) \tag{4}$$

Rather than representing a true singularity in power at the lowest frequencies, in the Selecta [17] he described the apparent infrared catastrophe in the power spectral density in the FRP as a "mirage" resulting from the fact that the moments of the model varied in time in a step-like fashion, a property he called "conditional covariance stationarity".

In [15] Mandelbrot noted a clear contrast between his conditionally stationary, non-Gaussian fractional renewal $1/f$ model and his stationary Gaussian fGn model (the 1968 paper about which, with Van Ness, was then in press at SIAM Review):

> Section VI showed that some $f^{D-2}L(f)$ noises have very erratic sampling behavior. Some *other* f^{D-2} noises, however, are Gaussian, which means that they are perfectly "well-behaved". An example is provided by "fractional white noise" which is the formal differential of the random process described in Mandelbrot and Van Ness, 1968 [i.e. fBm]

He identified the origin of the erratic sampling behaviour in the non-ergodicity of the FRP. Niemann et al. [7] have recently given a very precise analysis of the behaviour of the random prefactor $S(T)$, obtaining its Mittag-Leffler distribution and checking it by simulations.

4 The Hurst Effect Versus 1/f Versus LRD

Informed in part by the above historical investigations, the purpose of this section is now to distinguish conceptually between 3 things which are still frequently, and mistakenly, regarded as the same.

To recap, the phenomena are:

- The Hurst effect: the *observation* of "anomalous" growth of range in a time series using a diagnostic such as Hurst and Mandelbrot's R/S or detrended fluctuation analysis (DFA) (e.g. [1, 16]).
- $1/f$ noise: the *observation* of singular low frequency behaviour in the empirical periodogram of a time series.
- Long range dependence (LRD): a property of a stationary model *by construction*. This can only be *inferred* to be a property of an empirical time series if certain

additional conditions are known to be met, including the important one of stationarity.

The reason why it is necessary to unpick the relationship between these ideas is that there are three commonly held misperceptions about them.

The first is that observation of the Hurst effect in a time series necessarily implies stationary LRD. This is "well known" to be erroneous, see e.g. the work of [27] who showed the Hurst effect arising from an imposed trend rather than from stationary LRD, but is nonetheless in practice still not very widely appreciated.

The second is that observation of the Hurst effect in a time series necessarily implies a periodogram of power law form. Although less "well known", [28], for example, have shown an example where the Hurst effect arose in the Lorenz model which has an exponential power spectrum rather than 1/*f*.

The third is the idea that observation of a 1/f periodogram necessarily implies stationary LRD. As noted above, this is a more subtle issue, and although little appreciated since the pioneering work of [13–15] it has now become central to the investigation of weak ergodicity breaking in physics.

4.1 The Hurst Effect

The Hurst effect was originally observed as the growth of range in a time series, at first the Nile. The original diagnostic for this effect was rescaled range, or R/S. Using the notation J (not H) for the Joseph (i.e. Hurst) exponent that Mandelbrot latterly advocated [4], the Hurst effect is seen when the R/S [1, 16] grows with time as

$$\frac{R}{S} \sim \tau^J \tag{5}$$

in the case that $J \neq 1/2$. During the period between Feller's proof that an iid stationary process had $J = 1/2$, and Mandelbrot's papers of 1965–68 on long range dependence in fGn, there was a controversy [1] about whether the Hurst effect was a consequence of nonstationarity and/or a pre-asymptotic effect. The controversy has never fully subsided [1] because Occam's Razor frequently favours at least the possibility of change points in an empirically measured time series (e.g. [29]), and because of the (at first sight surprising) non-Markovian property of fGn.

A key point to appreciate is that it is easier to generate the Hurst effect over a finite scaling range, as measured for example by R/S, than it is to generate a true 1/*f* spectrum over many decades. [28] for example shows how a Hurst effect can appear over a finite range even when the power spectrum *is known a priori* to not be 1/*f*, e.g. in the Lorenz attractor case where the low frequency spectrum is in fact exponential.

4.2 1/f Spectra

The term 1/f spectrum is usually used to denote periodograms where the spectral density $S'(f)$ has an inverse power law form, e.g. the definition used in [14, 15]

$$S'(f) \sim f^{D-2} \tag{6}$$

where D runs between 0 and 2.

One needs to distinguish here between bounded and unbounded processes. Brownian, and fractional Brownian, motions are unbounded, nonstationary random walks and one can view their $1/f^{1+2H}$ spectral densities as a direct consequence of non-stationarity, as Mandelbrot did (see pp. 78–79 of [17]). In many physical contexts however, such as the on-off blinking quantum dot process [7] or the river Nile minima studied by Hurst [1] the signal amplitude is always bounded and does not grow in time, requiring a different explanation that is either stationary like fGn or "conditionally stationary" like the FRP.

Mandelbrot's best known model for 1/f noise remains the stationary, ergodic, fractional Gaussian noise (fGn) that he advocated so energetically in the 1960s. But, evidently aware that this had had received a disproportionate amount of attention, he was at pains late in his life (e.g. Selecta Volume N [17] p. 207, introducing the reprinted [14, 15]) to stress that:

> Self-affinity and an 1/f spectrum can reveal themselves in several *quite distinct* fashions ... forms of 1/f behaviour that are predominantly due to the fact that a process does not vary in "clock time" but in an "intrinsic time" that is fractal. Those 1/f noises are called "sporadic" or "absolutely intermittent", and can also be said to be "dustborne" and "acting in fractal time".

He thus clearly distinguished LRD stationary ergodic Gaussian models like fGn from from his "conditionally stationary" FRP, noting also that:

> There is a sharp contrast between a highly anomalous ("non-white") noise that proceeds in ordinary clock time and a noise whose principal anomaly is that it is restricted to fractal time.

In practise the main importance of this is to caution that, used on its own, even a very sophisticated approach to the periodogram like the GPH method [16] cannot tell the difference between a time series being stationary LRD and "just" a 1/f noise, unless independent information about stationarity is also available.

One route to reducing the ambiguity in future studies of 1/f is to develop non-stationary extensions to the Wiener-Khinchine theorem. An important step [26] has been to distinguish between one which relates the spectrum and the ensemble averaged correlation function, and a second relating the spectrum to the time averaged correlation function. The importance of this distinction can be seen by considering Fourier inverting the power spectrum-i.e. does inversion yield the time or the ensemble average? [E. Barkai, personal communication]. Another is to increase the emphasis on statistical hypothesis testing, where the degree of support for different models like ARFIMA and its seasonal or heavy tailed variants is compared (e.g. [30]).

4.3 LRD

Readers will, I hope, now be able to see why I believe that the commonly used spectral definition of LRD has caused misunderstandings. The problem has been that on its own a 1/*f* behaviour is necessary but not sufficient, and stationarity is also essential for LRD in the sense so widely studied in statistics community (e.g. in [16, 20]). One may in fact argue that the more crucial aspect of LRD is thus the "loose" one embodied in its name, rather than the formal one embodied in the spectral definition, because *a 1/f spectrum can only be synonymous with LRD when there is an infinitely long past*. The fact that fGn exhibits LRD *by construction* because the stationarity property is assumed, and *also* shows 1/*f* noise, and the Hurst effect has led to the widespread misconception that the converse is true, and that observing 1/*f* spectra and/or the Hurst effect *must imply* LRD.

5 Conclusions

Unfortunately [15] received far less contemporary attention than did Mandelbrot's papers on heavy tails in finance in the early 1960s or the series with van Ness and Wallis in 1968–69 on stationary fractional Gaussian models for LRD, gaining only about 20 citations in its first 20 years. The fact that his work on the AFRP was communicated primarily in the (IEEE) journals and conferences of telecommunications and computer science concealed it from to the contemporary audience that encountered fGn and fBm first in SIAM Review and Water Resources Research. In any event, it was so invisible that one of his most articulate critics, hydrologist, Vit Klemeš [31] actually proposed an AFRP model as a paradigm for the *absence* of the type of LRD seen in the stationary fGn model, clearly unaware of Mandelbrot's work. Sadly Klemeš and Mandelbrot seem not to have subsequently debated FRP and fGn approaches either, as with the advantage of historical distance one can see the importance of both as non-ergodic and ergodic solutions to the 1/*f* question.

Although he revisited the 1963–67 fractional renewal papers with new commentaries in the volume of his Selecta [17] that dealt with multifractals and 1/*f* noise, Mandelbrot himself did not mention them explicitly in his popular historical account of the genesis of LRD [32]. It is clear that he saw the FRP and FGn as a representing two different strands from the way each was allocated a separate Selecta volume [4, 17]. Despite the Selecta, the relatively low visibility has remained to the present day. Mandelbrot's fractional renewal papers are for example not cited or discussed even in encyclopedic books on LRD such as Beran et al. [16].

The long term consequence of this in the physics and statistics literatures may have been to emphasise ergodic solutions to the 1/*f* problem at the expense of non-ergodic ones. This seems to me to be important, because, for example, Per Bak's paradigm of Self-Organised Criticality, in which stationary spectra and correlation functions play an essential role, could not surely have been positioned as the unique solution to the $1/f$ problem [3] if it had been widely recognised how different Mandelbrot's two existing routes to 1/f already were.

Acknowledgements I would like to thank Rebecca Killick for inviting me to talk at ITISE 2016, and helpful comments on an earlier version from Eli Barkai. I also gratefully acknowledge many valuable discussions about the history of LRD and weak ergodicity breaking with Nick Moloney, Christian Franzke, Ralf Metzler, Holger Kantz, Igor Sokolov, Rainer Klages, Tim Graves, Bobby Gramacy, Andrey Cherstvy, Aljaz Godec, Sandra Chapman, Thordis Thorarinsdottir, Kristoffer Rypdal, Martin Rypdal, Bogdan Hnat, Daniela Froemberg, and Igor Goychuk among many others. I acknowledge travel support from KLIMAFORSK project number 229754 and the London Mathematical Laboratory, a senior visiting fellowship from the Max Planck Society in Dresden, and Office of Naval Research NICOP grant NICOP-N62909-15-1-N143 at Warwick and Potsdam.

References

1. Graves, T., Gramacy, R., Watkins, N.W., Franzke, C.L.E.: A brief history of long memory. (http://arxiv.org/abs/1406.6018)
2. Grigolini, P., Aquino, G., Bologna, M., Lukovic, M., West, B.J.: A theory of $1/f$ noise in human cognition. Physica A **388**, 4192 (2009)
3. Watkins, N.W., Pruessner, G., Chapman, S.C., Crosby, N.B., Jensen, H.J.: 25 years of self-organised criticality: concepts and controversies. Space Sci. Rev. **198**, 3–44 (2016). doi:10.1007/s11214-015-0155-x
4. Mandelbrot, B.B.: Gaussian self-affinity and fractals: globality, the earth, $1/f$ noise, and R/S, Selecta volume H, Springer (2002)
5. Margolin, G., Barkai, E.: Nonergodicity of a time series obeying Lévy statistics. J. Stat. Phys. **122**(1), 137–167 (2006)
6. Goychuk, I.: Life and death of stationary linear response in anomalous continuous random walk dynamics. Commun. Theor. Phys. **62**, 497 (2014)
7. Niemann, M., Barkai, E., Kantz, H.: Fluctuations of $1/f$ noise and the low frequency cutoff paradox. Phys. Rev. Lett. **110**, 140603 (2013)
8. Sadegh, S., Barkai, E., Krapf, D.: $1/f$ noise for intermittent quantum dots exhibits non-stationarity and critical exponents. New J. Phys. **16**, 113054 (2015)
9. Stefani, F.D., Hoogenboom, J.P., Barkai, E.: Beyond quantum jumps: Blinking nanoscale light emitters. Phys. Today **62**(2), 34–39 (2009)
10. Rodriguez, M.A: Complete spectral scaling of time series: toward a classification of $1/f$ noise. Phys. Rev. E **90**, 042122 (2014)
11. Rodriguez, M.A.: Class of perfect $1/f$ noise and the low frequency cutoff paradox. Phys. Rev. E **92**, 012112 (2015)
12. Berger, M., Mandelbrot, B.B.: A new model for error clustering in telephone circuits. IBM. J. Res. Dev. 224–236 (1963) [N6 in Mandelbrot, 1999]
13. Mandelbrot, B.B.: Self-similar error clusters in communications systems, and the concept of conditional stationarity. IEEE Trans. Commun. Technol. COM-13, 71–90 (1965a) [N7inMandelbrot, 1999]
14. Mandelbrot, B.B.: Time varying channels, $1/f$ noises, and the infrared catastrophe: or why does the low frequency energy sometimes seem infinite? In: IEEE Communication Convention, Boulder, Colorado (1965b) [N8 in Mandelbrot, 1999]
15. Mandelbrot, B.B.: Some noises with $1/f$ spectrum, a bridge between direct current and white noise. IEEE Trans. Inf. Theory, **13**(2), 289 (1967) [N9 in Mandelbrot, 1999]
16. Beran, J. et al.: Long memory processes. Springer (2013)
17. Mandelbrot, B.B.: Multifractals and $1/f$ noise: wild self-affinity in physics (1963–1976), Selecta volume N, Springer (1999)

18. Mertz, P.: Model of Impulsive Noise for Data Transmission. IRE Trans. Commun. Syst. 130–137 (1961)
19. Mertz, P.: Impulse noise and error performance in data transmission. Memorandum RM-4526-PR, RAND Santa Monica (April 1965)
20. Beran, J.: Statistics for long-range memory processes. Chapman and Hall (1994)
21. Bendat, J.: Principles and applications of random noise theory. Wiley (1958)
22. Bouchaud, J.-P.: Weak ergodicity breaking and aging in disordered systems. J. Phys. I France **2**, 1705–1713 (1992)
23. Lowen, S.B., Teich, M.C.: Fractal renewal processes generate 1/*f* noise. Phys. Rev. E **47**(2), 992 (1993)
24. Lowen, S.B., Teich, M.C.: Fractal-based point processes. Wiley (2005)
25. Dechant, A., Lutz, E.: Wiener-Khinchin theorem for nonstationary scale invariant processes. Phys. Rev. Lett. **115**, 080603 (2015)
26. Leibowich, N., Barkai, E.: Aging Wiener-Khinchin theorem. Phys. Rev. Lett. **115**, 080602 (2015)
27. Bhattacharya, R.N., Gupta, V.K., Waymire, E.: The Hurst effect under trends. J. Appl. Prob. **20**, 649–662 (1983)
28. Franzke, C.L.E., Osprey, S.M., Davini, P., Watkins, N.W.: A dynamical systems explanation of the Hurst effect and atmospheric low-frequency variability. Sci. Rep. **5**, 9068 (2015). doi:10.1038/srep09068
29. Mikosch, T., Starica, C.: Change of structure in financial time series and the GARCH Model. REVSTAT Stat. J. **2**(1), 41–73 (2004)
30. Graves, T.: PhD. Thesis, Statistics Laboratory, Cambridge University (2013)
31. Klemes, V.: The Hurst phenomenon: a puzzle? Water Resour. Res. **10**(4), 675 (1974)
32. Mandelbrot, B.B., Hudson, R.L.: The (mis)behaviour of markets: a fractal view of risk, ruin and reward. Profile books (2008)

Detection of Outlier in Time Series Count Data

Vassiliki Karioti and Polychronis Economou

Abstract Outlier detection for time series data is a fundamental issue in time series analysis. In this work we develop statistical methods in order to detect outliers in time series of counts. More specifically we are interesting on detection of an Innovation Outlier (IO). Models for time series count data were originally proposed by Zeger (Biometrika 75(4):621–629, 1988) [28] and have subsequently generalized into GARMA family. The Maximum Likelihood Estimators of the parameters are discussed and the procedure of detecting an outlier is described. Finally, the proposed method is applied to a real data set.

Keywords GARMA · Estimation · Likelihood ratio test · AIC

1 Introduction

In the last decades analysis of time series of counts has attracted the interest of many researchers. Davis et al. [8] there is a considerable current interest in the study of integer-valued time series models and particular for time series of counts. This kind of time series arise very often in many different fields such as public health and epidemiology [7, 24, 27], environmental processes [25], traffic management [15, 23], economics and finance [12, 13] and industrial processes [6].

Each observation in such applications represents the number of events occurring at a given time point or in a given time interval. For the analysis of such time series an integer valued distribution belonging to the exponential family is usually adopted. The Poisson and the Negative Binomial distributions are two of the most frequently choices.

V. Karioti (✉)
Department of Accounting, Technological Educational Institution of Western Greece (Patras), Patras, Greece
e-mail: vaskar@teiwest.gr

P. Economou
Department of Civil Engineering, University of Patras, Patras, Greece
e-mail: peconom@upatras.gr

I. Rojas et al. (eds.), *Advances in Time Series Analysis and Forecasting*, Contributions to Statistics, DOI 10.1007/978-3-319-55789-2_15

Moreover, in the last twenty or so years there has been a substantial development of a series of observation-driven models for time series. As mentioned in [9], the main models discussed include the autoregressive conditional Poisson model (ACP), the integer-valued autoregressive model (INAR), the integer-valued generalized autoregressive conditional heteroscedastic model (INGARCH), the conditional linear autoregressive process and the dynamic ordered probit model.

There is no single model class that covers all of these models. But the Generalized Autoregression Moving Average (GARMA) model described by Benjamin et al. [4] forms a quite general class which not only includes important models as special cases (see for example the Generalized linear autoregressive moving average models-GLARMA) but provides also a simple and flexible representation of the underlying process.

In this paper, we will consider the GARMA model as a regression model for time series count data. These models, originally proposed by Zeger [28], have been considered subsequently by several other authors (see, in particular, [19]) and extended by Benjamin et al. [5]. In these models, each observation y_t in the series is represented as an integer value variate Y which is conditionally independent of previous observations, given its mean, but whose mean depends on the previous observations $y_{t-1}, \dots, y_1$ and possibly on covariates.

One of the main challenges in the analysis of time series data is the detection of outliers since the presence of an outlier in time series may have significant affect on their form and on the estimation of the parameters. The existing literature focus on detecting individual outliers in a single time series [1, 2, 14, 21] and more recently on automatic procedure on outlier detection [3, 10], or examining the problem of detecting an outlying series in a set of time series [16, 17]. Regarding the detection of outliers in time series of counts very little work has been done (see for example [18] in which the outlier detection is based on the most extreme values of the randomized or Pearson residuals).

There are two basic types of outliers in time series, namely the so called Additive Outlier (AO) and the so called Innovative Outlier (IO). The first type is an outlier that affect a single observation while the second one acts as an addition to the noise term at a particular series point and affects the subsequent observations. The basic aim of this paper is to develop statistical methods in order to detect outliers in a time series of counts and in particular to detect an IO.

The rest of the paper is organized as follows. In Sect. 2, the model for the analysis of time series of counts is presented and it is extended in the case that an outlier of type IO appears. Section 3 describes the model fitting algorithm and in Sect. 4 the model inference on the presence of an outlier is considered. One typical example, the number of campylobacterosis cases, will be used as an illustration in Sect. 5 of this paper. Finally, the conclusions are represented in Sect. 6.

2 GARMA Models

Zeger [28] introduce the Generalized Autoregression Moving Average (GARMA) models in order to model time series of counts. Under the GARMA models the expected value μ_t of a variate is assumed to be related to past outcomes and possible on the past and present values of covariates $\boldsymbol{x}$ (i.e. it is assumed that $\mu_t = E(y_t|D_t)$ where $D_t = \{\boldsymbol{x}_t, \boldsymbol{x}_{t-1}, \ldots, \boldsymbol{x}_1, y_{t-1}, \ldots, y_1\})$ and is given by

$$g(\mu_t) = \eta_t = \boldsymbol{x}_t'\boldsymbol{\beta} + \sum_{j=1}^{p} \phi_j(g(y_{t-j}) - \boldsymbol{x}_{t-j}'\boldsymbol{\beta}) + \sum_{j=1}^{q} \theta_j(g(y_{t-j}) - \eta_{t-j}) \tag{1}$$

where g is a link function, $\boldsymbol{\beta}' = (\beta_0, \beta_!, \ldots, \beta_r)$ are the coefficients of the covariates and $\boldsymbol{\phi}' = (\phi_1, \phi_2, \ldots, \phi_p), \boldsymbol{\theta}' = (\theta_1, \theta_2, \ldots, \theta_q)$ are the autoregressive and moving average parameters that are to be estimated. We will denote the above model by $GARMA(p, q)$. In case that $p = 0$ or $q = 0$ the above sums is to be interpreted as a zero.

It is worth mentioning that because the link function is applied to the lagged observations y_{t-j} this model goes beyond standard generalized linear models (GLM) with independent data [22].

Seasonal AR and MA terms can be included in the model using data values and errors at times with lags that are multiples of S (the span of the seasonality). In such cases model (1) is expressed as

$$\begin{aligned} g(\mu_t) = \eta_t = \boldsymbol{x}_t'\boldsymbol{\beta} &+ \sum_{j=1}^{p} \phi_j(g(y_{t-j}) - \boldsymbol{x}_{t-j}'\boldsymbol{\beta}) + \sum_{j=1}^{q} \theta_j(g(y_{t-j}) - \eta_{t-j}) \\ &+ \sum_{j=1}^{P} \Phi_j(g(y_{t-Sj}) - \boldsymbol{x}_{t-Sj}'\boldsymbol{\beta}) + \sum_{j=1}^{Q} \Theta_j(g(y_{t-Sj}) - \eta_{t-Sj}) \end{aligned} \tag{2}$$

and will denote it as $GARMA(p, q) \times (P, Q)_s$, where P and Q are the seasonal AR and MA orders.

A special case of GARMA series arises when the conditional distribution for y_t (given D_t) is Poisson and the link function g is the logarithm, i.e. the canonical link function as in standard GLM. In this case relation (1) is expressed as

$$\begin{aligned} \log(\mu_t) = \eta_t = \boldsymbol{x}_t'\boldsymbol{\beta} &+ \sum_{j=1}^{p} \phi_j(\log(y_{t-j}) - \boldsymbol{x}_{t-j}'\boldsymbol{\beta}) + \sum_{j=1}^{q} \theta_j(\log(y_{t-j}) - \eta_{t-j}) \\ &+ \sum_{j=1}^{P} \Phi_j(\log(y_{t-Sj}) - \boldsymbol{x}_{t-Sj}'\boldsymbol{\beta}) + \sum_{j=1}^{Q} \Theta_j(\log(y_{t-Sj}) - \eta_{t-Sj}) \end{aligned}$$

To avoid the nonexistence of $\log(y_{t-j})$ for zero values of y_{t-j}, y_{t-j} can be replaced by $y^*_{t-j} = max(y_{t-j}, c)$, where $0 < c < 1$ or by $y^*_{t-j} = y_{t-j} + c$, where in this case $c \in \mathbb{N}$ (see for example [20, 29]).

In what follows we will assume that the conditional distribution for y_t (given D_t) is Poisson and the link function g is the logarithm and we will replace any zero values with $y^*_{t-j} = max(y_{t-j}, c)$.

2.1 Poisson GARMA Models with an Outlier

As already mentioned the basic aim of this paper is to develop statistical methods in order to detect outliers in a time series of counts and in particular to detect an IO. In order to do that we need first to describe the *GARMA(p, q)* in the presence of an IO.

A Poisson GARMA model with an IO at time point $t = t_0$ (which is assumed for the moment to be known) can be describe as

$$\begin{cases} y_t|D_{t-1} \sim P(\mu_t) & t \neq t_0 \\ y_q|D_{t_0-1} \sim P(\alpha\mu_{t_0}) & t = t_0, \alpha > 0 \end{cases}$$

This implies that the expected value μ_t is given by

$$\begin{aligned} \log(\mu_t) = \eta_t = & \delta_{t-t_0}\log(\alpha) + \boldsymbol{x}_t'\boldsymbol{\beta} \\ & + \sum_{j=1}^{p} \phi_j(\log(y^*_{t-j}) - \boldsymbol{x}_{t-j}'\boldsymbol{\beta}) + \sum_{j=1}^{q} \theta_j(\log(y^*_{t-j}) - \eta_{t-j}) \\ & + \sum_{j=1}^{P} \Phi_j(\log(y^*_{t-Sj}) - \boldsymbol{x}_{t-Sj}'\boldsymbol{\beta}) + \sum_{j=1}^{Q} \Theta_j(\log(y^*_{t-Sj}) - \eta_{t-Sj}) \end{aligned} \tag{3}$$

where δ_{t-t_0} is the Kronecker delta function. Relation (3) introduces an additional parameter α to the $GARMA(p,q) \times (P,Q)_s$. model. We will denote this model by $GARMA_{t_0,\alpha}(p,q) \times (P,Q)_s$. Note that $GARMA_{t_0,\alpha}(p,q) \times (P,Q)_s$ collapses to the $GARMA(p,q) \times (P,Q)_s$ model for $\alpha = 1$, i.e. when no outliers are present. This remark will be very useful in Sect. 4 in which the model inference on the presence or not of an outlier will be considered.

3 Estimation

Under the $GARMA_{t_0,\alpha}(p,q) \times (P,Q)_s$ model the likelihood function of the data $\{y_{m+1}, \dots, y_n\}$ conditional on the first m observations $\mathbf{H_m} = \{y_1, \dots, y_m\}$ where $m \geq max(p, q, SP, SQ)$ is given by

$$L\left(y_{m+1},\ldots,y_n|\mathbf{H_m}\right)=\prod_{t=m+1}^{n}P\left(Y_t=y_t|\mathbf{H_m}\right)\prod_{t=m+1}^{n}\frac{e^{-\mu_t}\mu_t^{y_t}}{y_t!}. \tag{4}$$

Since μ_t is a function of $\boldsymbol{\beta}$, $\boldsymbol{\phi}'=(\phi_1,\phi_2,\ldots,\phi_p),\boldsymbol{\theta}'=(\theta_1,\theta_2,\ldots,\theta_q)$, $\boldsymbol{\Phi}'=(\Phi_1,\Phi_2,\ldots,\Phi_P),\boldsymbol{\Theta}'=(\Theta_1,\Theta_2,\ldots,\Theta_Q)$ and α these parameters are to be estimated by maximizing the likelihood function.

Unfortunately, closed-form expressions are not available for the estimation of the parameters (an exception is the parameter α given the rest of the parameters). Additionally , the direct maximization of the likelihood, or equivalently of the log-likelihood, is not always possible, mainly due to the large number of parameters and the present of recursive equations in case in which moving average terms are present ($q>0$ or/and $Q>0$).

For these reasons the Poisson $GARMA_{t_0,\alpha}(p,q)\times(P,Q)_s$ model fitting procedure is heavily related to the maximum likelihood estimation (MLE) using iteratively reweighted least squares (IRLS) for the $GARMA(p,q)$ model developed by Benjamin et al. [4, 5]. In particular, we extend the Fisher Scoring Algorithm procedure to maximize the conditional log-likelihood function presented by Benjamin et al. [5] in order to estimate and the additional parameter α.

The used algorithm can be described by the following steps.

Step 0. Set $k=0$ and give initial values $\boldsymbol{\beta}^{(0)}$, $\boldsymbol{\phi}^{(0)}$, $\boldsymbol{\theta}^{(0)}$, $\boldsymbol{\Phi}^{(0)}$, $\boldsymbol{\Theta}^{(0)}$ and $\alpha^{(0)}$ for the parameters $\boldsymbol{\beta},\boldsymbol{\phi},\boldsymbol{\theta},\boldsymbol{\Phi},\boldsymbol{\Theta}$ and α.
If no previous information is available the constant term of $\boldsymbol{\beta}^{(0)}$ can be set equal to $\log(\bar{y})$ and the rest coefficients equal to zero, $\boldsymbol{\phi}^{(0)}=\mathbf{0}$, $\boldsymbol{\theta}^{(0)}=\mathbf{0}$, $\boldsymbol{\Phi}^{(0)}=\mathbf{0}$, $\boldsymbol{\Theta}^{(0)}=\mathbf{0}$ and $\alpha^{(0)}=1$.

Step 1. Set $k=k+1$ and calculate

(a) $\boldsymbol{\eta}^{(k)}$, $\boldsymbol{\mu}^{(k)}=e^{\boldsymbol{\eta}^{(k)}}$, $\left(\frac{\partial\eta}{\partial\mu}\right)^{(k)}=\frac{1}{\boldsymbol{\mu}^{(k)}}$, the variance $\boldsymbol{V}^{(k)}$ which for the Poisson distribution is equal to $\boldsymbol{\mu}^{(k)}$.

When fitting moving average components (i.e. when $q>0$) is necessary the initial values of $\boldsymbol{\eta}^{(k)}$ to be fixed. In the present work to avoid further complication in the algorithm we fix all values of $\boldsymbol{\eta}^{(k)}$ that do not contribute directly to the likelihood. In particular, for all $t\le m$ we use $\eta_t^{(k)}=g(y_t^*)=\log(y_t^*)$.

(b) the derivatives

$$\left(\frac{\partial\eta}{\partial\beta}\right)^{(k)}=x_{tm}-\sum_{j=1}^{p}\phi_j^{(k)}x_{t-j,m}-\sum_{j=1}^{P}\Phi_j^{(k)}x_{t-Sj,m}$$
$$-\sum_{j=1}^{q}\theta_j^{(k)}\left(\frac{\partial\eta_{t-m}}{\partial\beta_m}\right)^{(k)}-\sum_{j=1}^{Q}\Theta_j\left(\frac{\partial\eta_{t-Sm}}{\partial\beta_m}\right)^{(k)},\ \text{for } m=0,1,\ldots,r$$

$$\left(\frac{\partial \boldsymbol{\eta}}{\partial \boldsymbol{\phi}}\right)^{(k)} = g(y^*_{t-m}) - x'_{t-m}\boldsymbol{\beta}^{(k)} - \sum_{j=1}^{q} \theta_j \left(\frac{\partial \eta_{t-m}}{\partial \phi_m}\right)^{(k)}, \text{ for } m = 1, 2, \ldots, p$$

$$\left(\frac{\partial \boldsymbol{\eta}}{\partial \boldsymbol{\theta}}\right)^{(k)} = g(y^*_{t-m}) - \eta_{t-m} - \sum_{j=1}^{q} \theta_j \left(\frac{\partial \eta_{t-m}}{\partial \theta_m}\right)^{(k)}, \text{ for } m = 1, 2, \ldots, q$$

$$\left(\frac{\partial \boldsymbol{\eta}}{\partial \boldsymbol{\Phi}}\right)^{(k)} = g(y^*_{t-m}) - x'_{t-Sm}\boldsymbol{\beta}^{(k)} - \sum_{j=1}^{Q} \Theta_j \left(\frac{\partial \eta_{t-Sm}}{\partial \Phi_m}\right)^{(k)}, \text{ for } m = 1, 2, \ldots, P$$

$$\left(\frac{\partial \boldsymbol{\eta}}{\partial \boldsymbol{\Theta}}\right)^{(k)} = g(y^*_{t-Sm}) - \eta_{t-Sm} - \sum_{j=1}^{Q} \Theta_j \left(\frac{\partial \eta_{t-Sm}}{\partial \Theta_m}\right)^{(k)}, \text{ for } m = 1, 2, \ldots, Q$$

$$\left(\frac{\partial \boldsymbol{\eta}}{\partial \alpha}\right)^{(k)} = \delta_{t-t_0} \frac{1}{\alpha}.$$

Again to avoid further complication in the algorithm all the values for the derivatives for the observations that are not contributing directly in the likelihood are taken to be zero.

(c) the adjusted dependent variable $\boldsymbol{z}^{(k)}$, needed for the iteratively reweighted least squares approach (see Step 2) (Green 1984) where

$$z_t^{(k)} = \left(\frac{\partial \eta_t}{\partial \boldsymbol{\beta}'}\right)^{(k)} \boldsymbol{\beta}^{(k)} + \left(\frac{\partial \eta_t}{\partial \boldsymbol{\phi}'}\right)^{(k)} \boldsymbol{\phi}^{(k)} + \left(\frac{\partial \eta_t}{\partial \boldsymbol{\theta}'}\right)^{(k)} \boldsymbol{\theta}^{(k)}$$
$$+ \left(\frac{\partial \eta_t}{\partial \boldsymbol{\Phi}'}\right)^{(k)} \boldsymbol{\Phi}^{(k)} + \left(\frac{\partial \eta_t}{\partial \boldsymbol{\Theta}'}\right)^{(k)} \boldsymbol{\Theta}^{(k)} + \left(\frac{\partial \eta_t}{\partial \alpha}\right)^{(k)} \alpha^{(k)}$$
$$+ h(y_t - \mu_t{}^{(k)})\mu_t{}^{(k)}$$

where $h, 0 < h \leq 1$ is the step length of the algorithm (smaller values ensure better estimates at each repetition but slower convergence, in this paper s was set equal to 0.5) and the weights $\boldsymbol{w}^{(k)} = \boldsymbol{\mu}^{(k)}$.

Step 2. Update the parameters $\boldsymbol{\beta}^{(k)}$, $\boldsymbol{\phi}^{(k)}$, $\boldsymbol{\theta}^{(k)}$, $\boldsymbol{\Phi}^{(k)}$, $\boldsymbol{\Theta}^{(k)}$ and $\alpha^{(k)}$ by fitting to $\boldsymbol{z}^{(k)}$ a weighted least squares linear model on $\left(\frac{\partial \eta}{\partial \beta}\right)^{(k)}$, $\left(\frac{\partial \eta}{\partial \boldsymbol{\phi}}\right)^{(k)}$, $\left(\frac{\partial \eta}{\partial \theta}\right)^{(k)}$, $\left(\frac{\partial \eta}{\partial \boldsymbol{\Phi}}\right)^{(k)}$, $\left(\frac{\partial \eta}{\partial \boldsymbol{\Theta}}\right)^{(k)}$ and $\left(\frac{\partial \eta}{\partial \alpha}\right)^{(k)}$ with weights $\boldsymbol{w}^{(k)}$.

Step 3. Repeat steps 1 and 2 until the parameters estimates converge or the value of the likelihood (4) does not increase any further.

By fixing $\alpha = 1$, i.e. by assuming that no outlier is present at $t = t_0$, and so by setting $\left(\frac{\partial \eta_t}{\partial \alpha}\right)^{(k)} = 0$ the above algorithm can be used to fit a $GARMA(p, q) \times (P, Q)_s$ model.

4 Model Inference—Outlier Detection

We have already mentioned that the $GARMA_{t_0,\alpha}(p,q) \times (P,Q)_s$ model collapses to the $GARMA(p,q) \times (P,Q)_s$ model for $\alpha = 1$, i.e. when no outliers are present. As a consequence a test for detecting if an outlier is present or not at time point $t = t_0$, which is assumed for the moment to be known, can be conducted by testing the hypotheses $H_0 : \alpha = 1$ versus $H_1 : \alpha \neq 1$ using a Likelihood Ratio (LR) test .

In order to conduct a LR test, the null $GARMA(p,q) \times (P,Q)_s$ model and the alternative $GARMA_{t_0,\alpha}(p,q) \times (P,Q)_s$ model, are fitted to the data and the log-likelihood is recorded in each case. The LR test, for a given t_0, is given by $T = -2(\ell_1 - \ell_0)$, where ℓ_0 and ℓ_1 are the log likelihood under the null and the alternative model respectively.

Since the time point $t = t_0$ at which an outlier may have occurred is not known or at least a specific choice can not be fully justified, the direct application of the above mentioned LR can not be applied. As a consequence, we present an algorithm in order to identify firstly the time point which is more likely an outlier to have occurred and secondly a modified LR test.

4.1 Determination of t_0 and LR Test

Since in most of the applications the time point $t = t_0$ at which an outlier may have occur is not known an algorithm is needed in order to identify the most likely time point for an outlier to have occurred. This can be done by successively fitting a $GARMA_{t_0}(p,q,\alpha)$ model with $m \leq t_0 \leq n$ and selecting the t_0 with the largest value of log likelihood. For that t_0 a LR test can be performed for the hypotheses $H_0 : \alpha = 1$ versus $H_1 : \alpha \neq 1$.

Unfortunately, in this case the LR statistic, denoted as T_{t_0}, for an outlier at the selected t_0 does not has an asymptotic χ^2 distribution with one degree of freedom, since it is the maximum of correlated χ^2 distributions with one degree of freedom.

In order to overcome the problem of the unknown distribution of the T_{t_0} test statistic one could perform a simulation study in order to investigate the distribution of the T_{t_0} under the null hypothesis and compute the critical values. Although, this will required an extended simulation since the distribution of T_{t_0} does not only depend on the sample size and the order of the *GARMA* model, i.e. on the values of p, q, P and Q but probably and on the values of the parameters $\boldsymbol{\phi}' = (\phi_1, \phi_2, \ldots, \phi_p)$, $\boldsymbol{\theta}' = (\theta_1, \theta_2, \ldots, \theta_q)$;, $\boldsymbol{\Phi}' = (\Phi_1, \Phi_2, \ldots, \Phi_P)$, $\boldsymbol{\Theta}' = (\Theta_1, \Theta_2, \ldots, \Theta_Q)$. Moreover, the possible presence of the covariates increase even more the complexity of the simulation study.

A solution to this problem is to perform a (parametric) bootstrap by generating samples from the fitted model under the null hypothesis with the same sample size as the original time series using the original values for the first m observations. For every bootstrap sample the T_{t_0} is calculated and its distribution (and so the critical values) are computed.

This bootstrap procedure can be very time consuming since the alternative model has to be fitted $(n - m)N$ times, where n is the sample size of the original time series and N the number of bootstrap samples (usually N = 1000–10000). This procedure can speed up by fitting the alternative model not $(n - m)$ times but in a significant smaller number time points.

From a small simulation study it was observed that the time point t_0 with the largest value of log likelihood was always among the 5% of the observations with the largest (in absolute value) Pearson residuals [22] under the null model given by

$$r_t^P = \frac{y_t - \mu_t}{\sqrt{\mu_t}}.$$

The proposed method can be summarized as follows

Step 1. Fit the null model for the original time series and calculate the Pearson residuals.
Step 2. Select the 5% largest, in absolute value, Pearson residuals and fit the alternative model for these observations. Select the time point with the largest value of log likelihood and calculate the corresponding LR test T_{t_0}.
Step 3. Generate N bootstrap samples from the fitted model under the null hypothesis; For each sample repeat Steps 1 and 2.
Step 4. Calculate the critical values for the LR test based on the bootstrap samples and compared them with the T_{t_0} of the original time series.

5 Application

Campylobacter poisoning is one of the most common cause of bacterial foodborne illness. Campylobacter is found most often in food, particularly in chicken. The infection is usually self-limiting and, in most cases, symptomatic treatment by liquid and electrolyte replacement is enough. For that reason is common only the most severe cases to be reported and not all the cases.

Ferland et al. [11] report the number of campylobacterosis cases over a period of more than 10 years starting from January 1990 to October 2000 in the north of Quebec in Canada. The number of campylobacterosis cases was reported every 28 days (13 times a year) and are presented in Fig. 1 of [11]. From the time series plot it is clear that a possible outlier may have occurred at the time point $t = 100$. The following observations remain in relative large values making the presence of an IO at time $t = 100$ a possibly event. Although, since [11] give no information on this value (actually, they are not even comment on this observation) we prefer to apply the algorithm presented in the previous section in order to identify the most likely time point for an outlier to have occurred. Another reason for this is to demonstrate that the time point with the largest value of log likelihood belongs to the observations with 5% largest, in absolute value, Pearson residuals under the null model.

Ferland et al. [11] assumed that Y_t given the past is Poisson distributed and used the identity link function to model the expected value μ_t. For taking into account serial dependence they included a regression on the previous observation. Additionally, seasonality was captured by regressing on μ_{t-13}.

In the present work the canonical link function, i.e. the logarithm, is adopted. Additionally, different *GARMA* models of order p and q were fitted to the data in order to determine the optimum model. Next, given the optimum *GARMA*(p,q) model, different seasonal models were also fitted in order to estimate the best *GARMA*$(p,q) \times (P,Q)_{13}$ model. For all p, q, P and Q the values 0, 1 and 2 were used since firstly models of small order are preferable and secondly since models of high order turned out to be very sensitive to initial values and the algorithm did not always converge.

The choice of the optimum *GARMA* model was made using a modified Akaike information criterion (AIC) suitable for time series. More specifically, since the AIC values are computed using conditional likelihoods they may not be comparable because the conditioning may be different for different models. This is reflected on the number of observations that have been used to estimate the model which is different for *GARMA* models of different order.

For this reason the criterion is normalized by dividing it by the number of observations $(n-m)$ that have been used to estimate the model (see for example [26]). The used modified AIC is given by

$$AIC_m = 2\frac{k}{n-m} - 2\frac{\ell}{n-m}$$

where k is the number of estimated parameters in the model $(k = (r+1) + p + q)$ and ℓ is the maximum value of the likelihood function for the model. As with the classical AIC, the preferred model is the one with the minimum AIC_m value.

In Tables 1 and 2 are presented the log-likelihood and the AIC_m (in parenthesis) for different *GARMA*(p,q) and *GARMA*$(p,q) \times (P,Q)_{13}$ models respectively for the campylobacterosis case data (n = 140) with no covariates (i.e. only the intercept β_0 is included in the model). Based on the results presented in Table 1 the preferred *GARMA*(p,q) model based on the AIC_m value is the *GARMA*$(2,1)$. Given the *GARMA*$(2,1)$ model the preferred seasonal *GARMA*$(2,1) \times (P,Q)_{13}$ model is the *GARMA*$(2,1) \times (2,0)_{13}$ model (see results in Table 2).

In Fig. 1 are presented the μ_t for the fitted *GARMA*$(2,1) \times (2,0)_{13}$ model for the campylobacterosis case data (Red line in the left plot) and the corresponding Pearson residuals (right plot). From the residuals plot is clear that the largest, in absolute value, Pearson residual is obtained at $t = 100$. This is also the time point with the largest value of log likelihood under the alternative hypothesis. In Table 3 are presented the estimates of the parameters of the *GARMA*$(2,1) \times (2,0)_{13}$ (upper half) model, the log-likelihood and the AIC_m (given also in Table 1).

Table 1 The log-likelihood and the AIC_m (in parenthesis) for different $GARMA(p, q)$ models for the campylobacterosis case data (n = 140). In bold the model with the smallest AIC_m

		MA		
		q = 0	q = 1	q = 2
AR	p = 0	−550.305 (7.87579)	−471.948 (6.81939)	−447.179 (6.52433)
	p = 1	−433.975 (6.27302)	−430.899 (6.24315)	−428.523 (6.26844)
	p = 2	−430.111 (6.27697)	**−425.545 (6.22529)**	−425.322 (6.23655)

Table 2 The log-likelihood and the AIC_m (in parenthesis) for different $GARMA(2, 1) \times (P, Q)_{13}$ models for the campylobacterosis case data (n = 140). In bold the model with the smallest AIC_m

		S. MA		
		Q = 0	Q = 1	Q = 2
S. AR	P = 0	−425.545 (6.22529)	−388.082 (5.69684)	−358.321 (5.28001)
	P = 1	−392.415 (5.75964)	−389.176 (5.72719)	−358.129 (5.29173)
	P = 2	**−358.285 (5.27949)**	−358.283 (5.29396)	−358.131 (5.30625)

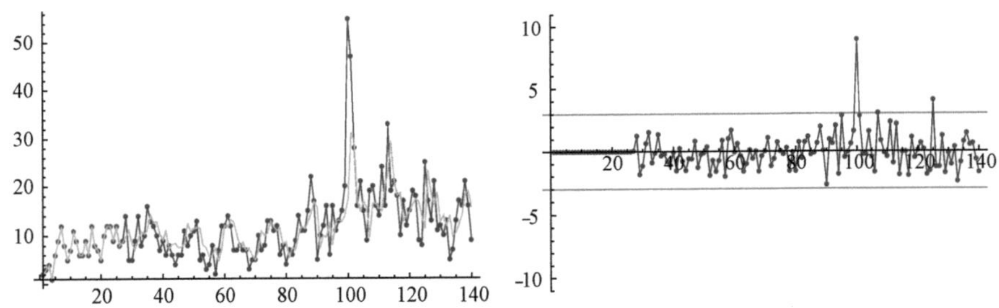

Fig. 1 The fitted $GARMA(2, 1) \times (2, 0)_{13}$model for the campylobacterosis case data (*Red line* in the *left plot*) and the corresponding Pearson residuals (*right plot*)

In Table 3 (bottom half) are presented the estimates of the parameters of the $GARMA_{100,\alpha}(2, 1) \times (2, 0)_{13}$, the log-likelihood and the AIC_m. In Fig. 2 are presented the μ_t for the fitted $GARMA_{100,\alpha}(2, 1) \times (2, 0)_{13}$ model for the data (Red line in the left plot) and the corresponding Pearson residuals (right plot).

From the log-likelihoods of the $GARMA(2, 1) \times (2, 0)_{13}$ and $GARMA_{100,\alpha}(2, 1) \times (2, 0)_{13}$ we can calculate the LR test $T_{100} = -2(\ell_1 - \ell_0) = 64.4498$.

In order to calculate the critical values, we have generated N = 10000 bootstrap samples under from the fitted model under the null hypothesis in order to determine

Table 3 The estimates of the parameters, the log-likelihood and the AIC_m of the $GARMA(2,1) \times (2,0)_{13}$ (upper half) and the $GARMA_{100,\alpha}(2,1) \times (2,0)_{13}$ (bottom half) for the campylobacterosis case data

$GARMA(2,1) \times (2,0)_{13}$				
$\hat{\beta}_0 = 2.83283$	$\hat{\phi}_1 = 0.434945$	$\hat{\theta}_1 = 0.0229244$	$\hat{\Phi}_1 = 0.244766$	
	$\hat{\phi}_2 = 0.0885754$		$\hat{\Phi}_2 = 0.022835$	
	$\ell_1 = -425.545$	$AIC_m = 6.22529$		
$GARMA_{100,\alpha}(2,1) \times (2,0)_{13}$				
$\hat{\beta}_0 = 2.63206$	$\hat{\phi}_1 = 0.702666$	$\hat{\theta}_1 = -0.398191$	$\hat{\alpha} = 3.59439$	$\hat{\Phi}_1 = 0.21262$
	$\hat{\phi}_2 = -0.0724158$			$\hat{\Phi}_2 = -0.00325188$
	$\ell_1 = -326.06$	$AIC_m = 4.82696$		

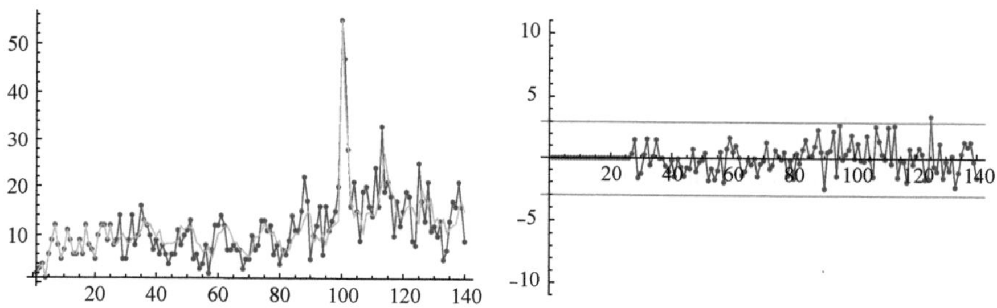

Fig. 2 The fitted $GARMA_{100,\alpha}(2,1) \times (2,0)_{13}$ model for the campylobacterosis case data (*Red line* in the *left plot*) and the corresponding Pearson residuals (*right plot*)

Table 4 The critical values obtained by the bootstrap procedure for different significant levels by generating N = 10000 bootstrap samples from the fitted model under the null hypothesis

Significant level	0.05	0.025	0.01
Critical value	11.6678	13.4697	17.7944

the critical values of the T_{100}, i.e. the critical values for the distribution of the maximum a series of correlated χ^2 distributions with one degree of freedom.

In Table 4 are presented the critical values obtained by the bootstrap procedure for different significant levels. From the critical values given in the Table we conclude again that the null hypothesis $H_0 : \alpha = 1$ can be rejected. Actually, the parametric bootstrap p-value was found equal to 0.0015. As a consequence, we can conclude that at time point $t = 100$, which correspond to September 1997 an IO occurred. More specifically, the (expected) number of the campylobacterosis cases reported that time was 3.59439 times larger than it would have been if this outlier did not occurred.

6 Conclusions

A method of detection of an IO in time series of count data was presented assuming a Poisson GARMA model. The proposed method includes a heuristic approach on identifying the time point which is more likely an outlier to have occurred, the estimation of the parameters in the presence of an outlier and finally the inference on if or not an outlier is actually presented in the data.

Challenges that remain include the development of a similar method in order to detect an AO in a time series of counts and to investigate possible extensions of the proposed method in order to detect multiple outliers.

References

1. Abraham, B., Chuang, A.: Outlier detection and time series modeling. Technometrics **31**(2), 241–248 (1989)
2. Barnett, V., Lewis, T.: Outliers in Statistical Data, 3rd edn. Wiley, Chichester (1994)
3. Basu, S., Meckesheimer, M.: Automatic outlier detection for time series: an application to sensor data. Knowl. Inf. Syst. **11**(2), 137–154 (2007)
4. Benjamin, M.A., Rigby, R.A., Stasinopoulos, D.M.: Generalized autoregressive moving average models. J. Am. Stat. Assoc. **98**(461), 214–223 (2003)
5. Benjamin, M.A., Rigby, R.A., Stasinopoulos, M.D.: Fitting Non-Gaussian Time Series Models, pp. 191–196. Physica-Verlag HD, Heidelberg (1998)
6. Blundell, R., Griffithand, R., Van Reenen, J.: Dynamic count data models of technological innovation. Econ. J. **105**(429), 333–344 (1995)
7. Cardinal, M., Roy, R., Lambert, J.: On the application of integer-valued time series models for the analysis of disease incidence. Stat. Med. **18**(15), 2025–2039 (1999)
8. Davis, R., Holan, S., Lund, R., Ravishanker, N.: Handbook of Discrete-Valued Time Series. Chapman & Hall/CRC Handbooks of Modern Statistical Methods. Taylor & Francis (2015)
9. Dunsmuir, W., Scott, D.: The glarma package for observation-driven time series regression of counts. J. Stat. Softw. **067**(i07) (2015)
10. Ferdousi, Z., Maeda, A.: Unsupervised outlier detection in time series data. In: 22nd International Conference on Data Engineering Workshops (ICDEW'06), pp. x121–x121 (2006)
11. Ferland, R., Latour, A., Oraichi, D.: Integer-valued GARCH process. J. Time Ser. Anal. **27**(6), 923–942 (2006)
12. Freeland, R.K., McCabe, B.P.M.: Analysis of low count time series data by Poisson autoregression. J. Time Ser. Anal. **25**(5), 701–722 (2004)
13. Heinen, A., Rengifo, E.: Multivariate autoregressive modeling of time series count data using copulas. J. Empirical Finan. **14**(4), 564–583 (2007)
14. Hotta, L., Neves, M.: A brief review on tests for detection of time series outliers. Estadistica **44**(142, 143), 103–148 (1992)
15. Johansson, P.: Speed limitation and motorway casualties: a time series count data regression approach. Accid. Anal. Prev. **28**(1), 73–87 (1996)
16. Karioti, V., Caroni, C.: Detecting outlying series in sets of short time series. Comput. Stat. Data Anal. **39**(3), 351–364 (2002)
17. Karioti, V., Caroni, C.: Simple detection of outlying short time series. Stat. Pap. **45**(2), 267–278 (2004)
18. Karioti, V., Caroni, C.: Properties of the GAR(1) model for time series of counts. J. Modern Appl. Stat. Methods **5**(1), 140–151 (2006)

19. Kedem, B., Fokianos, K.: Regression Models for Time Series Analysis. Wiley Series in Probability and Statistics. Wiley, New York (2005)
20. Li, W.K.: Time series models based on generalized linear models: some further results. Biometrics **50**(2), 506–511 (1994)
21. Ljung, G.: On outlier detection in time series. J. R. Stat. Soc. Ser. B (Methodological) **55**(2), 559–567 (1993)
22. McCullagh, P., Nelder, J.: Generalized Linear Models, Second Edition. Chapman & Hall/CRC Monographs on Statistics & Applied Probability. Taylor & Francis (1989)
23. Quddus, M.A.: Time series count data models: an empirical application to traffic accidents. Accid. Anal. Prev. **40**(5), 1732–1741 (2008)
24. Schmidt, A.M., Pereira, J.B.M.: Modelling time series of counts in epidemiology. Int. Stat. Rev. **79**(1), 48–69 (2011)
25. Thyregod, P., Carstensen, J., Madsen, H., Arnbjerg-Nielsen, K.: Integer valued autoregressive models for tipping bucket rainfall measurements. Environmetrics **10**(4), 395–411 (1999)
26. Vogelvang, B.: Econometrics: Theory and Applications with EViews. Financial Times. Pearson/Addison Wesley (2005)
27. Yu, X., Baron, M., Choudhary, P.K.: Change-point detection in binomial thinning processes, with applications in epidemiology. Sequential Anal. **32**(3), 350–367 (2013)
28. Zeger, S.L.: A regression model for time series of counts. Biometrika **75**(4), 621–629 (1988)
29. Zeger, S.L., Qaqish, B.: Markov regression models for time series: a quasi-likelihood approach. Biometrics **44**(4), 1019–1031 (1988)

Ratio Tests of a Change in Panel Means with Small Fixed Panel Size

Barbora Peštová and Michal Pešta

Abstract The aim of this paper is to develop stochastic methods for detection whether a change in panel data occurred at some unknown time or not. Panel data of our interest consist of a moderate or relatively large number of panels, while the panels contain a small number of observations. Testing procedures to detect a possible common change in means of the panels are established. To this end, we consider several competing ratio type test statistics and derive their asymptotic distributions under the no change null hypothesis. Moreover, we prove the consistency of the tests under the alternative. The main advantage of the proposed approaches is that the variance of the observations neither has to be known nor estimated. The results are illustrated through a simulation study. An application of the procedure to actuarial data is presented.

Keywords Change point · Panel data · Change in mean · Hypothesis testing · Structural change · Fixed panel size · Short panels · Ratio type statistics

1 Introduction

The problem of an unknown common change in means of the panels is studied here, where the panel data consist of N panels and each panel contains T observations over time. Various values of the change are possible for each panel at some unknown common time $\tau = 1, \ldots, N$. The panels are considered to be independent, but this restriction can be weakened. In spite of that, observations within the panel are usually not independent. It is supposed that a common unknown dependence structure is present over the panels.

B. Peštová
Institute of Computer Science, The Czech Academy of Sciences, Prague, Czech Republic
e-mail: pestova@cs.cas.cz

M. Pešta (✉)
Faculty of Mathematics and Physics, Charles University, Prague, Czech Republic
e-mail: michal.pesta@mff.cuni.cz

I. Rojas et al. (eds.), *Advances in Time Series Analysis and Forecasting*, Contributions to Statistics, DOI 10.1007/978-3-319-55789-2_16

1.1 State of Art

Tests for change point detection in the panel data have been proposed only in case when the panel size T is sufficiently large, i.e., T increases over all limits from an asymptotic point of view, cf. [3] or [5]. However, the change point estimation has already been studied for finite T not depending on the number of panels N, see [2] or [14]. The remaining task is to develop testing procedures to decide whether a common change occurs or not in the panels, while taking into account that the length T of each observation regime is fixed and can be relatively small.

1.2 Motivation

Structural changes in panel data—especially *common breaks in means*—are wide spread phenomena. Our primary motivation comes from non-life insurance business, where associations in many countries uniting several insurance companies collect claim amounts paid by every insurance company each year. Such a database of cumulative claim payments can be viewed as panel data, where insurance company $i = 1, \ldots, N$ provides the total claim amount $Y_{i,t}$ paid in year $t = 1, \ldots, T$ into the common database. The members of the association can consequently profit from the joint database.

For the whole association it is important to know, whether a possible change in the claim amounts occurred during the observed time horizon. Usually, the time period is relatively short, e.g., 10–15 years. To be more specific, a widely used and very standard actuarial method for predicting future claim amounts—called chain ladder—assumes a kind of stability of the historical claim amounts. The formal necessary and sufficient condition is derived in [12]. This paper shows a way how to test for a possible historical instability.

1.3 Structure of the Paper

The remainder of this paper is organized as follows. Section 2 introduces an abrupt change point model for panel data together with stochastic assumptions. Various ratio type test statistics for the abrupt change in panel means are proposed in Sect. 3. Consequently, asymptotic behavior of the considered change point test statistics under the null as well as under the alternatives is derived, which covers the main theoretical contribution. As a by-product of the developed tests, we provide estimation of the correlation structure in Sect. 4. Section 5 contains a simulation study that illustrates finite sample performance of the test statistics. It numerically emphasizes the advantages and disadvantages of the proposed approach. A practical application of the developed approach to an actuarial problem is presented in Sect. 6. Proofs are given in the Appendix.

2 Panel Change Point Model

Let us consider the panel change point model

$$Y_{i,t} = \mu_i + \delta_i \mathcal{I}\{t > \tau\} + \sigma\varepsilon_{i,t}, \quad 1 \leq i \leq N,\ 1 \leq t \leq T; \tag{1}$$

where $\sigma > 0$ is an unknown variance-scaling parameter and T is fixed, not depending on N. The possible *common change point time* is denoted by $\tau \in \{1, \dots, T\}$. A situation where $\tau = T$ corresponds to *no change* in means of the panels. The means μ_i are panel-individual. The amount of the break in mean, which can also differ for every panel, is denoted by δ_i. Furthermore, it is assumed that the sequences of panel disturbances $\{\varepsilon_{i,t}\}_t$ are independent and within each panel the errors form a weakly stationary sequence with a common correlation structure. This can be formalized in the following assumption.

Assumption A1 The vectors $[\varepsilon_{i,1}, \dots, \varepsilon_{i,T}]^\top$ existing on a probability space $(\Omega, \mathcal{F}, \mathsf{P})$ are *iid* for $i = 1, \dots, N$ with $\mathsf{E}\varepsilon_{i,t} = 0$ and $\mathsf{Var}\,\varepsilon_{i,t} = 1$, having the autocorrelation function

$$\rho_t = \mathsf{Corr}\left(\varepsilon_{i,s}, \varepsilon_{i,s+t}\right) = \mathsf{Cov}\left(\varepsilon_{i,s}, \varepsilon_{i,s+t}\right), \quad \forall s \in \{1, \dots, T-t\},$$

which is independent of the lag s, the cumulative autocorrelation function

$$r(t) = \mathsf{Var} \sum_{s=1}^{t} \varepsilon_{i,s} = \sum_{|s|<t} (t - |s|)\rho_s,$$

and the shifted cumulative correlation function

$$R(t, v) = \mathsf{Cov}\left(\sum_{s=1}^{t} \varepsilon_{i,s}, \sum_{u=t+1}^{v} \varepsilon_{i,u}\right) = \sum_{s=1}^{t} \sum_{u=t+1}^{v} \rho_{u-s}, \quad t < v$$

for all $i = 1, \dots, N$ and $t, v = 1, \dots, T$.

The sequence $\{\varepsilon_{i,t}\}_{t=1}^{T}$ can be viewed as a part of a *weakly stationary* process. Note that the dependent errors within each panel do not necessarily need to be linear processes. For example, GARCH processes as error sequences are allowed as well. The assumption of independent panels can indeed be relaxed, but it would make the setup much more complex. Consequently, probabilistic tools for dependent data need to be used (e.g., suitable versions of the central limit theorem). Nevertheless, assuming, that the claim amounts for different insurance companies are independent, is reasonable. Moreover, the assumption of a common homoscedastic variance parameter σ can be generalized by introducing weights $w_{i,t}$, which are supposed to be known. Being particular in actuarial practice, it would mean to normalize the total claim amount by the premium received, since bigger insurance companies are expected to have higher variability in total claim amounts paid.

It is required to test the *null hypothesis* of no change in the means

$$H_0 \,:\, \tau = T$$

against the *alternative* that at least one panel has a change in mean

$$H_1 \,:\, \tau < T \quad \text{and} \quad \exists i \in \{1,\ldots,N\} \,:\, \delta_i \neq 0.$$

3 Test Statistics and Asymptotic Results

We propose several *ratio type statistics* to test H_0 against H_1, because this type of statistics does not require estimation of the nuisance parameter for the variance. Generally, this is due to the fact that the variance parameter simply cancels out from the numerator and denominator of the statistic. For surveys on ratio type test statistics, we refer to [6], [8], and [10]. Our first particular panel change point test statistic of the ratio type is

$$\mathcal{S}_N(T) = \max_{t=2,\ldots,T-2} \frac{\sum_{s=1}^{t} \left[\overline{\mathcal{E}}_{N,T,s,t}\right]^2}{\sum_{s=t}^{T-1} \left[\widetilde{\mathcal{E}}_{N,T,s,t}\right]^2}$$

where

$$\overline{\mathcal{E}}_{N,T,s,t} = \sum_{i=1}^{N} \sum_{r=1}^{s} \left(Y_{i,r} - \overline{Y}_{i,t}\right) \quad \text{and} \quad \widetilde{\mathcal{E}}_{N,T,s,t} = \sum_{i=1}^{N} \sum_{r=s+1}^{T} \left(Y_{i,r} - \widetilde{Y}_{i,t}\right),$$

such that $\overline{Y}_{i,t}$ is the average of the first t observations in panel i, $\widetilde{Y}_{i,t}$ is the average of the last $T-t$ observations in panel i, i.e.,

$$\overline{Y}_{i,t} = \frac{1}{t}\sum_{s=1}^{t} Y_{i,s} \quad \text{and} \quad \widetilde{Y}_{i,t} = \frac{1}{T-t}\sum_{s=t+1}^{T} Y_{i,s}.$$

The second panel change point test statistic is deduced from [4] and modified for the panel setup

$$\mathcal{T}_N(T) = \max_{t=2,\ldots,T-2} \frac{\sum_{s=1}^{t} \left[\overline{\mathcal{E}}_{N,T,s,t}\right]^2 - \frac{1}{t}\left[\sum_{s=1}^{t} \overline{\mathcal{E}}_{N,T,s,t}\right]^2}{\sum_{s=t}^{T-1} \left[\widetilde{\mathcal{E}}_{N,T,s,t}\right]^2 - \frac{1}{T-t}\left[\sum_{s=t}^{T-1} \widetilde{\mathcal{E}}_{N,T,s,t}\right]^2}$$

and the third one is motivated by [7]

$$\mathcal{U}_N(T) = \max_{t=2,\ldots,T-2} \frac{\max_{s=1,\ldots,t} \overline{\mathcal{E}}_{N,T,s,t} - \min_{s=1,\ldots,t} \overline{\mathcal{E}}_{N,T,s,t}}{\max_{s=t,\ldots,T-1} \widetilde{\mathcal{E}}_{N,T,s,t} - \min_{s=t,\ldots,T-1} \widetilde{\mathcal{E}}_{N,T,s,t}}.$$

The above defined ratio type test statistics can be compared to test statistic

$$\mathcal{R}_N(T) = \max_{t=2,\ldots,T-2} \frac{\max_{s=1,\ldots,t} \left|\overline{\mathcal{E}}_{N,T,s,t}\right|}{\max_{s=t,\ldots,T-1} \left|\widetilde{\mathcal{E}}_{N,T,s,t}\right|},$$

elaborated in [13]. Although, it will be demonstrated by simulations that $\mathcal{S}_N(T)$ and $\mathcal{T}_N(T)$ provide higher power of the test than $\mathcal{R}_N(T)$.

Firstly, we derive the behavior of the test statistics under the null hypothesis.

Theorem 1 (Under Null) *Under hypothesis H_0 and Assumption A1*

$$\mathcal{S}_N(T) \xrightarrow[N\to\infty]{\mathscr{D}} \max_{t=2,\ldots,T-2} \frac{\sum_{s=1}^{t} \left(X_s - \frac{s}{t}X_t\right)^2}{\sum_{s=t}^{T-1} \left(Z_s - \frac{T-s}{T-t}Z_t\right)^2},$$

$$\mathcal{T}_N(T) \xrightarrow[N\to\infty]{\mathscr{D}} \max_{t=2,\ldots,T-2} \frac{\sum_{s=1}^{t} \left(X_s - \frac{s}{t}X_t\right)^2 - \frac{1}{t}\left[\sum_{s=1}^{t} \left(X_s - \frac{s}{t}X_t\right)\right]^2}{\sum_{s=t}^{T-1} \left(Z_s - \frac{T-s}{T-t}Z_t\right)^2 - \frac{1}{T-t}\left[\sum_{s=t}^{T-1} \left(Z_s - \frac{T-s}{T-t}Z_t\right)\right]^2},$$

$$\mathcal{U}_N(T) \xrightarrow[N\to\infty]{\mathscr{D}} \max_{t=2,\ldots,T-2} \frac{\max_{s=1,\ldots,t} \left(X_s - \frac{s}{t}X_t\right) - \min_{s=1,\ldots,t} \left(X_s - \frac{s}{t}X_t\right)}{\max_{s=t,\ldots,T-1} \left(Z_s - \frac{T-s}{T-t}Z_t\right) - \min_{s=t,\ldots,T-1} \left(Z_s - \frac{T-s}{T-t}Z_t\right)};$$

where $Z_t := X_T - X_t$ and $[X_1, \ldots, X_T]^\top$ is a multivariate normal random vector with zero mean and covariance matrix $\boldsymbol{\Lambda} = \{\lambda_{t,v}\}_{t,v=1}^{T,T}$ such that

$$\lambda_{t,t} = r(t) \quad \textit{and} \quad \lambda_{t,v} = r(t) + R(t,v),\ t < v.$$

The limiting distributions do not depend on the variance nuisance parameter σ, but they depend on the unknown correlation structure of the panel change point model, which has to be estimated for testing purposes. The way of its estimation is shown in Sect. 4. Note that in case of independent observations within the panel, the correlation structure and, hence, the covariance matrix $\boldsymbol{\Lambda}$ is simplified such that $r(t) = t$ and $R(t, v) = 0$.

Next, we show how the test statistic behaves under the alternative.

Assumption A2 $\lim_{N\to\infty} \frac{1}{\sqrt{N}} \left|\sum_{i=1}^{N} \delta_i\right| = \infty.$

Theorem 2 (Under Alternative) *If $\tau \leq T - 3$, then under Assumptions A1, A2 and alternative H_1*

$$\mathcal{S}_N(T) \xrightarrow[N\to\infty]{\mathsf{P}} \infty, \quad \mathcal{T}_N(T) \xrightarrow[N\to\infty]{\mathsf{P}} \infty, \quad \mathcal{U}_N(T) \xrightarrow[N\to\infty]{\mathsf{P}} \infty.$$

Assumption A2 is satisfied, for instance, if $0 < \delta \le \delta_i \, \forall i$ (a common lower change point threshold) and $\delta\sqrt{N} \to \infty$, $N \to \infty$. Another suitable example of δ_is for the condition in Assumption A2, can be $0 < \delta_i = KN^{-1/2+\eta}$ for some $K > 0$ and $\eta > 0$. Or $\delta_i = Ci^{\alpha-1}\sqrt{N}$ may be used as well, where $\alpha \ge 0$ and $C > 0$. The assumption $\tau \le T - 3$ means that there are at least three observations in the panel after the change point. It is also possible to redefine the ratio type test statistic by interchanging the numerator and the denominator. Afterwards, Theorem 2 for the modified test statistic would require three observations before the change point, i.e., $\tau \ge 3$.

Theorem 2 says that in presence of a structural change in the panel means, the test statistics explode above all bounds. Hence, the procedures are consistent and the asymptotic distributions from Theorem 1 can be used to construct the tests.

4 Estimation of the Correlation Structure

The estimation of the covariance matrix $\boldsymbol{\Lambda}$ from Theorem 1 requires panels as vectors with elements having common mean (i.e., without a jump). Therefore, it is necessary to construct an estimate for a possible change point. A *consistent estimate* of the change point τ in the panel data is proposed in [14] as

$$\widehat{\tau}_N := \arg\min_{t=2,\ldots,T} \frac{1}{w(t)} \sum_{i=1}^{N} \sum_{s=1}^{t} (Y_{i,s} - \bar{Y}_{i,t})^2, \tag{2}$$

where $\{w(t)\}_{t=2}^{T}$ is a sequence of weights specified in [14].

Since the panels are considered to be independent and the number of panels may be sufficiently large, one can estimate the correlation structure of the errors $[\varepsilon_{1,1}, \ldots, \varepsilon_{1,T}]^\top$ empirically. We base the errors' estimates on *residuals*

$$\widehat{e}_{i,t} := \begin{cases} Y_{i,t} - \bar{Y}_{i,\widehat{\tau}_N}, & t \le \widehat{\tau}_N, \\ Y_{i,t} - \widetilde{Y}_{i,\widehat{\tau}_N}, & t > \widehat{\tau}_N. \end{cases} \tag{3}$$

Then, the autocorrelation function can be estimated by its empirical counterpart $\widehat{\rho}_t := \frac{1}{\widehat{\sigma}^2 NT} \sum_{i=1}^{N} \sum_{s=1}^{T-t} \widehat{e}_{i,s}\widehat{e}_{i,s+t}$. Consequently, the kernel estimation of the cumulative autocorrelation function and shifted cumulative correlation function is adopted in lines with [1]:

$$\widehat{r}(t) = \sum_{|s|<t} (t - |s|)\kappa\left(\frac{s}{h}\right)\widehat{\rho}_s, \quad \widehat{R}(t,v) = \sum_{s=1}^{t} \sum_{u=t+1}^{v} \kappa\left(\frac{u-s}{h}\right)\widehat{\rho}_{u-s}, \quad t < v;$$

where $h > 0$ stands for the window size and κ belongs to a class of kernels

$$\Big\{\kappa(\cdot) : \mathbb{R} \to [-1, 1] \,|\, \kappa(0) = 1,\ \kappa(x) = \kappa(-x),\ \forall x,\ \int_{-\infty}^{+\infty} \kappa^2(x)\mathrm{d}x < \infty,$$
$$\kappa(\cdot) \text{ is continuos at 0 and at all but a finite number of other points}\Big\}.$$

Since the variance parameter σ simply cancels out from the limiting distributions of Theorem 1, it neither has to be estimated nor known. Nevertheless, one can use $\widehat{\sigma}^2 := \frac{1}{NT}\sum_{i=1}^{N}\sum_{s=1}^{T}\widehat{e}_{i,s}^2$.

5 Simulation Study

A simulation experiment was performed to study the *finite sample* properties of the test statistics for a common change in panel means. In particular, the interest lies in the empirical *sizes* of the proposed tests (i.e., based on $\mathcal{R}_N(T)$, $\mathcal{S}_N(T)$, $\mathcal{T}_N(T)$, and $\mathcal{U}_N(T)$) under the null hypothesis and in the empirical *rejection* rate (power) under the alternatives. Random samples of panel data (5000 each time) are generated from the panel change point model (1). The panel size is set to $T = 10$ and $T = 25$ in order to demonstrate the performance of the testing approaches in case of small and intermediate panel length. The number of panels considered is $N = 50$ and $N = 200$.

The correlation structure within each panel is modeled via random vectors generated from iid, AR(1), and GARCH(1,1) sequences. The considered AR(1) process has coefficient $\phi = 0.3$. In case of GARCH(1,1) process, we use coefficients $\alpha_0 = 1$, $\alpha_1 = 0.1$, and $\beta_1 = 0.2$, which according to [11, Example 1] gives a strictly stationary process. In all three sequences, the innovations are obtained as iid random variables from a standard normal $\mathsf{N}(0, 1)$ or Student t_5 distribution. Simulation scenarios are produced as all possible combinations of the above mentioned settings.

When using the asymptotic distributions from Theorem 1, the covariance matrix is estimated as proposed in Sect. 4 using the Parzen kernel

$$\kappa_P(x) = \begin{cases} 1 - 6x^2 + 6|x|^3, & 0 \le |x| \le 1/2; \\ 2(1 - |x|)^3, & 1/2 \le |x| \le 1; \\ 0, & \text{otherwise}. \end{cases}$$

Several values of the smoothing window width h are tried from the interval $[2, 5]$ and all of them work fine providing comparable results. To simulate the asymptotic distribution of the test statistics, 2000 multivariate random vectors are generated using the pre-estimated covariance matrix. To access the theoretical results under H_0 numerically, Table 1 provides the empirical size (one minus specificity) of the asymptotic tests based on $\mathcal{R}_N(T)$, $\mathcal{S}_N(T)$, $\mathcal{T}_N(T)$, and $\mathcal{U}_N(T)$, where the significance level is $\alpha = 5\%$.

Table 1 Empirical size (1-specificity) of the test under H_0 for test statistics $\mathcal{R}_N(T)$, $\mathcal{S}_N(T)$, $\mathcal{T}_N(T)$, $\mathcal{U}_N(T)$, and $\mathcal{C}_N(T)$, considering a significance level of 5%, $w(t) = t^2$, $h = 2$

T	N	innovations	IID					AR(1)					GARCH(1,1)				
10	50	N(0,1)	.058	.056	.057	.058	.066	.067	.067	.068	.066	.176	.054	.051	.050	.048	.065
		t_5	.049	.050	.049	.062	.075	.068	.069	.070	.066	.178	.054	.051	.052	.053	.071
	200	N(0,1)	.050	.050	.051	.051	.067	.061	.060	.611	.059	.175	.050	.050	.049	.049	.062
		t_5	.052	.051	.049	.051	.073	.065	.064	.067	.064	.179	.052	.053	.058	.053	.063
25	50	N(0,1)	.054	.053	.055	.056	.060	.068	.069	.072	.068	.220	.053	.052	.051	.053	.055
		t_5	.052	.048	.051	.051	.055	.068	.067	.069	.068	.210	.054	.053	.055	.054	.057
	200	N(0,1)	.051	.050	.050	.045	.061	.070	.069	.750	.067	.199	.049	.050	.050	.049	.061
		t_5	.047	.048	.046	.048	.059	.069	.065	.068	.065	.187	.048	.049	.047	.048	.054

For a comparison, the procedure based on *non-ratio* (CUSUM) statistic

$$\mathcal{C}_N(T) = \frac{1}{\widehat{\sigma}\sqrt{N}} \max_{t=1,\ldots,T-1} \left| \sum_{i=1}^{N} \sum_{s=1}^{t} \left(Y_{i,s} - \overline{Y}_{i,T} \right) \right|$$

does not firmly keep the theoretical significance level (Table 1). Although, it may give higher power under some alternatives, because for the ratio type test statistics the data are loosely speaking split into two parts, where the first one is used for the numerator and the second one for the denominator.

It may be seen that all approaches based on the ratio type test statistics are close to the theoretical value of size 0.05. As expected, the best results are achieved in case of independence within the panel, because there is no information overlap between two consecutive observations. The precision of not rejecting the null is increasing as the number of panels is getting higher and the panel is getting longer as well.

The performance of the testing procedures under H_1 in terms of the empirical rejection rates is shown in Table 2, where the change point is set to $\tau = \lfloor T/2 \rfloor$ and the change sizes δ_i are independently uniform on $[1, 3]$ in 33%, 66% or in all panels.

One can conclude that the power of all four tests increases as the panel size and the number of panels increase, which is straightforward and expected. Moreover, higher power is obtained when a larger portion of panels is subject to have a change in mean. The test power drops when switching from independent observations within the panel to dependent ones. Innovations with heavier tails (i.e., t_5) yield smaller power than innovations with lighter tails. Generally, the newly defined test statistics $\mathcal{S}_N(T)$ and $\mathcal{T}_N(T)$ *outperform* $\mathcal{R}_N(T)$ in all scenarios with respect to the power. The highest powers are reached in case of $\mathcal{S}_N(T)$, the second highest are in case of $\mathcal{T}_N(T)$. On the other hand, the test statistic $\mathcal{U}_N(T)$ gives the lowest powers among four considered test statistics. Our simulation study also reveals that the proposed approaches can be used even for panel data of a small panel length ($T = 10$) with relatively small number of panels ($N = 50$).

Finally, an early change is discussed very briefly. We stay with standard normal innovations, iid observations within the panel, the size of changes δ_i being independently uniform on $[1, 3]$ in all panels, and the change point is $\tau = 3$ in case of $T = 10$ and $\tau = 5$ for $T = 25$. The empirical sensitivities of all four tests for small values of τ are shown in Table 3.

When the change point is not in the middle of the panel, the power of the test generally falls down. The source of such decrease is that the left or right part of the panel possesses less observations with constant mean, which leads to a decrease of precision in the correlation estimation. Nevertheless, $\mathcal{S}_N(T)$ and $\mathcal{T}_N(T)$ again outperform $\mathcal{R}_N(T)$ even for early or late changes (the late change points are not numerically demonstrated here). The test statistic $\mathcal{S}_N(T)$ still seems to be the most powerful one from four considered ratio type test statistics according to our simulation study.

Table 2 Empirical sensitivity (power) of the test under H_1 for test statistics $\mathcal{R}_N(T)$, $\mathcal{S}_N(T)$, $\mathcal{T}_N(T)$, and $\mathcal{U}_N(T)$ considering a significance level of 5%, $w(t) = t^2$, $h = 2$

H_1	T	N	innovations	IID				AR(1)				GARCH(1,1)			
33%	10	50	N(0, 1)	.235	.267	.249	.189	.256	.286	.265	.209	.193	.212	.206	.152
			t_5	.174	.345	.190	.180	.202	.232	.209	.164	.201	.239	.213	.166
		200	N(0, 1)	.453	.474	.473	.367	.486	.532	.503	.398	.387	.427	.404	.312
			t_5	.360	.433	.382	.288	.393	.433	.409	.322	.389	.430	.412	.321
	25	50	N(0, 1)	.376	.436	.418	.279	.394	.459	.432	.293	.312	.368	.348	.243
			t_5	.294	.328	.331	.215	.301	.355	.337	.292	.312	.369	.348	.249
		200	N(0, 1)	.685	.776	.726	.543	.699	.758	.739	.542	.584	.654	.635	.431
			t_5	.561	.603	.604	.414	.565	.645	.610	.430	.590	.656	.631	.438
66%	10	50	N(0, 1)	.450	.529	.472	.365	.491	.515	.504	.393	.386	.424	.409	.302
			t_5	.360	.477	.378	.356	.377	.420	.395	.298	.390	.431	.407	.310
		200	N(0, 1)	.774	.794	.796	.663	.807	.849	.823	.697	.677	.729	.701	.572
			t_5	.642	.781	.679	.547	.692	.737	.711	.585	.688	.728	.709	.572
	25	50	N(0, 1)	.688	.769	.731	.526	.694	.756	.735	.547	.581	.661	.631	.514
			t_5	.558	.614	.601	.419	.570	.640	.614	.547	.594	.652	.635	.520
		200	N(0, 1)	.951	.972	.971	.865	.959	.982	.972	.854	.905	.941	.934	.761
			t_5	.874	.933	.903	.723	.888	.934	.917	.739	.906	.944	.932	.762
100%	10	50	N(0, 1)	.641	.741	.668	.531	.667	.701	.686	.570	.563	.599	.583	.454
			t_5	.519	.597	.542	.517	.547	.591	.568	.448	.546	.593	.570	.450
		200	N(0, 1)	.928	.941	.940	.850	.945	.968	.957	.878	.868	.902	.889	.760
			t_5	.844	.887	.863	.734	.869	.904	.882	.780	.872	.911	.887	.772
	25	50	N(0, 1)	.873	.889	.905	.725	.884	.929	.913	.734	.792	.847	.829	.694
			t_5	.760	.832	.795	.592	.771	.834	.813	.733	.789	.845	.827	.696
		200	N(0, 1)	.997	.999	.999	.975	.997	.999	.999	.975	.985	.994	.992	.925
			t_5	.977	.984	.989	.895	.982	.993	.991	.918	.986	.996	.994	.927

Table 3 Empirical sensitivity of the test for small values of τ under H_1 for test statistics $\mathcal{R}_N(T)$, $\mathcal{S}_N(T)$, $\mathcal{T}_N(T)$, and $\mathcal{U}_N(T)$ considering a significance level of 5%, $w(t) = t^2$, $h = 2$

T	τ	N	H_1, iid, N(0, 1)				T	τ	N	H_1, iid, N(0, 1)			
10	3	50	0.551	0.582	0.560	0.436	25	5	50	0.629	0.681	0.670	0.464
		200	0.867	0.871	0.895	0.749			200	0.927	0.948	0.941	0.783

6 Real Data Analysis

As mentioned in the introduction, our primary motivation for testing the panel mean change comes from the *insurance business*. The data set is provided by the National Association of Insurance Commissioners (NAIC) database, see [9]. We concentrate on the 'Commercial auto/truck liability/medical' insurance line of business. The data collect records from $N = 157$ insurance companies (one extreme insurance company was omitted from the analysis). Each insurance company provides $T = 10$ yearly total claim amounts starting from year 1988 up to year 1997. One can consider normalizing the claim amounts by the premium received by company i in year t. That is thinking of panel data $Y_{i,t}/p_{i,t}$, where $p_{i,t}$ is the mentioned premium. This may yield a stabilization of series' variability, which corresponds to the assumption of a common variance. Figure 1 graphically shows series of normalized claim amounts and their logarithmic versions.

The data are considered as panel data in the way that each insurance company corresponds to one panel, which is formed by the company's yearly total claim amounts normalized by the earned premium. The length of the panel is quite short. This is very typical in insurance business, because considering longer panels may invoke incomparability between the early claim amounts and the late ones due to changing market or policies' conditions over time.

We want to test whether or not a change in the normalized claim amounts occurred in a common year, assuming that the normalized claim amounts are approximately constant in the years before and after the possible change for every insurance company. Our ratio type test statistic gives $\mathcal{S}_{157}(10) = 10{,}544$. The asymptotic critical

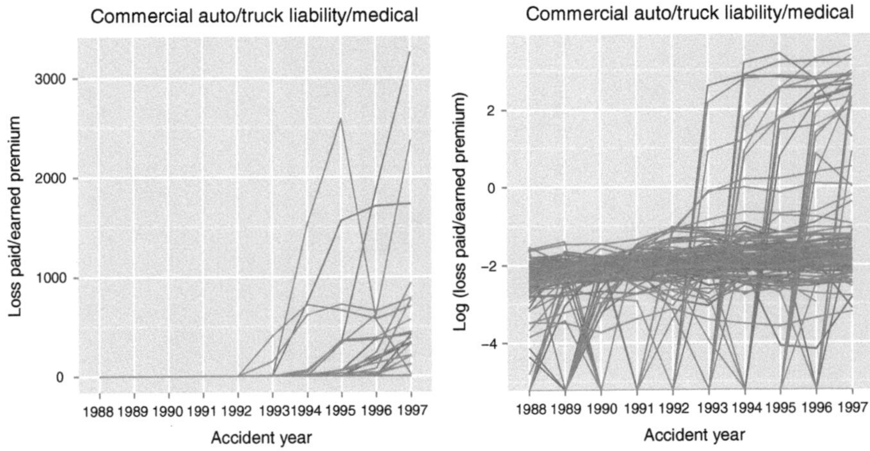

Fig. 1 Development of yearly total claim amounts normalized by earned premium (*left*) together with the log normalized amounts (*right*)

Table 4 Ratio type test statistics with critical values for the 'Commercial auto/truck liability/medical' insurance, considering a significance level of 5%, $w(t) = t^2$, $h = 2$

T	N	$\mathcal{R}_N(T)$	$\mathcal{S}_N(T)$	$\mathcal{T}_N(T)$	$\mathcal{U}_N(T)$	Critical values			
10	157	39.9	10,544	4,414	52.8	52.4	8,698	8,564	75.9

value is 8,698. These values mean that we do reject the hypothesis of no change in panel means. However, the null hypothesis is not rejected using the asymptotic tests based on $\mathcal{R}_{157}(10)$, $\mathcal{T}_{157}(10)$, and $\mathcal{U}_{157}(10)$, which can be explained by lower power of these three tests compared to the one based on $\mathcal{S}_N(T)$, see Table 4. We also try to take the decadic logarithms of claim amounts normalized by the earned premium and to consider log normalized amounts as the panel data observations. Nevertheless, we reject the hypothesis of no change in the panel means (i.e., means of $\log_{10}$ normalized amounts) again.

7 Conclusions

We consider the change point problem in panel data with fixed panel size. Occurrence of common breaks in panel means is tested. We introduce ratio type test statistics and derive their asymptotic properties. Under the null hypothesis of no change, the test statistics weakly converge to functionals of the multivariate normal random vector with zero mean and covariance structure depending on the intra-panel covariances. These covariances can be estimated and, consequently, used for testing whether a change in means occurred or not. This is indeed feasible, because the test statistics under the alternatives converge to infinity in probability. Furthermore, the whole stochastic theory behind requires relatively simple assumptions, which are not too restrictive.

A simulation study illustrates that even for small panel size, all four investigated approaches—the newly derived ones based on $\mathcal{S}_N(T)$, $\mathcal{T}_N(T)$, and $\mathcal{U}_N(T)$ and the older one proposed in [13]—work fine. One may judge that all four methods keep the significance level under the null, while various simulation scenarios are considered. Besides that, the highest power of the test is reached in case of $\mathcal{S}_N(T)$. The proposed ratio statistics outperform the non-ratio one by keeping the significance level under the null, mainly when stronger dependence within the panel is present. Finally, the proposed methods are applied to insurance data, for which the panel change point analysis provides an appealing approach.

7.1 Discussion

Our setup can be modified by considering large panel size, i.e., $T \to \infty$. Consequently, the whole theory leads to convergences to functionals of Gaussian processes with a covariance structure derived in a similar fashion as for fixed T. However, our motivation is to develop tests for fixed and small panel size.

Dependent panels may be taken into account and the presented work might be generalized for some kind of asymptotic independence over the panels or prescribed dependence among the panels. Nevertheless, our incentive is determined by a problem from non-life insurance, where the association of insurance companies consists of a relatively high number of insurance companies. Thus, the portfolio of yearly claims is so diversified, that the panels corresponding to insurance companies' yearly claims may be viewed as independent and neither natural ordering nor clustering has to be assumed.

Acknowledgements With institutional support RVO:67985807. Supported by the Czech Science Foundation project No. P402/12/G097.

Appendix: Proofs

Proof (of Theorem 1*)* Let us define

$$U_N(t) := \frac{1}{\sigma\sqrt{N}} \sum_{i=1}^{N} \sum_{s=1}^{t} (Y_{i,s} - \mu_i).$$

Using the multivariate Lindeberg-Lévy CLT for a sequence of T-dimensional iid random vectors $\{[\sum_{s=1}^{1} \varepsilon_{i,s}, \ldots, \sum_{s=1}^{T} \varepsilon_{i,s}]^\top\}_{i\in}$, we have under H_0

$$[U_N(1), \ldots, U_N(T)]^\top \xrightarrow[N\to\infty]{\mathscr{D}} [X_1, \ldots, X_T]^\top,$$

since $\mathsf{Var}\,[\sum_{s=1}^{1} \varepsilon_{1,s}, \ldots, \sum_{s=1}^{T} \varepsilon_{1,s}]^\top = \boldsymbol{\Lambda}$. Indeed, the t-th diagonal element of the covariance matrix $\boldsymbol{\Lambda}$ is $\mathsf{Var}\,\sum_{s=1}^{t} \varepsilon_{1,s} = r(t)$ and the upper off-diagonal element on position (t, v) is

$$\begin{aligned}\mathsf{Cov}\left(\sum_{s=1}^{t} \varepsilon_{1,s}, \sum_{u=1}^{v} \varepsilon_{1,u}\right) &= \mathsf{Var}\sum_{s=1}^{t} \varepsilon_{1,s} + \mathsf{Cov}\left(\sum_{s=1}^{t} \varepsilon_{1,s}, \sum_{u=t+1}^{v} \varepsilon_{1,u}\right)\\ &= r(t) + R(t, v),\end{aligned}$$

for $t < v$. Moreover, let us define the reverse analogue to $U_N(t)$, i.e.,

$$V_N(t) := \frac{1}{\sigma\sqrt{N}} \sum_{i=1}^{N} \sum_{s=t+1}^{T} (Y_{i,s} - \mu_i) = U_N(T) - U_N(t).$$

Hence,

$$\begin{aligned} U_N(s) - \frac{s}{t} U_N(t) &= \frac{1}{\sigma\sqrt{N}} \sum_{i=1}^{N} \left\{ \sum_{r=1}^{s} \left[\left(Y_{i,r} - \mu_i\right) - \frac{\sum_{v=1}^{t} \left(Y_{i,v} - \mu_i\right)}{t} \right] \right\} \\ &= \frac{1}{\sigma\sqrt{N}} \sum_{i=1}^{N} \sum_{r=1}^{s} \left(Y_{i,r} - \overline{Y}_{i,t}\right) \end{aligned}$$

and, consequently,

$$\begin{aligned} V_N(s) - \frac{T-s}{T-t} V_N(t) &= \frac{1}{\sigma\sqrt{N}} \sum_{i=1}^{N} \left\{ \sum_{r=s+1}^{T} \left[\left(Y_{i,r} - \mu_i\right) - \frac{\sum_{v=t+1}^{T} \left(Y_{i,v} - \mu_i\right)}{T-t} \right] \right\} \\ &= \frac{1}{\sigma\sqrt{N}} \sum_{i=1}^{N} \sum_{r=s+1}^{T} \left(Y_{i,r} - \widetilde{Y}_{i,t}\right). \end{aligned}$$

Using the Cramér-Wold device, we end up with

$$\max_{t=2,\ldots,T-2} \frac{\sum_{s=1}^{t} \left[U_N(s) - \frac{s}{t} U_N(t)\right]^2}{\sum_{s=t}^{T-1} \left[V_N(s) - \frac{T-s}{T-t} V_N(t)\right]^2} \xrightarrow[N\to\infty]{\mathscr{D}} \max_{t=2,\ldots,T-2} \frac{\sum_{s=1}^{t} \left[X_s - \frac{s}{t} X_t\right]^2}{\sum_{s=t}^{T-1} \left[(X_T - X_s) - \frac{T-s}{T-t}(X_T - X_t)\right]^2},$$

$$\max_{t=2,\ldots,T-2} \frac{\sum_{s=1}^{t} \left[U_N(s) - \frac{s}{t} U_N(t)\right]^2 - \frac{1}{t}\left[\sum_{s=1}^{t} \left(U_N(s) - \frac{s}{t} U_N(t)\right)\right]^2}{\sum_{s=t}^{T-1} \left[V_N(s) - \frac{T-s}{T-t} V_N(t)\right]^2 - \frac{1}{T-t}\left[\sum_{s=t}^{T-1} \left(V_N(s) - \frac{T-s}{T-t} V_N(t)\right)\right]^2}$$

$$\xrightarrow[N\to\infty]{\mathscr{D}} \max_{t=2,\ldots,T-2} \frac{\sum_{s=1}^{t} \left[X_s - \frac{s}{t} X_t\right]^2 - \frac{1}{t}\left[\sum_{s=1}^{t} \left(X_s - \frac{s}{t} X_t\right)\right]^2}{\sum_{s=t}^{T-1} \left[Z_s - \frac{T-s}{T-t} Z_t\right]^2 - \frac{1}{T-t}\left[\sum_{s=t}^{T-1} \left(Z_s - \frac{T-s}{T-t} Z_t\right)\right]^2},$$

and

$$\max_{t=2,\ldots,T-2} \frac{\max_{s=1,\ldots,t}\left\{U_N(s)-\frac{s}{t}U_N(t)\right\}-\min_{s=1,\ldots,t}\left\{U_N(s)-\frac{s}{t}U_N(t)\right\}}{\max_{s=t,\ldots,T-1}\left\{V_N(s)-\frac{T-s}{T-t}V_N(t)\right\}-\min_{s=t,\ldots,T-1}\left\{V_N(s)-\frac{T-s}{T-t}V_N(t)\right\}}$$

$$\xrightarrow[N\to\infty]{\mathscr{D}} \max_{t=2,\ldots,T-2} \frac{\max_{s=1,\ldots,t}\left\{X_s-\frac{s}{t}X_t\right\}-\min_{s=1,\ldots,t}\left\{X_s-\frac{s}{t}X_t\right\}}{\max_{s=t,\ldots,T-1}\left\{Z_s-\frac{T-s}{T-t}Z_t\right\}-\min_{s=t,\ldots,T-1}\left\{Z_s-\frac{T-s}{T-t}Z_t\right\}}.$$

□

Proof (of Theorem 2) Let $t=\tau+1$. Then under alternative H_1, for the numerator of $\mathcal{S}_N(T)$ it holds

$$\sum_{s=1}^{\tau+1}\left(\frac{1}{\sigma\sqrt{N}}\overline{\mathcal{E}}_{N,T,s,\tau+1}\right)^2 \geq \left(\frac{1}{\sigma\sqrt{N}}\overline{\mathcal{E}}_{N,T,\tau,\tau+1}\right)^2$$

$$=\left[\frac{1}{\sigma\sqrt{N}}\sum_{i=1}^{N}\sum_{r=1}^{\tau}\left(\mu_i+\sigma\varepsilon_{i,r}-\frac{1}{\tau+1}\sum_{v=1}^{\tau+1}(\mu_i+\sigma\varepsilon_{i,v})-\frac{1}{\tau+1}\delta_i\right)\right]^2$$

$$=\left[\frac{1}{\sqrt{N}}\sum_{i=1}^{N}\sum_{r=1}^{\tau}\left(\varepsilon_{i,r}-\bar{\varepsilon}_{i,\tau+1}\right)-\frac{\tau}{\sigma(\tau+1)\sqrt{N}}\sum_{i=1}^{N}\delta_i\right]^2 \xrightarrow{\mathsf{P}} \infty, \quad N\to\infty,$$

where $\bar{\varepsilon}_{i,\tau+1}=\frac{1}{\tau+1}\sum_{v=1}^{\tau+1}\varepsilon_{i,v}$. The latter convergence holds due to Assumption A2 and

$$\frac{1}{\sqrt{N}}\sum_{i=1}^{N}\sum_{r=1}^{\tau}\left(\varepsilon_{i,r}-\bar{\varepsilon}_{i,\tau+1}\right)=\mathcal{O}_{\mathsf{P}}(1), \quad N\to\infty.$$

In case of $\mathcal{T}_N(T)$ under H_1, we get

$$\sum_{s=1}^{\tau+1}\left(\frac{1}{\sigma\sqrt{N}}\overline{\mathcal{E}}_{N,T,s,\tau+1}\right)^2-\frac{1}{\tau+1}\left(\sum_{s=1}^{\tau+1}\frac{1}{\sigma\sqrt{N}}\overline{\mathcal{E}}_{N,T,s,\tau+1}\right)^2$$

$$=\sum_{s=1}^{\tau+1}\left(\frac{1}{\sigma\sqrt{N}}\overline{\mathcal{E}}_{N,T,s,\tau+1}-\frac{1}{\tau+1}\sum_{r=1}^{\tau+1}\frac{1}{\sigma\sqrt{N}}\overline{\mathcal{E}}_{N,T,r,\tau+1}\right)^2$$

$$=\sum_{s=1}^{\tau+1}\left[\frac{1}{\sqrt{N}}\sum_{i=1}^{N}\sum_{r=1}^{s}\left(\varepsilon_{i,r}-\bar{\varepsilon}_{i,\tau+1}\right)-\frac{s}{\sigma(\tau+1)\sqrt{N}}\sum_{i=1}^{N}\delta_i\right.$$

$$\left.-\frac{1}{\tau+1}\sum_{u=1}^{\tau+1}\frac{1}{\sqrt{N}}\sum_{i=1}^{N}\sum_{r=1}^{u}\left(\varepsilon_{i,r}-\bar{\varepsilon}_{i,\tau+1}\right)+\frac{1}{\tau+1}\sum_{u=1}^{\tau+1}\frac{u}{\sigma(\tau+1)\sqrt{N}}\sum_{i=1}^{N}\delta_i\right]^2$$

$$= \sum_{s=1}^{\tau+1} \left[\frac{1}{\sqrt{N}} \sum_{i=1}^{N} \left\{ \sum_{r=1}^{s} \left(\varepsilon_{i,r} - \bar{\varepsilon}_{i,\tau+1} \right) - \frac{1}{\tau+1} \sum_{u=1}^{\tau+1} \sum_{r=1}^{u} \left(\varepsilon_{i,r} - \bar{\varepsilon}_{i,\tau+1} \right) \right\} \right.$$

$$\left. - \frac{2s - \tau - 2}{2\sigma(\tau+1)\sqrt{N}} \sum_{i=1}^{N} \delta_i \right]^2 \xrightarrow{\mathsf{P}} \infty, \quad N \to \infty.$$

The numerator of $\mathcal{U}_N(T)$ under H_1 can be treated as

$$\max_{s=1,\ldots,\tau+1} \frac{1}{\sigma\sqrt{N}} \bar{\mathcal{E}}_{N,T,s,\tau+1} - \min_{s=1,\ldots,\tau+1} \frac{1}{\sigma\sqrt{N}} \bar{\mathcal{E}}_{N,T,s,\tau+1}$$

$$\geq \left| \frac{1}{\sigma\sqrt{N}} \bar{\mathcal{E}}_{N,T,1,\tau+1} - \frac{1}{\sigma\sqrt{N}} \bar{\mathcal{E}}_{N,T,\tau+1,\tau+1} \right|$$

$$= \left| \frac{1}{\sigma\sqrt{N}} \sum_{i=1}^{N} \left(\mu_i + \sigma\varepsilon_{i,1} - \frac{1}{\tau+1} \sum_{v=1}^{\tau+1} (\mu_i + \sigma\varepsilon_{i,v}) - \frac{1}{\tau+1} \delta_i \right) \right|$$

$$= \left| \frac{1}{\sqrt{N}} \sum_{i=1}^{N} \left(\varepsilon_{i,1} - \bar{\varepsilon}_{i,\tau+1} \right) - \frac{1}{\sigma(\tau+1)\sqrt{N}} \sum_{i=1}^{N} \delta_i \right| \xrightarrow{\mathsf{P}} \infty, \quad N \to \infty,$$

because $\bar{\mathcal{E}}_{N,T,\tau+1,\tau+1} = 0$.

Since there is no change after $\tau + 1$ and $\tau \leq T - 3$, then by Theorem 1 we have for the denominators of $\mathcal{S}_N(T)$, $\mathcal{T}_N(T)$, and $\mathcal{U}_N(T)$ the following

$$\sum_{s=\tau+1}^{T-1} \left[\frac{1}{\sigma\sqrt{N}} \tilde{\mathcal{E}}_{N,T,s,\tau+1} \right]^2 \xrightarrow[N\to\infty]{\mathscr{D}} \sum_{s=\tau+1}^{T-1} \left(Z_s - \frac{T-s}{T-\tau} Z_{\tau+1} \right)^2,$$

$$\sum_{s=\tau+1}^{T-1} \left[\frac{1}{\sigma\sqrt{N}} \tilde{\mathcal{E}}_{N,T,s,\tau+1} \right]^2 - \frac{\left[\sum_{s=\tau+1}^{T-1} \frac{1}{\sigma\sqrt{N}} \tilde{\mathcal{E}}_{N,T,s,\tau+1} \right]^2}{T - \tau - 1}$$

$$\xrightarrow[N\to\infty]{\mathscr{D}} \sum_{s=\tau+1}^{T-1} \left(Z_s - \frac{T-s}{T-\tau} Z_{\tau+1} \right)^2 - \frac{\left[\sum_{s=\tau+1}^{T-1} \left(Z_s - \frac{T-s}{T-\tau} Z_{\tau+1} \right) \right]^2}{T - \tau - 1},$$

and

$$\max_{s=\tau+1,\ldots,T-1} \frac{1}{\sigma\sqrt{N}} \tilde{\mathcal{E}}_{N,T,s,\tau+1} - \min_{s=\tau+1,\ldots,T-1} \frac{1}{\sigma\sqrt{N}} \tilde{\mathcal{E}}_{N,T,s,\tau+1}$$

$$\xrightarrow[N\to\infty]{\mathscr{D}} \max_{s=\tau+1,\ldots,T-1} \left(Z_s - \frac{T-s}{T-\tau} Z_{\tau+1} \right) - \min_{s=\tau+1,\ldots,T-1} \left(Z_s - \frac{T-s}{T-\tau} Z_{\tau+1} \right).$$

□

References

1. Andrews, D.W.K.: Heteroskedasticity and autocorrelation consistent covariance matrix estimation. Econometrica **59**(3), 817–858 (1991)
2. Bai, J.: Common breaks in means and variances for panel data. J. Econom. **157**(1), 78–92 (2010)
3. Chan, J., Horváth, L., Hušková, M.: Darling-Erdős limit results for change-point detection in panel data. J. Stat. Plan. Infer. **143**(5), 955–970 (2013)
4. Giraitis, L., Kokoszka, P., Leipus, R., Teyssière, G.: Rescaled variance and related tests for long memory in volatility and levels. J. Econom. **112**(2), 265–294 (2003)
5. Horváth, L., Hušková, M.: Change-point detection in panel data. J. Time Ser. Anal. **33**(4), 631–648 (2012)
6. Csörgő, M., Horváth, L.: Limit Theorems in Change-Point Analysis. Wiley, Chichester (1997)
7. Lo, A.: Long-term memory in stock market prices. Econometrica **59**(5), 1279–1313 (1991)
8. Madurkayová, B.: Ratio type statistics for detection of changes in mean. Acta Universitatis Carolinae: Mathematica et Physica **52**(1), 47–58 (2011)
9. Meyers, G.G., Shi, P.: Loss Reserving Data Pulled from NAIC Schedule P. http://www.casact.org/research/index.cfm?fa=loss_reserves_data (2011). Updated 01 Sept 2011. Accessed 10 June 2014
10. Horváth, L., Horváth, Z., Hušková, M.: Ratio tests for change point detection. In: Balakrishnan, N., Peña, E.A., Silvapulle, M.J. (eds.) Beyond Parametrics in Interdisciplinary Research: Festschrift in Honor of Professor Pranab K. Sen, vol. 1, pp. 293–304. IMS Collections, Beachwood, Ohio (2009)
11. Lindner, A.M.: Stationarity, mixing, distributional properties and moments of GARCH(p, q)-processes. In: Andersen, T.G., Davis, R.A., Kreiss, J.P., Mikosch, T. (eds.) Handbook of Financial Time Series, pp. 481–496. Springer, Berlin (2009)
12. Pešta, M., Hudecová, Š.: Asymptotic consistency and inconsistency of the chain ladder. Insur. Math. Econ. **51**(2), 472–479 (2012)
13. Peštová, B., Pešta, M.: Testing structural changes in panel data with small fixed panel size and bootstrap. Metrika **78**(6), 665–689 (2015)
14. Peštová, B., Pešta, M.: Erratum to: testing structural changes in panel data with small fixed panel size and bootstrap. Metrika **79**(2), 237–238 (2016)

Part IV
Advanced Time Series Forecasting Methods

Operational Turbidity Forecast Using Both Recurrent and Feed-Forward Based Multilayer Perceptrons

Michaël Savary, Anne Johannet, Nicolas Massei, Jean-Paul Dupont and Emmanuel Hauchard

Abstract Approximately 25% of the world population drinking water depends on karst aquifers. Nevertheless, due to their poor filtration properties, karst aquifers are very sensitive to pollutant transport and specifically to turbidity. As physical processes involved in solid transport (advection, diffusion, deposit…) are complicated and badly known in underground conditions, a black-box modelling approach using neural networks is promising. Despite the well-known ability of universal approximation of multilayer perceptron, it appears difficult to efficiently take into account hydrological conditions of the basin. Indeed these conditions depend both on the initial state of the basin (schematically wet or dry), and on the intensity of rainfalls. To this end, an original architecture has been proposed in previous works to take into account phenomenon at large temporal scale (moisture state), coupled with small temporal scale variations (rainfall). This architecture, called hereafter as "two-branches" multilayer perceptron is compared with the classical two layers perceptron for both kinds of modelling: recurrent and non-recurrent. Applied in this way to the *Yport* pumping well (*Normandie*, France) with 12 h lag time, it appears that both models proved crucial information: amplitude and synchronization are

M. Savary · N. Massei · J.-P. Dupont
M2C Laboratory, Rouen University, Place E. Blondel, 76821
Mont-Saint-Aignan, France
e-mail: michael.savary@univ-rouen.fr

N. Massei
e-mail: nicolas.massei@univ-rouen.fr

J.-P. Dupont
e-mail: jean-paul.dupont@univ-rouen.fr

M. Savary · A. Johannet (✉)
LGEI, Ecole des mines d'Alès, 6 avenue de Clavières,
30 319 Alès Cedex, France
e-mail: anne.johannet@mines-ales.fr

E. Hauchard
Communauté d'Agglomération Havraise, 19 Rue Georges Braque,
76600 Le Havre, France
e-mail: emmanuel.hauchard@codah.fr

I. Rojas et al. (eds.), *Advances in Time Series Analysis and Forecasting*,
Contributions to Statistics, DOI 10.1007/978-3-319-55789-2_17

better with "two-branches" feed forward model when thresholds surpassing prediction is better using classical feed forward perceptron.

Keywords Neural networks · Recurrent · Feed-forward · Turbidity · Karst

1 Introduction

Turbidity is crucial for water quality because it is generally the indicator of the contamination of the underground water by surface water, potentially polluted by phytosanitary products or biological organisms. When turbid water is pumped, complex and expensive treatments are engaged. Predicting turbid events allows thus optimising treatment processes in order to provide drinking water satisfying standards. Nevertheless, both complexity of the hydrologic system and difficulty to quantify physical behaviours, prevent designing operational physical models; statistical framework and specifically machine learning appear thus as a complementary solutions. In this context, the present study is one of the early studies devoted to prediction of rainfall-turbidity relation.

The paper is organized following six parts: after the introduction, turbidity and the state of the art are described. Presentation of neural networks follows in Sect. 3, and the *Yport* (*Normandie*—France) watershed and database are presented. Section 5 presents results and discussion, and the conclusion shows in Sect. 6 that an original coupling between both recurrent and non-recurrent models allows significant anticipation on the occurrence of turbid event.

2 Estimating Turbidity by Machine Learning

2.1 Definition of the Turbidity

Turbidity is the cloudiness of a fluid caused by suspended particles. The unit of turbidity is the Nephelometer Turbidity Unit (or NTU) and water is considered as potable when turbidity is inferior to 1 NTU. Various measurement methods develop analysis of the interactions between light beams and suspend mater. All of them need to be carefully calibrated in the proper range of: pH, temperature, conductivity… and suspended particles (size, shape, color, number…). Due to complex suspended particles composition, a direct relation between NTU and suspended sediment mass is not possible. This complicates modeling of turbidity and makes it especially difficult to perform.

2.2 *State of the Art*

At present, due to the lack of knowledge about physical properties of underground circulations, physical modeling of turbidity can't be successfully performed. For this reasons other strategies were developed using statistical approaches in the framework of systemic modeling. Amongst them, one can note first the exploration of the causal relation between velocity of water and turbidity. The relation between discharge and turbidity, called sediment-rating curve, is thus established using various tools and strategies: SVM [1], multi linear regression [2], correlation analysis [3], neural networks… Because of their flexibility, neural networks were applied to various kinds of relations: sediment-rating curve, chemistry-turbidity relation (conductivity, temperature, pH, ammonium concentration…). Neural networks were proved better than others methods by [4, 5]. When discharge measurements are not available, rainfall-turbidity relation could be investigated using a rainfall-runoff model [6]. Synthetically, it appears that modeling the direct relation between rainfall and turbidity is little published, for the best of our knowledge, due to the complexity of the relation.

2.3 *Turbidity, Uncertainty and Water Production*

At *Yport* Plant, turbidity is measured with a nephelometer (which analyses the light scattered at 90° by the suspended particles). The nephelometer is considered as well calibrated, thus the estimation of uncertainty is the one given by the manufacturer: 2% for turbidity between 0 NTU and 40 NTU and 5% for turbidity superior to 40 NTU. Regarding the production process, when turbidity exceeds the threshold of 100 NTU it is necessary to make water longer decanting. This diminishes the output flow of 20% to 30%. Being able to anticipate the 100 NTU threshold thus would allow to anticipate by: (*i*) stock more water, (*ii*) assessing the quality of treatment chain.

3 Design of the Model

3.1 *Multilayer Perceptron*

The multilayer perceptron was chosen due to its property of universal approximation [7] and of parsimony [8]. The model is shown in Fig. 1. It is fed by exogenous variables, in this study: rainfalls ($\mathbf{u}_r$), evapotranspiration ($\mathbf{u}_e$) and observed turbidity (y_o), and delivers, as output, the estimated variable of interest (y), k is the discrete time step. As this model is very famous it is not detailed herein; for more information on multilayer perceptron, the reader can refer to [9].

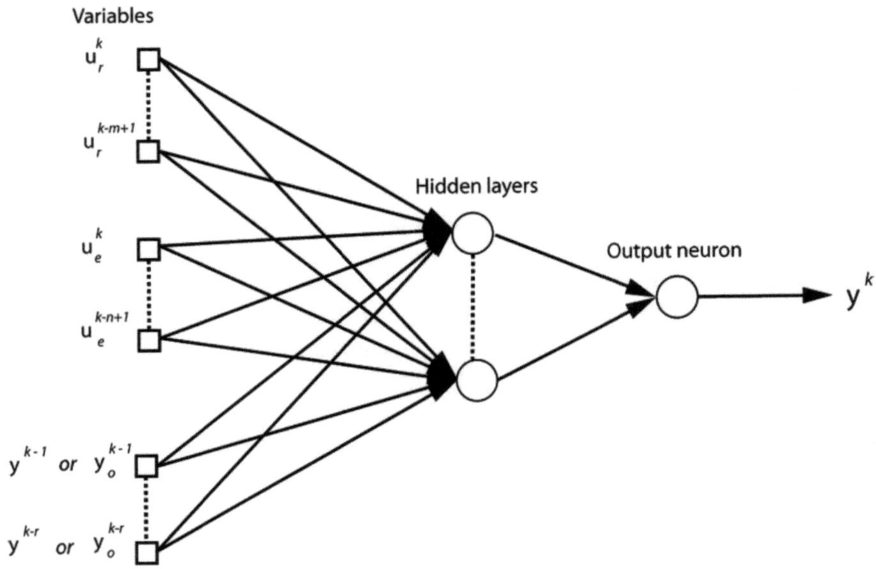

Fig. 1 Standard multilayer perceptron

3.2 *Specific Architectures*

As the behavior of the rainfall-turbidity relation is dynamic, it is important to take into account information about the state of the basin, this can be done usually using two kinds of models: feed-forward and recurrent models [10].

Feed-Forward/Recurrent Models

The feed-forward model is a multilayer perceptron fed by only exogenous inputs. Specifically, added to exogenous variables (rainfall, temperature, evapotranspiration…), this model receives variables of the measured output, here the turbidity, at previous time steps ($k - 1, \ldots k - r$). In automatic control, this information can be considered as providing the state of the system (position, speed, acceleration). The feed-forward model can be mathematically explained as:

$$y^k(\mathbf{u}^k, \mathbf{w}) = g_{NN}(y_o^{k-1}, \ldots, y_o^{k-r}, \mathbf{u}^k, \ldots, \mathbf{u}^{k-m+1}, \mathbf{w}), \quad (1)$$

where y^k is the estimated turbidity, g_{NN} is the non-linear function implemented by the neural network, k is the discrete time step, y_o^k is the measured (or observed) turbidity, $\mathbf{u}^k$ is the vector of exogenous variables (rainfalls, evapotranspiration, etc.), r is the order of the model, m is the width of the sliding time window of exogenous variables, $\mathbf{w}$ is the matrix of parameters.

When turbidity measurements are corrupted by noise, these data can be replaced by turbidity estimations calculated by the model, at previous times steps. The advantage of this model is that it takes better into account the dynamics of the system. Nevertheless, it is generally less effective for predicting the future as illustrated by [10].

With the same notations, the recurrent model can be stated mathematically as:

$$y^k(\mathbf{u}^k, \mathbf{w}) = g_{NN}(y^{k-1}, \ldots, y^{k-r}, \mathbf{u}^{k-1}, \ldots, \mathbf{u}^{k-m+1}, \mathbf{w}) \tag{2}$$

One-branch/Two-branches

A specific ad hoc model was built in order to represent a conceptual hypothesis about the role of evapotranspiration and rainfalls on the hydrogeological basin [11].

In this view, the process is split in, (*i*) the rainfall-turbidity relation and (*ii*) the evapotranspiration influence on the previous relation. The rainfall-turbidity relation is fast and controlled by recent rainfalls while the potential evapotranspiration (ETP) has slower dynamics. Because of the different dynamics it could be advantageous to calculate a nonlinear transformation for each of the processes (ETP or rainfalls) before taking them into account in a coupled model. The model presented in Fig. 2 aims at implement this strategy; it is composed of two branches: one for the rainfall-turbidity relation (upper branch), the other for the evapotranspiration (lower branch). Finally both branches are connected in a supplementary non-linear hidden layer. Hidden layers are composed of non-linear neurons based on arctg function.

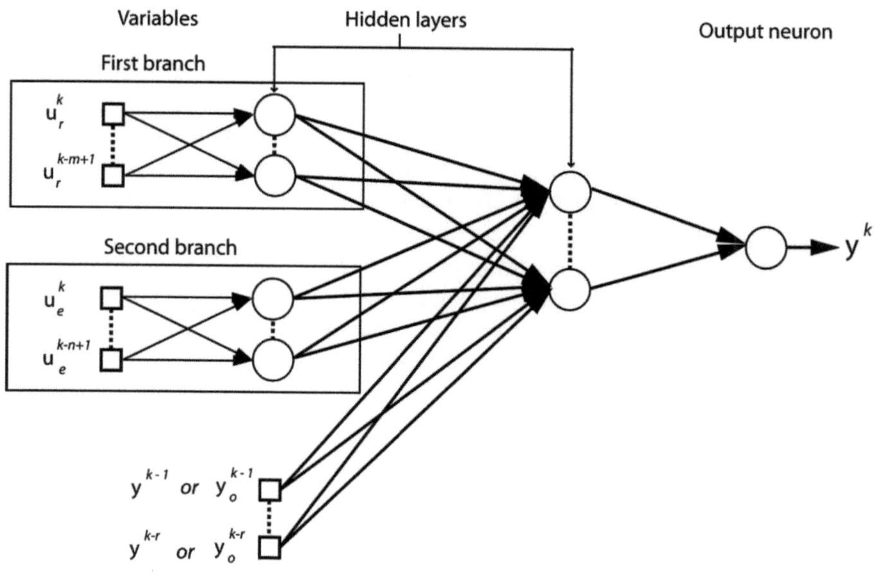

Fig. 2 Two-branches multilayer perceptron

3.3 Bias Variance and Regularization Methods

Being statistical models, neural networks are designed in relation to a database. This database is usually divided into three sets: a training set, a stop set, and a test set. The training set is used to calculate parameters through a training procedure that minimizes the mean quadratic error calculated on output neurons. In this study the Levenberg-Marquardt training rule was chosen [9]. The training is stopped thanks to the stop set (usually called validation set), and the model quality is measured on the remaining part of the database: the test set, which is separate from the previous sets. The choice of the stop set is crucial as it influences the model in a very important way. For this reason we proposed in [12] to choose the stop set, for each model, in taking the set having the best score in validation. This choice guarantees that there is a strong coherence between the training set and the stop set.

The model's ability to be efficient on the test set is called *generalisation*. One has to underline that the training error is not an efficient estimator of the generalisation error because the efficiency of the training algorithm makes the model specific to the training set. This specialisation of the model over the training set is called overtraining. Overtraining is exacerbated by large errors and uncertainties in field measurements; the model then learns the specific realization of noise in the training set. This major issue in neural network modelling is called bias-variance trade-off [13]. This trap can be avoided using regularization methods, particularly cross-validation [14, 15].

3.4 Model Selection

References [15, 16] showed that overfitting can be avoided thanks to a rigorous model selection. This consists to choose not only the number of neurons in the hidden layers but also the order of the model and the dimension of input variables vectors using cross-validation. By this way, numerous combinations of variables are tried, and the one minimizing the variance is chosen. Another hyper-parameter to choose is the initialization of the parameters. This can be done thanks to cross-validation in general case. Nevertheless it was shown by [17] that a more robust model could be designed using an ensemble strategy. Ten models are thus trained and the median of the output, at each time step, is taken. The model design is thus made as the following: first the hidden neurons number, then the number of input variables (selecting m), and lastly the order (r).

3.5 Quality Criteria

In order to assess the performance of models, several quality criteria are used: R^2, persistency and the percentage of pic discharge (P_{PD}).

The Nash-Sutcliff efficiency, or R^2 [18], is the most commonly used criterion in hydrology.

$$R^2 = 1 - \frac{\sum_{k=1}^{n} (y_o^k - y^k)^2}{\sum_{k=1}^{n} (y_o^k - \overline{y_o})^2} \tag{3}$$

The nearest than 1 the R^2 is, the best the results are. Nevertheless this criterion can reach good values even if the model proposes bad forecasts [19]. To avoid this problem, the persistency is used.

The Persistency: C_p [20], provides information on the prediction capability of the model compared to the naive forecast. The naive forecast postulates that the output of the process at time step $k + l$ (where l is the lead-time) is the same than the value at time k. The nearest than 1 the persistence efficiency is, the best the results are. A positive result means that model prediction is better than the naive prediction.

$$C_p = 1 - \frac{\sum_{k=1}^{n} (y_o^{k+1} - y^{k+1})^2}{\sum_{k=1}^{n} (y_o^{k+1} - y_o^k)^2} \tag{4}$$

The percentage of the Turbidity Peak: P_{TP}, inspired from [10], assesses performance of a model at the time of the peak. It calculates the ratio between the observed and forecast peak values. Calculation is visualized in Fig. 3; $k_{\max}$ is the instant of the peak. In Fig. 3, as there is two curves, there is consequently two different instants for the turbidity peak (one for observed, one for simulated peak).

$$P_{TP} = 100 \frac{y^{k_{max}}}{y_o^{k_{max}}} \tag{5}$$

4 Site of Study: *Yport* Pumping Well

4.1 Overview of the Basin

Yport pumping well is situated in *Normandie* (North-West of France). Managed by the CODAH (*Communauté d'agglomération Havraise*), it delivers roughly half of

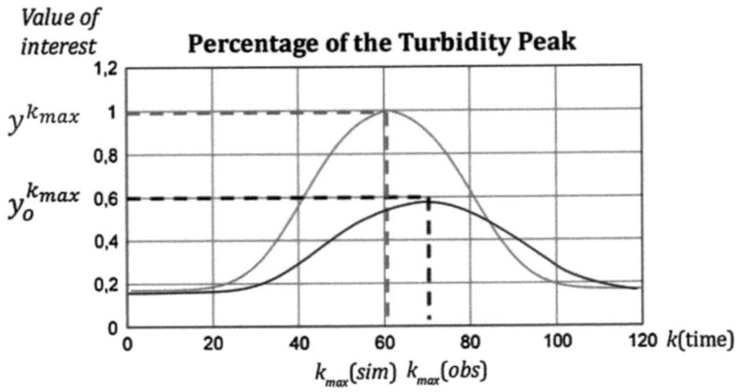

Fig. 3 Definition of the P_{TP}

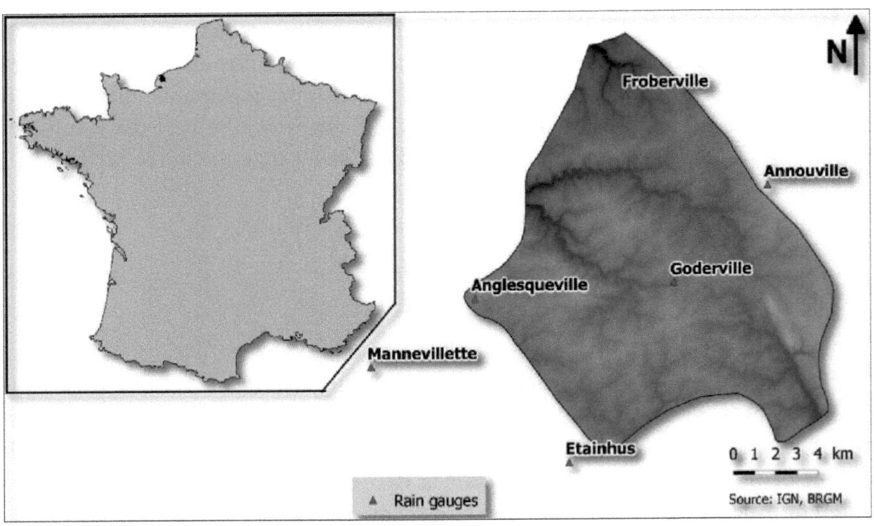

Fig. 4 Yport Basin: Location of rain gauges

Le Havre conurbation drinking water (236 000 inhabitants). The area of the alimentation basin is estimated to 320 km^2 and it is essentially devoted to agriculture. Rain falling on the basin is measured by six rain gauges (*Froberville*, *Annouville*, *Goderville*, *Anglesqueville*, *Manevillette* and *Etainhus*) as shown in Fig. 4. Drinking water is pumped in a well, dug in a natural underground conduit. The turbidity is recorded at the entry of the *Yport* treatment plant.

4.2 Database

Rainfalls are measured by the six previously cited stations between 01/07/2009 and 28/04/2015. Turbidity was measured at *Yport* Plant between 23/10/1993 and 06/02/2015.

Database was hourly re-sampled from the original five minutes period for both turbidity and rainfall. Hourly rainfalls were obtained by addition, and turbidity by picking the maximum hourly value. Because of gaps in turbidity measurements, an event-based modelling approach was chosen. Events whose cumulative rainfalls exceeded 3.5 mm in 24 h were extracted. This way of selection intended to avoid false positive (induced by a heavy rain without turbidity peaks). Finally, 22 events were extracted. The Table 1 presents these events. Amongst them, 10 events (events 2, 3, 6, 7, 10, 11, 16, 17, 18 and 22) present peaks of turbidity.

As explained in Sect. 3, three sets were distinguished: test set (event 11), stop set (7, 10, 13, 16 and 17) and training set (the rest of the database). Event 11 is chosen for test set as it contains high and double peak of turbidity.

5 Results

5.1 Selected Architecture

Based on MLP, the four selected architectures are presented in Table 2. One can note that the two-branches feed-forward model is more parsimonious than the recurrent two-branches model, specifically regarding the number of hidden neurons.

Table 1 Database composed of 22 events

"Event" without turbidity peak					Event with turbidity peak				
Event	Duration (h)	Turbidity (NTU)		Rain (mm)	Event	Duration (h)	Turbidity (NTU)		Rain (mm)
		Max	Min	Cumul			Max	Min	Cumul
1	288	7.07	1	15.9	2	624	302.48	1.54	41.3
4	384	9.82	0.91	14.1	3	1008	135.03	0	26.7
5	336	7.71	1.52	17.5	6	720	245.38	1.53	42.0
8	360	26.87	0.97	22.5	7	744	84.67	0.05	19.2
9	384	9	1.00	20.2	10	576	256.15	0.92	24.9
12	456	12	0.84	28.7	11	744	307.89	0.87	54.8
13	576	13	0.86	30.8	16	648	405.25	0.81	53.8
14	384	14	0.86	23.3	17	744	157.45	0.49	50.7
15	600	15	0.85	31.7	18	744	86.67	2.18	42.8
19	504	19	1.50	30.3	22	623	53.91	0.80	44.2
20	576	20	0.89	40.8					
21	600	48.44	0.93	48.5					

Table 2 Models architectures

Parameters		Recurrent		Feed-forward	
		MLP	Two branches	MLP	Two branches
Hidden layer	Rainfall layer	X	15	X	10
	Evapotranspiration layer	X	1	X	1
	Global layer	5	15	5	10
Input windows widths	Rainfall	30	50	50	50
	Evapotranspiration	3	3	3	3
	Order	1	1	1	10

After selection of the best model and training, the test set was run (event 11), and forecasts for each model are shown in Fig. 5. They correspond to the prediction of an ensemble of ten models differing by random initialization of parameters. The grey line corresponds to the median of the output of the ensemble. The grey area around this line shows the "uncertainty" provided by the model (max and min of the prediction at each time step).

It appears on Fig. 5 that two-branches models seem working better that the standard multilayer perceptron and that the feed-forward models provide a good amplitude prediction of the maximum amplitude when the recurrent model delivers a good synchronization of the peaks. Quality criteria are provided in Table 3.

After this first step of validation, a kind of *cross-test* was performed in order to assess the quality of prediction on the whole database. The test is thus performed on each event of turbidity of the database each one its turn; "events" without turbidity, but high rainfall were not tried. We can note on Table 4 that, satisfactorily, the forecasting behavior is quite stable on the whole database. Moreover it appears clearly that the best model appears to be the feed-forward two branches model.

As suggested in Sect. 2, another way to assess the quality of the modeling approach for operational end users, is to focus on operational stakes. Regarding the *Yport* Plant it is important to be able to detect the occurrence of turbid events exceeding 100 NTU. Counting the number of false predictions can assess this. To this end, Table 5 presents the number of errors in warning: false positive and false negative. One can note that these false warnings are marginal for the threshold of 100 NTU. Regarding Table 5, it appears that the multilayer feed-forward and two-branches recurrent models are globally the best models to predict the thresholds surpassing.

Synthetically, an operational tool based on a multi approach: multilayer feed-forward, two-branches feed-forward and recurrent models will be of great interest.

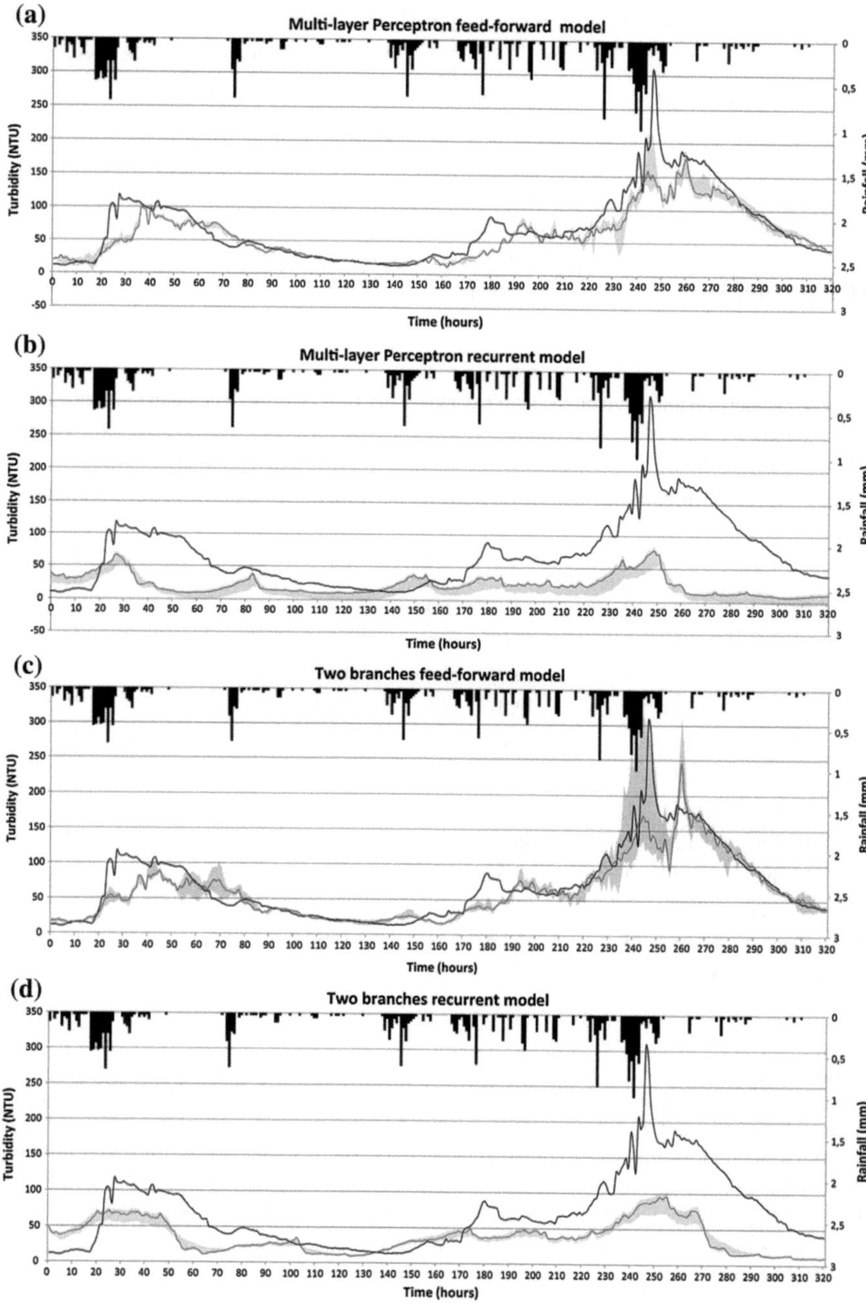

Fig. 5 Measured (black) and forecast (grey) turbidity with a lag time of 12 h. Test on the event 11. Uncertainty is shown in grey area

Table 3 Quality criteria for the 10-ensemble model and the four architectures. Test on ev. 11

		P_{TP}	Nash	Persistency
Two branches recurrent	Maximum	33.73	0.28	0.12
	Minimum	23.57	−0.17	−2.11
Two branches feed-forward	Maximum	103.40	0.81	0.31
	Minimum	70.37	−1.20	−1.20
MLP recurrent	Maximum	27.71	0.37	0.37
	Minimum	18.18	−0.23	−3.74
MLP feed-forward	Maximum	70.40	0.85	0.34
	Minimum	50.17	−0.67	−0.66

Table 4 Models performance on the whole database for 12 h lag time. The model named T_n is the model designed with the event n in test. Best results are highlighted in bold. The median calculated over all events is shown in last row

		Two branches feed-forward		Two branches recurrent		MLP feed-forward		MLP recurrent	
		P_{TP}	Peak delay (h)	P_{TP}	Peak delay (h)	P_{TP}	Peak delay (h)	P_{TP}	Peak delay (h)
T_2	Median	79.95	**4**	38.01	18	**80.62**	19	39.58	21
T_3	Median	112.95	**4**	60.99	15	**108.67**	**4**	70.63	3
T_6	Median	217.41	**8**	79.97	10	**93.3**	14	69.7	10.5
T_7	Median	Stop set		75.06	**2**	**82.89**	14	75.44	3
T_{10}	Median	**86.04**	7	40.92	**4**	50.38	14.5	Stop set	
T_{11}	Median	**80.36**	15	28.48	10	59.46	8.5	25.69	**3**
T_{16}	Median	**106.39**	**2**	Stop set		65.52	14	44.16	10
T_{17}	Median	**105.55**	**5**	87.55	101	Stop set		50.21	9
T_{18}	Median	**98.17**	97	152.86	3	89.84	102	79.53	**0**
T_{22}	Median	**143.54**	**3**	152.33	14	169.18	4	149.17	6
Median		**105**	**5**	75	10	83	14	70	6

Table 5 Prediction of the 100 NTU threshold surpassing. All events of the database are investigated, successively in test. The model designed with the event n in test is called T_n. F_p means *false positive* and F_n *false negative;* d_r is the delay for the rising part of the curve, and d_d for the decreasing part. Best values are highlighted in bold. The X means that no 100 NTU threshold surpassing in observed in simulated or observed data. In the last row, M means the average for F_p and F_n, and means the median for the delays d_r and d_d. Best results are highlighted in bold

	Two branches feed-forward				Two branches recurrent				Multilayer perceptron feed-forward				Multilayer perceptron recurrent			
T	F_p	F_n	d_r (h)	d_d (h)	F_p	F_n	d_r (h)	d_d (h)	F_p	F_n	d_r (h)	d_d (h)	F_p	F_n	d_r (h)	d_d (h)
2	0	0	6	11	0	0	6	−76	**0**	**0**	**6**	**8**	0	0	6	−53
3	**1**	**0**	**0**	**3**	1	0	−11	−4	1	0	7	4	0	0	0	−22

(continued)

Table 5 (continued)

	Two branches feed-forward				Two branches recurrent				Multilayer perceptron feed-forward				Multilayer perceptron recurrent			
T	F_p	F_n	d_r (h)	d_d (h)	F_p	F_n	d_r (h)	d_d (h)	F_p	F_n	d_r (h)	d_d (h)	F_p	F_n	d_r (h)	d_d (h)
6	0	0	−4	10	0	0	−5	6	0	0	−4	9	**0**	**0**	**−4**	**4**
7	Stop set				*X*	*X*	*X*	*X*	*X*	*X*	*X*	*X*	*X*	*X*	*X*	*X*
10	0	1	−2	10	**0**	**0**	**−1**	**9**	0	0	−9	11	Stop set			
11	0	1	3	6	0	1	−17	−31	**0**	**0**	**−12/8**	**−3/4**	0	2	X	X
16	0	0	14	18	Stop set				**0**	**0**	**1**	**12**	0	0	6	−19
17	1	1	X	X	0	1	X	X	Stop set				0	1	X	X
18	**0**	**0**	**X**	**X**	**0**	**0**	**X**	**X**	**0**	**0**	**X**	**X**	1	0	X	X
22	1	0	X	X	**0**	**0**	**X**	**X**	**0**	**0**	**X**	**X**	1	0	X	X
M	0.2	0.3	1.5	10	**0.1**	0.3	−5	**−4**	0.2	**0**	**−1.5**	8	0.3	0.4	3	−20

6 Conclusion

Due to the complex phenomena involved in turbidity, prediction is a very difficult task seldom investigated in bibliography. Nevertheless, as water policy imposes norms on turbidity, and because turbidity is usually associated with pollutants transport, end users must take this aspect into account. In this context this study aims to predict peaks of turbidity with 12 h lag time.

Recurrent and feed-forward models were run and it was shown that, thanks to the design of a new architecture, taking into account explicitly the role of evapotranspiration, called two-branches network, and to a rigorous selection of the model, it will be possible to anticipate the instant and the amplitude of the peak of turbidity as well as 100 NTU thresholds surpassing. A synthesis of several MLP-based architectures will allow designing an efficient and operational tool for water managers. Future works will investigate longer horizon of prediction and the way to improve performances of the recurrent two branches model, which seems specifically promising.

Acknowledgements The authors would like to thank the CODAH for providing rainfall and turbidity data. The *Normandie* Region and *Seine-Normandie* Water Agency are thanked for the co-funding of the study. We are also very grateful to S. Lemarie and J. Ratiarson of for the very helpful discussions they helped organize. Our thanks are extended to D. Bertin for his extremely fruitful collaboration in the design and implementation of the Neural Network simulation tool: RnfPro.

References

1. Kisi, O., Dailr, A.H., Cimen, M., Shiri, J.: Suspended sediment modeling using genetic programming and soft computing techniques. J. Hydrol. **450**, 48–58 (2012)

2. Rajaee, T., Mirbagheri, S.A., Zounemat-Kermani, M., Nourani, V.: Daily suspended sediment concentration simulation using ANN and neuro-fuzzy models. Sci. Total Environ. **407**(17), 4916–4927 (2009)
3. Massei, N., Dupont, J.P., Mahler, B.J., Laignel, B., Fournier, M., Valdes, D., Ogier, S.: Investigating transport properties and turbidity dynamics of a karst aquifer using correlation, spectral, and wavelet analyses. J. Hydrol. **329**, 1–2, 244–25 (2006)
4. Nieto, P.G., García-Gonzalo, E., Fernández, J.A., Muñiz, C.D.: Hybrid PSO–SVM-based method for long-term forecasting of turbidity in the Nalón river basin: a case study in Northern Spain. Ecol. Eng. **73**, 192–200 (2014)
5. Iglesias, C., Torres, J.M., Nieto, P.G., Fernández, J.A., Muñiz, C.D., Piñeiro, J.I., Taboada, J.: Turbidity prediction in a river basin by using artificial neural networks: a case study in northern Spain. Water Resour. Manag. **28**(2), 319–331 (2014)
6. Beaudeau, P., Leboulanger, T., Lacroix, M., Hanneton, S., Wang, H.Q.: Forecasting of turbid floods in a coastal, chalk karstic drain using an artificial neural network. Ground Water **39**(1), 109–118 (2001)
7. Hornik, K., Stinchcombe, M., White, H.: Multilayer feedforward networks are universal approximators. Neural Netw. **2**(5), 359–366 (1989)
8. Barron, A.R.: Universal approximation bounds for superpositions of a sigmoidal function. IEEE Trans. Inf. Theory **39**(3), 930–945 (1993)
9. Dreyfus, G.: Neural Networks: Methodology and Applications, p. 497. Springer Science & Business Media (2005)
10. Artigue, G., Johannet, A., Borrell, V., Pistre, S.: Flash flood forecasting in poorly gauged basins using neural networks: case study of the Gardon de Mialet basin (southern France). Nat. Hazards Earth Syst. Sci. **12**(11), 3307–3324 (2012)
11. Johannet, A., Vayssade, B., Bertin, D.: Neural networks: from black box towards transparent box. Application to evapotranspiration modeling. Int. J. Comput. Intell. **4**(3), 163–170 (2008)
12. Toukourou, M., Johannet, A., Dreyfus, G., Ayral, P.A.: Rainfall-runoff modeling of flash floods in the absence of rainfall forecasts: the case of "Cévenol Flash Floods". Appl. Intell. **35** (2), 178–189 (2011)
13. Geman, S., Bienenstock, E., Doursat, R.: Neural networks and the bias/variance dilemma. Neural Comput. **4**(1), 1–58 (1992)
14. Stone, M.: Cross-validatory choice and assessment of statistical predictions. J. R. Stat. Soc. Ser. B (Methodological) 111–147 (1974)
15. Kong-A-Siou, L., Johannet, A., Valérie, B.E., Pistre, S.: Optimization of the generalization capability for rainfall–runoff modeling by neural networks: the case of the Lez aquifer (southern France). Environ. Earth Sci. **65**(8), 2365–2375 (2012)
16. Kong-A-Siou, L., Johannet, A., Borrell, V., Pistre, S.: Complexity selection of a neural network model for karst flood forecasting: the case of the Lez basin (southern France). J. Hydrol. **403**, 367–380 (2011)
17. Darras, T., Johannet, A., Vayssade, B., Kong-A-Siou, L., Pistre, S.: In: Garcia, G.R. (eds.) Influence of the Initialization of Multilayer Perceptron for Flash Floods Forecasting: How Designing a Robust Model, (ITISE 2014), pp. 687–698. Ruiz, IR (2014)
18. Nash, J.E., Sutcliffe, J.V.: River flow forecasting through conceptual models part I-A discussion of principles. J. Hydrol. **10**(3), 282–290 (1970)
19. Moussa, R.: When monstrosity can be beautiful while normality can be ugly: assessing the performance of event-based flood models. Hydrol. Sci. J. **55**(6) (2010). Special Issue: the court of miracles of hydrology, pp. 1074–1084
20. Kitanidis, P.K., Bras, R.L.: Real-time forecasting with a conceptual hydrologic model: 2 applications and results. Water Resour. Res. **16**(6), 1034–1044 (1980)

Productivity Convergence Across US States in the Public Sector. An Empirical Study

Miriam Scaglione and Brian W. Sloboda

Abstract This paper will examine the productivity of the public sectors in the US across the states. Because there is heterogeneity across states in terms of public services provided that could impact its productivity. In fact, there could be a convergence among the states. The services provided by the public sectors have come under increased scrutiny with the ongoing process of reform in recent years. The public sector unlike the private sector or in the absence of contestable markets, and the information and incentives provided by these markets, performance information, particularly measures of comparative performance, have been used to gauge the productivity of the public service sector. This paper will examine the productivity of the public sector across states throughout the United States. The research methodology marries exploratory (i.e. Kohonen clustering) and empirical techniques (panel model) via the Cobb-Douglas production function. Given that there is a homogeneity across states in terms of the use of a standard currency, it will be easy to identify the nature of the convergence process in the public sectors by states throughout the United States.

Keywords Productivity • Public capital • Clustering • Cobb-Douglas

1 Introduction

There is a great interest by policy-makers in the United States concerning the measurement of productivity in the public sector across states as many states are confronted with budget deficits. Consequently, policy-makers want to know if the

M. Scaglione
Institute of Tourism, University of Applied Sciences and Arts, Western Switzerland Valais, Sierre, Switzerland
e-mail: miriam.scaglione@hevs.ch

B.W. Sloboda (✉)
University of Maryland University College, Upper Marlboro, MD, USA
e-mail: brian.sloboda@faculty.umuc.edu

I. Rojas et al. (eds.), *Advances in Time Series Analysis and Forecasting*, Contributions to Statistics, DOI 10.1007/978-3-319-55789-2_18

state governments are using its limited resources efficiently and cost effective to the taxpayers. This focus on the public sector productivity within the United States is becoming important is very interesting because of the homogenization of the factor price caused by the adoption of the standard currency. There are greater pressures to provide optimal social outcomes and being accountable often leads an organization with a 'productivity paradox' and a service dilemma because state agencies often have years of spending on structure and infrastructure. However, such spending do not seem to have led to long-term gains in either productivity or effectiveness as desired by policy-makers and the public.

Because of this standardization, policy-makers can easily compare the same service across states; this latter fact makes clearer the factor price equalization and its influence on the convergence process [1]. It will further be interesting to observe if the rate of convergence is greater within the United States rather than outside of the United States. More important, when considering the impacts of the role of public capital on productivity and efficiency, policy-makers and public administrators want to be able to answer the following questions: "how much you have," "how you pay for it," and "how you use it."

The primary objective of this paper is to contribute towards an assessment of the evolution of public sector services across states throughout the United States from the year 2000–2014 using annual data by shedding some light on difference performance and possible convergence patterns. The secondary objective of this research is a methodological one. Its aim is showing the relevant of data mining technics to reduce heterogeneity across states by clustering on their dynamics. These techniques not only increase the accuracy of estimates of productivity estimates but also help in the determination of leaders and catcher-ups clubs between the productivity between the states.

2 Existing Research

Despite these well-known and historical difficulties to measure productivity in the public sector given limited resources, there recent pressures on public expenditures have made it essential that state administrators continue to search for ways to increase productivity of their operations while simultaneously enhancing responding to the publics' needs especially the use of information technology to provide services [2]. Despite intense criticism in the public administration literature, there is a strong focus upon the public as customer, with state agencies attempting to develop service or quality-based models that wisely employ current information technologies and simultaneously guarantee "effective, efficient, and responsive government" [3].

In the literature, there is great discussion on the effects of public capital on the national economy. Aschauer [4] started the close examination of the effects of public capital on the macroeconomy. That is, government spending on infrastructures such as roads, bridges, airports could improve economic productivity.

As infrastructure spending increased during the 1950s and 1960s, productivity also increased. However, as public investment declined from the 1970s to the early 1980s, productivity also declined.

Ever since the advent of Aschauer [4] the literature developed concerning the importance of the role of the public capital for economic growth. In general, the empirical results seem to indicate a positive role for public capital in the determination of economic growth of a nation. Some of these empirical models are simple extensions of the neoclassical growth model of Solow [5]. To examine more closely the role of public capital and productivity, researchers used different data sets to investigate the linkages between the role of public capital and the macroeconomy. Namely, many authors have made use of state level data to look at the importance of infrastructure to productivity [6–8] used state level data to assess spillover effects; and costs of production in manufacturing sectors were used on state level as espoused by Holtz-Eakin and Schwartz [9]. Subsequently, future empirical work as given in the preceding sentence using state level data removed the trends, taking into account any missing explanatory variables, such as oil price shocks and estimated an elasticity of close to zero. Munnell [10] and Aschauer [4] originally used the aggregate data at the national level that ignored trends of the time series. In fact, some of the research has revealed nonlinearity between the role of public capital and economic growth on the state level. Aschauer [7] provided one such explanation that the benefits of public capital rise at a diminishing rate but the costs of providing public capital (e.g., through distorting taxation) rise at a constant rate.

Part of the analysis of this paper will focus on convergence. Convergence in macroeconomics by countries even by states, or regions that are poorer (e.g., per capita income) grow faster than countries (states, region) that are richer. This is known as β convergence. The reasons for convergence include capital accumulation to steady state; labor migration; technology transfer; and other factors. Baumol [1] delves into the idea of conditional convergence. That is, how can nations join the "club"? There are several ways for the latter to occur: openness to trade, financial markets, educational attainment of the population. Convergence can also occur regionally or by state and the United States would be a good example. That is, convergence among U.S. states: "reversal of fortune" of the South because of the differences in the achieving of economic development.

3 Data and Methodology

3.1 Data

Data on public spending or public capital were obtained from the National Association of State Budget Officers (NASBO). The NASBO does not collect data for the District of Columbia (DC), so DC was omitted from the analysis. Compared with accounting data based on capital stock and depreciation schedules, these fiscal data have certain advantages, particularly reliability, because it represents actual spending by the state governments. Also these data are a more objective measure, which avoids the controversy that ensues from estimating state data following [10]. The largest spending function of most state governments is elementary and secondary education. The spending series are deflated using the price index for private fixed investment in structures from the Bureau of Economic Analysis because a deflator for public spending which includes infrastructure is currently not available because the Bureau of Economic Analysis does not have a complete series of public capital to develop such a deflator.

For the other inputs on labor, a common variable used in public sector productivity studies which is obtained the Bureau of the Census Employment and Payroll Survey. The Bureau of the Census conducts a Census of Governments of all state and local government organization units every 5 years, for years ending in 2 and 7 which incidentally coincide with the Economic Census, as required by law. Because of the infrequency of the Census of Governments, we used the Employment and Payroll Survey to provide the annual data. In this analysis we used full time equivalent (FTE) for labor and a measure of payroll for all workers including part-time. In general, in productivity studies it is best to use the number of hours of full-time labor. However, the Bureau of the Census does not collect such data, so we used the payroll data over the FTE of workers by state government.

3.2 Methodology and Models

Productivity analysis across US States

In order to the assess the variation of productivity level across US states, Cobb-Douglas production function [11, 12] was estimated using panel regression models [13, 14]. The model for the state *i* at time *t* is

$$\log(y_{it}) = \alpha + \beta_L \log(L_{it}) + \beta_K \log(K_{it}) + \mu_{it} \; \mathrm{i} = 1, \ldots, \mathrm{N}, \mathrm{t} = 1, \ldots, T \quad (1)$$

with *i* denotes the cross-section dimension of N = 50 US State and *t* time denotes the time series dimension where T = 15 (2000–2014).

For a given State *i* and the time *t*; y_{it} is the public service output per inhabitant, and the exogenous input variables are L_{it} is the labor input measure as the payroll over the number of full time employees (FTE) and K_{it} is public capital over inhabitants. Panels models used in this research have either one-way error component of disturbance, with

$$\mu_{it} = \mu_i + \nu_{it} \quad (2)$$

or two-way error component as shows in Eq. (3)

$$\mu_{it} = \mu_i + \gamma_t + \nu_{it} \tag{3}$$

where μ_{it} denotes the unobservable individual effect and the reminder disturbance $\nu_{it} \sim IID(0, \sigma_\upsilon^2)$ and it is assumed the independence of the exogenous variables with the disturbance component and μ_ι and γ_t are no random parameters.

In order to test the differences effects across US States or time (t), the test based on F-statistics which has as null hypothesis that $\mu_\iota = 0$ for $i = 1$ *to* $N-1$ was used, namely that the efficient estimator is the pooled least square [13].

Additionally, all the models used in the present research are fixed effect models. This latter restriction lays on Baltagi [14] recommendation about the goodness of fit of this kind of models when cross-section are US States. Even though the Hausman test which has a null hypothesis of no correlation between the individual effect and exogenous variables could turn out to be significant, for this first study, the authors decided to keep the fixed effect models. Further research will be carried taking into account the controversial discussion between distinguished scholars about fixed vs random effect models [cf. 14].

Convergence analysis

The convergence analysis was carried out using the classical model by Baumol [15] using panel models as shows in Eq. (4)

$$Growth\,(y_{it}) = \alpha + \beta_i \log\,(y_{it}) + \mu_{it},\ \mathrm{i} = 1,\ \ldots,\ \mathrm{N}, \mathrm{t} = 1,\ \ldots,\ T-1. \tag{4}$$

where $Growth(y_{it})$ is the slope unobservable component filtered using structural times series model [16, 17] and with i denotes the cross-section dimension of N = 50 US State and t time denotes the time series dimension where T = 14 (2000–2013). The authors calculated panel growth models in the same way as described in the precedent section.

Exploratory methods

These kind of methods could be useful in shedding some light on the heterogeneity of the evolution of output in public service across US States. As mentioned above, the data under study have two dimensions: cross-sectional and time. In order to inspect similar evolution dynamics across of US State, the authors proceeded in two steps. In order to grasp clusters of similar dynamics in an accurately way Self-Organizing Maps (SOM) by Kohonen [18] clustering methods were applied not only on the raw time series of productivity but also on the two unobservable components (trend and slope) filtered by structural times series (STS) model [16, 17].

SOM are a family of neural networks useful for data visualization which uses unsupervised training, that means that target output is provided and the process runs until stabilization [19, 20].

The authors first apply SOM clustering on the raw series, having as output eight different clusters. Then, a Structural Time Series multivariate model was adjusted using Stamp [17] within each of these clusters. The unobservable components, on the one hand, trends and on the other hand, slopes filtered in the process were pooled into a two respectively sets. Finally, the authors apply SOM to the set of trends and to the set of slope producing eight clusters for each of them.

4 Results

This section is organized as follows, first the explanatory analysis, then panels models of Cobb-Douglas equation and finally the convergence analysis.

4.1 Explanatory Analysis of Outputs

Figure 1 shows SOM clustering on the raw output per inhabitants (panel a) and the filtered rate of growth obtained using STS (panel b). The output of SOM for raw output series (panel a) can roughly be interpreted as showing by columns similar shapes and decreasing range by rows. The output of rate of growth (panel b) bear similar interpretation than the former one, but the names were casted taking into account the pooled mean of only the last 4 years rates of growth and not the overall. If the mean of rate of growth over those years is positive, the cluster is labeled as increasing ("Inc."), contrary decreasing ("Decr."). It is interesting to note the case of Oregon which is classified alone (Panel a, cluster 4 U4) and has a fixed rate of growth (Panel b, cluster 4 U4, Inc. U1). The results of the SOM of the another unobservable component filtered using STS, namely the level, are not shown here for the sake of the space.

Figure 2 (panel **a**) is the heat-map representing the cross table of US States, the SOM clustering for slope were decrease ordered following the pooled mean: rank 1 represented the series whose mean 2.9% to rank 8 with a mean of −2.1%. The dotted square shows series that fulfilled the convergence hypothesis; on the one hand, "Leaders slowing down" and on the other hand, "Catchers-up". Figure 2 (panel **b**) is a map of the US States. The SOM technic sheds some light on the members of Leaders and Catchers-Up and on the States that seems not to fulfill the convergence hypothesis. Finally, the χ^2 test with a null hypothesis indicates that there is no link between the evolution of output and rate of growth SOM clustering and is thus not significant ($\chi^2(1) = 2.38$, p-value = 0.123). Therefore, in this exploratory analysis, we don't find enough evidences that the convergence hypothesis is globally fulfilled, but at least, we have found some evidence of convergence clubs.

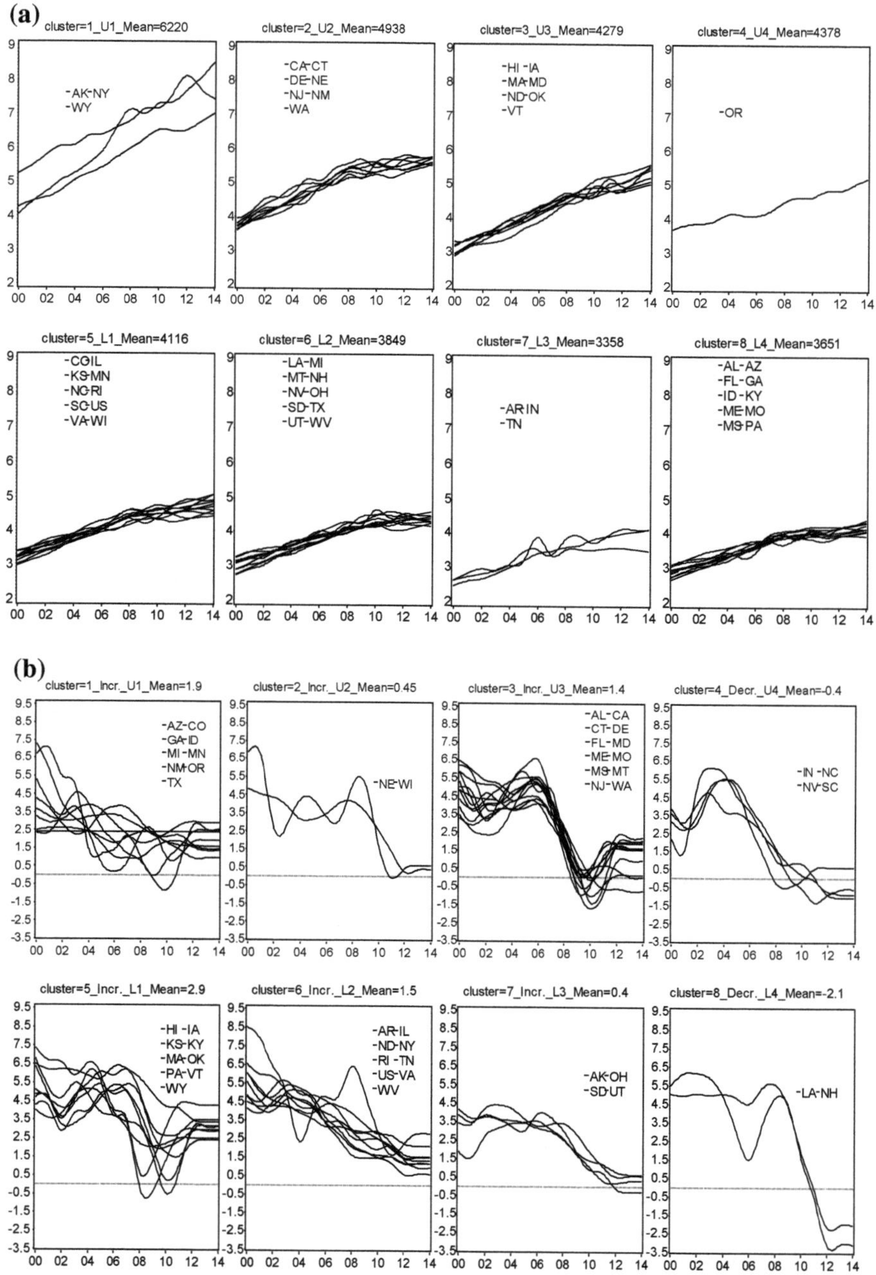

Fig. 1 SOM clusters. Panel **a** raw output data per inhabitants (in thousands of dollars) and Panel **b** rate of growth in %. Panel **a** Mean of the polled values on the series in each cluster (Ui = Upper cluster column i, Li = Lower upper cluster column i). Panel **b** the pooled mean of the last 4 years

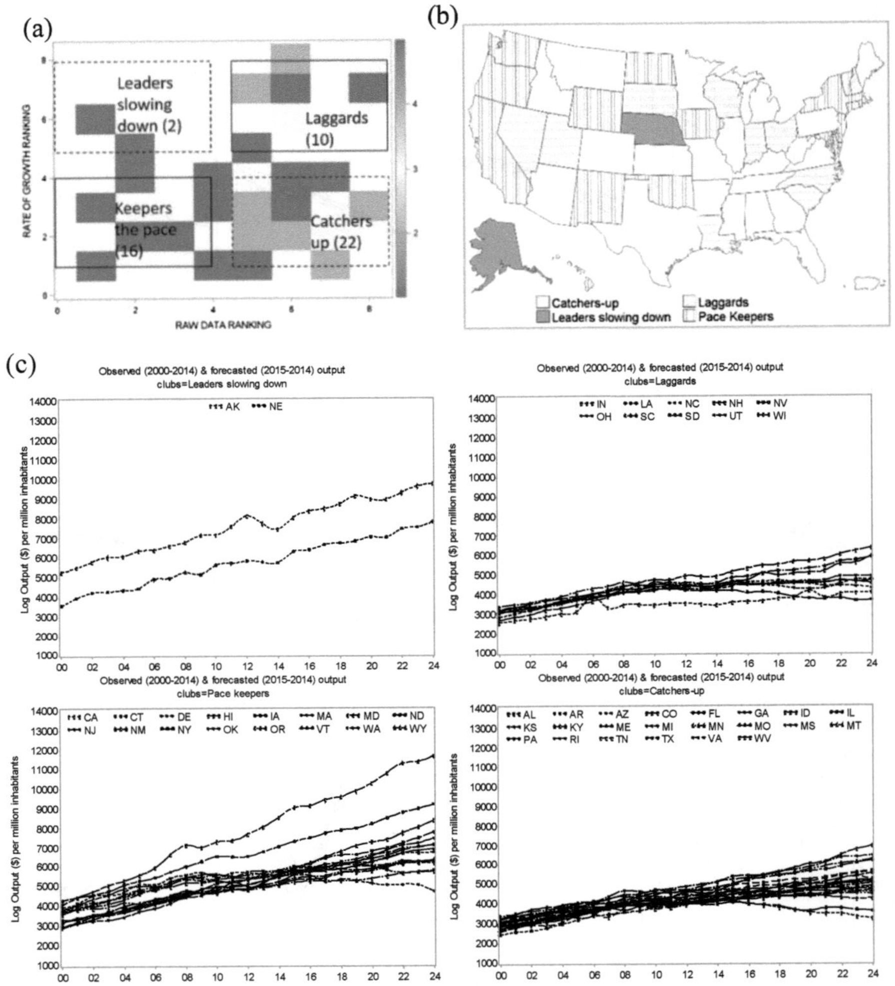

Fig. 2 In panel **a**, a heat-map of the cross-table of rate of growth and raw data SOM clustering, is ranked in decreasing order (highest rank = 1 lowest rank = 8), **b** map of the US States, **c** the 10-year forecasts using STS univariate

4.2 Productivity Analysis Across US States

Table 1 shows the estimates of Eq. (2) and (3) across the 50 US States. The fix model (Eq. 2) shows the estimates (labor and capital) are very significant and all the other states have fixed effects significantly lower that Wyoming. For the two-way model neither of the estimates (labor and capital) are significant, nor the intercept. The reason for the performance is similar to the models calculates within the SOM clusters of raw data shown in Fig. 1a and deserve further analysis beyond present one. (Table 2)

Table 1 Estimates of Eq. 2 of one-way cross sectional fixed US States effects and Eq. 3 two-way fixed effects (cross sectional and time). The last column, shows that WY is the reference state and all others are significantly lower than it. Note: L + C = 1 reports Wald test and column 'Bench' show that WY is the benchmark

Models	Int.	Labor	Capital	L + C = 1	MSE	R^2	Bench
FIXONE	0.3443***	1.045***	0.0143***	1.0598	0.0023	0.9481	**WY,** LW (all others)
(SE)/ p-val	(0.1224)	(0.014)	(0.0039)	<0.00			
FIXTWO	5.8428	0.37948	0.01348	0.3923	0.0014	0.9696	
(SE)/ p-val	(0.4358)	(0.0522)	(0.003)	<0.00			

Table 2 shows the estimates of Eq. 2 and 3 for each SOM clustering of raw data. Cluster 3 is omitted because it contains only one State. In the last column, in bold the reference state, between brackets after LW the list of US States which effects are sig. lower than the reference, HG the list of US States sig. higher. Interesting to note that the first SOM cluster, those situated in the upper row of panel (**a**) in Fig. 1 have significant estimates for labor and capital in the one-way fixed effect (in italic in Table 3). Probably this shows the benefit of the reduction of heterogeneity across States. Except for Cluster 3 and for Eq. 2 estimates, all mean square errors are lower are lower than the overall model in Table 1. Moreover, the inspection of the graph of series of estimates show a better adjustment than the estimates on the overall 50 states but the graphics are not included for the sake of space.

4.3 Convergence Analysis

Figure 2 (panel **b**) shows US states by convergence clubs. The aim of the forecasts graphs (panel **c**) is only a tool for evaluation the relevance of the converge clubs.

The two-way model of Eq. 4 that shows the error structure follows form of Eq. 3. Table 3 shows these estimates. The fit statistics of the $R^2 = 0.621$ and MSE = 1.498, the F-test is significant showing enough evidence that the effect coefficients are not simultaneal null (F(64,685) = 1.7.49). The coefficient of the log of output per inhabitant (−6.1687, std = 1.2004) is very significant, showing some evidence of a converge process across US States. All the same, time effects are significant for some period, and without claiming causal effect there are some events that are contemporaneous to those periods. Leaving out 2000 which could be under some side effects as the first observation of the sample, the 2001 and 2002 are significant and negative in coincidence with 9/11. The before the sub-prime (including 2008) crisis the (2005–2008) the effects are positive and significant, contrary the two following years after this events are negative and significant.

Table 2 Estimates of Eqs. 2 and 3 by SOM clustering of raw data. Note: L + C = 1 reports Wald test. In the last column, shows the reference State and those US states whose fixed effect are significant lower (LW) or higher (HG) than the latter

Clus (#)	Models	Int	Labor	Capital	L + C = 1	MSE	R^2e	Bench
1 (3)	FIXONE	−1.3382	*1.2205***	*−0.0212***	1.1992	0.0020	0.9504	**WY**, LW(Ak-NY)
	(SE)/p-val	(0.4689)	(0.0481)	(0.0273)	<0.00			
	FIXTWO	2.9353	0.6465*	−0.0863**	0.56023	0.0017	0.9728	
	(SE)/p-val	(2.4598)	(0.3218)	(0.0412)	0.2132			
2 (7)	FIXONE	0.4080	*0.984****	*0.0236***	1.0079	0.0018	0.9151	**WA**, LW(CA-CT-NJ), HG(DE-NE-NM)
	(SE)/p-val	(0.2761)	(0.0313)	(0.0091)	0.8064			
	FIXTWO	7.72636***	0.1250	0.029***	0.14684	0.001	0.9556	
	(SE)/p-val	(1.2488)	(0.1458)	(0.0067)	<0.00			
3 (7)	FIXONE	−0.0818	*1.084****	*0.047****	1.1318	0.0031	0.9069	**VT**, LW(HI-MA), HG(OK)
	(SE)/p-val	(0.4839)	(0.0494)	(0.0144)	<0.00			
	FIXTWO	7.7384***	0.1167	0.016*	0.13271	0.001	0.973	
	(SE)/p-val	(0.688)	(0.082)	(0.009)	<0.00			
5 (8)	FIXONE	0.1138***	1.00450	0.00700	1.0115	0.0018	0.9068	**WI**, LW(CO-IL-MN-RI), HG(KS-NC-SV-VA)
	(SE)/p-val	(0.2627)	(0.0296)	(0.0082)	0.6996			
	FIXTWO	6.8965***	0.1929	0.0039	0.19679	0.0009	0.9587	
	(SE)/p-val	(1.1512)	(0.1369)	(0.006)	<0.00			
6 (9)	FIXONE	0.2661	1.000***	0.00621	1.0064	0.0021	0.8925	**WV**, LW(MI-NH-NV-OH-TX-UT), HG(SD)
	(SE)/p-val	(0.2812)	(0.0312)	(0.0106)	0.8327			
	FIXTWO	6.0402***	0.2850**	0.0016	0.28667	0.0011	0.9494	
	(SE)/p-val	(1.0799)	(0.1312)	(0.0080)	p < 0.00			

(continued)

Table 2 (continued)

Clus (#)	Models	Int	Labor	Capital	L + C = 1	MSE	R^2e	Bench
7 (3)	FIXONE	−1.43852**	1.201***	.00908	1.2010	0.0029	0.8833	**TN**, LW(IN), HG(AR)
	(SE)/p-val	(0.568)	(0.0701)	(0.0115)	0.003			
	FIXONE	2.2133	0.7647**	0.0233**	0.78808	0.0015	0.9612	
	(SE)/p-val	(2.7566)	(0.3323)	(0.0096)	0.5187			
8 (9)	FIXONE	−0.1318	1.008***	0.00220	1.0091	0.0021	0.8774	**PA**, LW(nill), HG(all others)
	(SE)/p-val	(0.2726)	(0.0325)	(0.0079)	0.7722			
	FIXTWO	8.057***	0.03196	0.00095	0.03291	0.0011	0.9455	
	(SE)/p-val	(1.0153)	(0.1195)	(0.00577)	<0.00			

Table 3 Estimates of Eq. 4. Individual cross effects of states are omitted due to a lack of space

Variable and cross sectional effect	DF	Estimate	Std error	t Value	Pr > |t|
2000	1	57.0099	10.6766	5.34	<0.0001
2001	1	−2.2876	0.5655	−4.05	<0.0001
2002	1	−1.2382	0.5084	−2.44	0.0151
2003	1	−0.5369	0.4667	−1.15	0.2504
2004	1	−0.2280	0.4288	−0.53	0.5950
2005	1	0.8152	0.3844	2.12	0.0343
2006	1	1.8372	0.3474	5.29	<0.0001
2007	1	2.0346	0.3142	6.48	<0.0001
2008	1	2.6576	0.2885	9.21	<0.0001
2009	1	−0.4976	0.2642	−1.88	0.0600
2010	1	−0.4314	0.2590	−1.67	0.0963
2011	1	−0.3060	0.2518	−1.22	0.2248
2012	1	−0.2577	0.2498	−1.03	0.3027
2013	1	−0.1735	0.2471	−0.70	0.4829
2014	1	−0.0893	0.2454	−0.36	0.7160
Log output per inhabitants	1	−6.1687	1.2004	−5.14	<0.0001

Note: Cross effect of Wyoming is zero by calculation

Further analysis to models test the convergence hypothesis will be carried out inside each of the club of Fig. 2 (panel **b**). Moreover, dynamic panel models deserved to be tested.

5 Discussion and Conclusions

This paper attempts to understand the economic returns of public sector productivity versus other public capital in three ways: the use of fiscal data from the NASBO data, the consideration of convergence between the states, and the application of the panel data models along with clustering analysis to assess convergence. Employing state level data for output and public spending was used to assess productivity. We estimated a specification of the Cobb-Douglas production function. Recall that earlier studies, e.g., Munnell [10] and Aschauer [4] had reported a positive and significant effect of public capital on private sector output, which later was attributed to spurious estimates due to trends in the time series. In this analysis we reported smaller estimates of the effects of public capital on productivity which can be attributed to the use of the panel data models which accounted for the heteroscedasticity and autocorrelation corrections which commonly plague panel regression analysis.

6 Limitations and Future Research

This study has identified some empirical evidence about convergence in productivity by state government using Baumol's convergence theory. More specifically, Baumol's convergence is applied to each state government not the entire country of the United States. These results by state yielded more fruitful results. However, there could be a case of dynamic change to these models that may lead to the convergence hypothesis that could occur out inside each of the club as exhibited in Fig. 2 panel b. Consequently, dynamic panel models may need to be examined. That is, the productivity may entail a dynamic process rather than the static panel data model. Ignoring the dynamics in a model is an omitted variables problem and careful attention needs to be made to the number of lags to include. Thus, with the lagged productivity as an explanatory variable in a panel regression, the fixed effects (FE) and the random effects (RE) estimators will be biased. As a remedy for the biased FE and RE, the first difference can be applied to the panel data as well as the use of the instrumental variables. If the lagged dependent variable is used in a static panel data, there is a correlation between the error term, ε_{it} and the lag of productivity. That is, we want to be able to use additional lags that do not pose a problem in the use of a first difference model (FD) or have a FD estimator. In fact, these lags can also be used as instruments as a proxy for the lag of productivity. Arellano–Bond estimator [21] starts by transforming all regressors, usually by differencing to remove the fixed effects, and uses the generalized method of moments (GMM). Before commencing, a careful assessment will need to be undertaken to see if there is a dynamic relationship actually exists.

Acknowledgements An early version of this research was presented at the 36th International Symposium on Forecasting, Santander Spain, and the ITISE (International Work Conference on Time Series), Granada, Spain. The authors gratefully thank the audience at the ISF, ITISE and the anonymous referees from the ITISE for their feedback.

References

1. Baumol, W.J.: Multivariate growth patterns: contagion and common forces as possible sources of convergence. Converg. Prod. 62–85 (1994)
2. Lee, G., Perry, J.L.: Are computers boosting productivity? a test of the paradox in state governments. J. Public Adm. Res. Theor. **12**(1), 77–102 (2002)
3. Danziger, J.N., Andersen, K.V.: The impacts of information technology on public administration: an analysis of empirical research from the "golden age" of transformation [1]. Int. J. Public Adm. **25**(5), 591–627 (2002)
4. Aschauer, D.A.: Is public expenditure productive? J. Monet. Econ. **23**(2), 177–200 (1989)
5. Solow, R.M.: Technical change and the aggregate production function. Rev. Econ. Stat. 312–320 (1957)
6. Munnell, A.H.: Why has productivity growth declined? Productivity and public investment. N. Engl. Econ. Rev. 3–22 (1990)
7. Aschauer, D.A.: Dynamic Output and Employment Effects of Public Capital bv. (1997)

8. Sloboda, B.W., Yao, V.W.: Interstate spillovers of private capital and public spending. Ann. Reg. Sci. **42**(3), 505–518 (2008)
9. Holtz-Eakin, D., Schwartz, A.E.: Spatial productivity spillovers from public infrastructure: evidence from state highways. Int. Tax Public Finance **2**(3), 459–468 (1995)
10. Munnell, A.H., Cook, L.M.: How does public infrastructure affect regional economic performance? N. Engl. Econ. Rev. 11–33 (1990)
11. Nerlove, M.: Estimation and Identification of Cobb-Douglas Production Functions. Rand Menally & Co—Chicago, Chicago (1965)
12. Arrow, K.J., et al.: Capital labor substitution and economic efficiency. Rev. Econ. Stat. **18**(3), 225–250 (1961)
13. Greene, W.H.: Econometric analysis, vol. XXIII, 791 S, 2nd edn. Macmillan, New York (1993)
14. Baltagi, B.H.: Econometric analysis of panel data, vol. X, 293 S, 2nd edn. Wiley, Chichester (2001)
15. Baumol, W.J.: Productivity growth, convergence, and welfare: what the long-run data show. Am. Econ. Rev. 1072–1085 (1986)
16. Harvey, A.C.: Forecasting, structural time series models and the Kalman filter. Cambridge university press (1990)
17. Koopman, S.J. et al.: STAMP 6.0: Structural Time Series Analyser, Modeller and Predictor. Timberlake Consultants, London (2000)
18. Kohonen, T.: Self-organizing maps, volume 30 of Springer Series in Information Sciences. Springer, Berlin (2001)
19. Sarada, C., Alivelu, K., Prayaga, L.: Self-Organising Mapping Networks (SOM) with SAS E-Miner. (Unknown)
20. Mangiameli, P., Chen, S.K., West, D.: A comparison of SOM neural network and hierarchical clustering methods. Eur. J. Oper. Res. **93**(2), 402–417 (1996)
21. Arellano, M., Bond, S.: Some tests of specification for panel data: Monte Carlo evidence and an application to employment equations. Rev. Econ. Stud. **58**(2), 277–297 (1991)

Proposal of a New Similarity Measure Based on Delay Embedding for Time Series Classification

Basabi Chakraborty and Sho Yoshida

Abstract Time series data is abundant in many areas of practical life such as medical and health related applications, biometric or process industry, financial or economical analysis etc. The categorization of multivariate time series (MTS) data poses problem due to its dynamical nature and conventional machine learning algorithms for static data become unsuitable for time series data processing. For classification or clustering, a similarity measure to assess similarity between two MTS data is needed. Though various similarity measures have been developed so far, dynamic time warping (DTW) and its variants have been found to be the most popular. An approach of time series classification with a similarity measure (Cross Translational Error CTE) based on multidimensional delay vector (MDV) representation of time series has been proposed previously. In this work another new similarity measure (Dynamic Translational Error DTE), an improved version of CTE, and its two variants are proposed and the performance study of DTE(1) and DTE(2) in comparison to several other currently available similarity measures have been done using 43 publicly available bench mark data sets with simulation experiments. It has been found that the new measures produce the best recognition accuracy in larger number of data sets compared to other measures.

Keywords Similarity measure · Multivariate time series · Delay vector embedding · Embedding dimension · Cross translational error · Dynamic translational error

1 Introduction

Time series data is abundant in nature and also in real life events. It is being generated in various application domains by experimental observations of different state variables of a complex system over a period of time. When the number of variables is

B. Chakraborty (✉) · S. Yoshida
Faculty of Software and Information Science, Iwate Prefectural University, 152-52 Sugo, Takizawa 020-0693, Japan
e-mail: basabi@iwate-pu.ac.jp

I. Rojas et al. (eds.), *Advances in Time Series Analysis and Forecasting*, Contributions to Statistics, DOI 10.1007/978-3-319-55789-2_19

more than one, the time series is designated as multivariate time series (MTS). Some of the examples are on line handwritten signature data or human gait data in the area of biometric authentication, stock market or exchange rate fluctuation in the area of financial analysis, EEG or ECG data in medical domain or temperature, humidity time series in weather pattern recognition. The analysis of univariate or multi variate time series is essential for mining, prediction, classification or clustering of data in variety of domains as mentioned above. Though statisticians proposed a variety of techniques for time series analysis [1], linear statistical models are unsuitable for analyzing real life non-stationary time series while complex methodologies are needed for non linear models. Due to huge volume, random noise and dynamical nature of MTS data, their analysis is a challenging task. Traditional machine learning algorithms for classification of static data also seem to be unsuitable for analysis of MTS data.

Due to importance for classification of time series data, various approaches have been developed ranging from Neural and Bayesian networks to Genetic Algorithms, Support Vector Machines and Characteristic Pattern Extraction [2]. Traditional classification techniques like Bayesian classifier or decision tree are modified for MTS data and temporal naive Bayesian model (T-NB) and temporal decision tree (T-DT) are developed [3]. In [4] MTS data is transformed to a lower dimensional compact representation by extracting characteristic features to facilitate the use of classical machine learning algorithms for classification. Effective and efficient time series classification process has two main challanges, representation of the time series and similarity measure. Raw time series has large number of points leading to expensive processing and storage. So it is desirable to reduce the data while preserving the characteristics of the data. On the other hand, unlike static data where the distance measures are straightforward distance between two points, the distance between time series needs to be carefully defined in order to properly capture the dissimilarity between them.

Many techniques of representation for reducing time series data have been proposed, for example, Discrete Fourier Transform (DFT), Singular Value Decomposition (SVD), Discrete Cosine Transform (DCT), Discrete Wavelet Transform (DWT) are a few of them. Now for any classification task, pair wise similarity measure for grouping the time series is also important. There are a number of distance measures available in literature for evaluating similarity of two time series. Euclidean distance is widely used as the simple similarity measure. Dynamic time Warping (DTW) and its various variants are considered to be the most successful similarity measure. Among other measures, Longest Common Subsequence (LCSS) and edit distance are quite popular [5]. Most of the existing researches on the similarity measures of time series is focussed on univariate time series [6]. However, a few researches focus on multivariate time series [7, 8] in which similarity is calculated with the extended Frobenius norm and from information theoretic framework respectively.

An algorithm for multivariate time series classification based on multidimensional delay vector representation of time series has been proposed by author in [9]. A similarity measure for measuring similarity of two time series based on time delay embedding proposed in [10] is extended and used for the proposed classification algorithm. The proposed similarity measure has been shown to be computationally efficient but it was found that the recognition accuracy is poor compared to dynamic time warping. DTW is found to be the most effective regarding classification accuracy while having high computational cost. In this work our previously proposed measure has been integrated with DTW to propose a new measure. The classification efficiency has been verified with benchmark data set and is found to be better than DTW though computational cost is comparatively high. The next section describes the existing approaches for time series classification and popular similarity measures in brief.

2 Time Series Classification and Similarity Measures

In this section current approaches for time series classification and similarity measures used for grouping time series are discussed in brief [11].

2.1 Approaches for Time Series Classification

Existing approaches for time series classification can be broadly classified into following three categories [12]:

2.1.1 Feature Based Classification:

In feature based classification approaches, a multidimensional time series is transformed into a feature vector and then is classified by conventional classification algorithms such as artificial neural network or decision trees. The choice of appropriate features plays an important role in this approach. A number of techniques has been proposed for feature subset selection by using compact representation of high dimensional MTS into one row to facilitate application of traditional feature selection algorithms like recursive feature elimination (RFE), zero norm optimization etc. [3, 13]. Time series shapelets, characteristic subsequences of the original series, are recently proposed as the features for time series classification [14]. Lesh et al. [15] proposed a pattern based feature selection method in which short segments of time series are considered to be features when they appear frequently in a class. Ji et al. [16] introduced a pattern-extraction algorithm called Minimal Distinguishing Subsequence, more appropriate for classifying biological sequences.

Another group of techniques extract features from the original time series by using various transformation techniques like Fourier, Wavelet etc. In [17], a family of techniques have been introduced to perform unsupervised feature selection on MTS data based on common principal component analysis (CPCA), a generalization of PCA for multivariate data items where all the data items have the same number of dimensions. Kernel methods are also used in feature extraction, particularly in text time series with huge number of features. Any distance metric is used for classification of the feature based representation of the time series data.

2.1.2 Model Based Classification:

In model based classification approaches, a model is constructed from the data and the new data is classified according to the model that best fits it. Models used in time series classification problems are mainly statistical, such as Gaussian, Poisson, Markov and Hidden Markov Model (HMM) or based on neural network models. Naive Bayes is the simplest model and is used in text classification [18]. Hidden Markov models (HMM) are successfully used for biological sequence classification as they are able to handle variable length time series while neural network models require fixed length inputs. Some neural network models such as recurrent neural network (RNN) are suitable for temporal data classification. RNN models also does not require any knowledge of data in contrast to HMM models. Probabilistic distance measures are generally suitable for model based classification of time series.

2.1.3 Distance Based Classification:

In distance based classification, a distance function which measures the similarity between two time series is used for classification. Similarity or dissimilarity measures are the most important component of this approach. Euclidean distance is the most widely used measure with 1NN classifier for time series classification. Though computationally simple, it requires two series to be of equal length and is sensitive to time distortion. Elastic similarity measures such as Dynamic Time Warping (DTW) [19] and its variants overcome the above problems and seem to be the most successful similarity measure for time series classification in spite of high computational cost. There are some works [20] on speeding up of DTW techniques. In the next subsection some popular similarity measures are presented.

2.2 Time Series Similarity Measures

The list of time series similarity measures proposed so far is quite long and a comprehensive enumeration of all of them will take a lot of space. Here we present a several representative example which we use in our work for comparison. Similarity measures popularly used for multivariate time series analysis from different categories listed below are Euclidean distance (lock step measure), Fourier coefficient (feature based measure), DTW, EDR, TWED (elastic measure) and autoregressive AR as model based measure.

Euclidean Distance Euclidean measure is the simplest and the most popular dissimilarity measure.

The dissimilarity $D(x, y)$ between two time series x and y using any L_n norm is defined as

$$D_{ec}(x, y) = \left(\Sigma_{i=1}^{M}(x_i - y_i)^n\right)^{\frac{1}{n}} \tag{1}$$

where n is a positive integer, M is the length of the time series, x_i and y_i are the i th elements of x and y time series respectively. For $n = 2$, we obtain Euclidean distance. This measure is difficult to use for time series of different lengths and having a time lag.

Fourier Coefficient Measure Instead of comparing the raw time series, comparison can be done between the ith Fourier coefficients of the time series pair after the Discrete Fourier Transform. This measure falls under the category of feature based classification. The equation is given as

$$D_{fc}(x, y) = \left(\Sigma_{i=1}^{\theta}(\hat{x}_i - \hat{y}_i)^2\right)^{\frac{1}{2}} \tag{2}$$

where $\hat{x}_i$ and $\hat{y}_i$ represent ith Fourier coefficients of x and y time series and $\theta = \frac{M}{2}$, M is the length of the time series.

Auto Regression Coefficient Measure This distance measure falls under the category of model based classification and uses the model parameters for calculating similarity values. Auto regression coefficients of two time series are calculated beforehand from AR (Auto Regressive) models and the distance between corresponding coefficients is taken as the dissimilarity measure. The number of AR coefficients is controlled by a parameter in this model and directly affects the speed of the similarity calculation.

Dynamic Time Warping (DTW) Distance Measure Dynamic Time Warping (DTW) is a classic approach for computing dissimilarity between two time series. DTW belongs to the group of elastic measures and works by optimally aligning the time series in temporal domain so that the accumulated cost of the alignment is minimal. The accumulated cost can be calculated by dynamic programming, recursively applying

$$D_{i,j} = f(x_i, y_j) + \min\left(D_{i,j-1}, D_{i-1,j}, D_{i-1,j-1}\right) \tag{3}$$

for $i = 1 \ldots M$ and $j = 1 \ldots N$ where M and N are the length of the time series x and y respectively and $f(x_i, y_j) = \sqrt{(x_i - y_j)^2}$.

Currently DTW is the main benchmark against any promising new similarity measure, though its computational cost is quite high.

Edit Distance on Real Sequences Edit distance on real sequences or EDR is an extension of original edit or Levensthein [21] distance to real valued time series. Computation of EDR formalized by dynamic programming is similar to DTW but $f(x_i, y_j)$ is different as follows:

$$m(x_i, y_j) = \Theta(\epsilon - f(x_i, y_j)) \tag{4}$$

where Θ is the Heaviside step function, such that $\Theta(z) = 1$ if $z \geq 0$ and 0 otherwise.

Time-Warped Edit Distance Time-warped edit distance or TWED is an extension and combination of DTW and EDR [22]. Twed used a mismatch penalty λ and a stiffness parameter ν. For uniformly sampled time series, the formulation of TWED is as follows:

$$D_{i,j} = \min(D_{i,j} + \Gamma_{x,y}, D_{i-1,j} + \Gamma_x, D_{i,j-1} + \Gamma_y) \tag{5}$$

for $i = 1 \ldots M$ and $j = 1 \ldots N$ where
$\Gamma_{x,y} = f(x_i, y_j) + f(x_{i-1}, y_{j-1}) + 2\nu|i - j|$
$\Gamma_x = f(x_i, x_{i-1}) + \nu + \lambda$
$\Gamma_y = f(y_j, y_{j-1}) + \nu + \lambda$

3 Proposal for a New Similarity Measure

A similarity measure *Cross Translation Error (CTE)* based on time series representation by delay coordinate embedding , a standard approach for analysis and modeling of nonlinear time series [23], has been proposed by author in [25]. A deterministic time series signal $\{s_n(t)\}_{t=1}^{T_n} (n = 1, 2, \ldots, N)$ can be embedded as a sequence of time delay co-ordinate vector $v_{s_n}(t)$ known as experimental attractor, with an appropriate choice of embedding dimension m and delay time τ for reconstruction of the original dynamical system as follows:

$$v_{s_n}(t) \equiv \{s_n(t), s_n(t + \tau), \ldots, s_n(t + (m - 1)\tau)\}, \tag{6}$$

Now for correct reconstruction of the attractor, a fine estimation of embedding parameters (m and τ) is needed. There are variety of heuristic techniques for estimating those parameters [24]. The author proposed an approach for fine estimation of optimal embedding parameters in [10].

3.1 *Cross Translation Error: CTE*

Cross Translation Error (CTE) has been proposed in [25] for calculating similarity between two time series. The algorithm is described below, the details can be found in [25].

1. Multi-dimensional delay vector $v_{s_n}(t)$ can be generated from time series $\{s_n(t)\}_{t=1}^{T_n}$ ($n = 1, 2, \ldots, N$) based on Eq. (6). $(m + 1)$ dimensional vector $v_{s_n}(t)'$ including the normalized time index t/T_n is defined as follows;

$$v_{s_n}(t)' \equiv \{s_n(t), s_n(t+\tau), \ldots, s_n(t+(m-1)\tau), t/T_n\}. \tag{7}$$

2. Let $v_{s_i}(t)$ and $v_{s_e}(t)$ denote m-dimensional delay vectors generated from time series $s_i(t)$ and time series $s_e(t)$ respectively. $v_{s_i}(t)'$ and $v_{s_e}(t)'$ denote the corresponding $(m+1)$-dimensional vector including the normalized time index t/T_n.
3. A random vector $v_{s_i}(k)$ is picked up from $v_{s_i}(t)$. Let the nearest vector of $v_{s_i}(k)$ from $v_{s_e}(t)$ be $v_{s_e}(k')$. The index k' for the nearest vector is defined as follows;

$$k' \equiv \arg\min_t ||v_{s_i}(k)' - v_{s_e}(t)'|| \tag{8}$$

4. For the vectors $v_{s_i}(k)$ and $v_{s_e}(k')$, the transition in the each orbit after one step are calculated as follows;

$$V_{s_i}(k) = v_{s_i}(k+1) - v_{s_i}(k), \tag{9}$$

$$V_{s_e}(k') = v_{s_e}(k'+1) - v_{s_e}(k'). \tag{10}$$

5. Cross Translation Error (CTE) e_{cte} is calculated from $V_{s_i}(k)$ and $V_{s_e}(k')$ as

$$e_{cte} = \frac{1}{2}\left(\frac{|V_{s_i}(k) - \bar{V}|}{|\bar{V}|} + \frac{|V_{s_e}(k') - \bar{V}|}{|\bar{V}|}\right), \tag{11}$$

where $\bar{V}$ denotes average vector between $V_{s_i}(k)$ and $V_{s_e}(k')$.
6. e_{cte} is calculated for L times for different selection of random vector $v_{s_i}(k)$ and the median of e^i_{cte} $(i = 1, 2, \ldots, L)$ is calculated as

$$M(e_{cte}) = Median(e^1_{cte}, \ldots, e^L_{cte}). \tag{12}$$

The final cross translation error E_{cte} is calculated by taking the average, repeating the procedure Q times to suppress the statistical error generated by random sampling in the step (3).

$$E_{cte} = \frac{1}{Q}\sum_{i=1}^{Q} M_i(e_{cte}). \tag{13}$$

Cross translation error is a distance metric, so lower value of E_{cte} represents higher similarity. For multivariable time series, each dimension is considered separately as a single time series and represented by a multidimensional delay vector. So MTS data can be represented as a set of multidimensional delay vectors, each element corresponding to a single variable time series.

3.2 Dynamic Translational Error DTE

CTE is computationally light like Euclid distance but unlike Euclid distance CTE can be used to measure similarity between time series of unequal length. However it is found in one of our earlier works [26] that average classification accuracy (by 1NN classifier) using CTE is poor compared to that using DTW. In this work a modification of CTE has been proposed to enhance the classification accuracy. The new proposed measure Dynamic Translational Error is a combination of CTE and DTW. Here distance calculation of DTW is done according to the strategy of calculation of CTE. The measure considers the two time series represented by multidimensional delay vectors and aligns them in the phase space considering the nearest vectors so that the accumulated cost is minimum.

The algorithm in brief is as follows:

1. The time series is to be converted to multidimensional delay vector form

$$v_{s_n}(t) \equiv \{s_n(t), s_n(t+\tau), \ldots, s_n(t+(m-1)\tau)\}, \tag{14}$$

2. Calculate the similarity matrix as in DTW, but here $f(x_i, y_j)$ of Eq. 3 is CTE
$D_{i,j} = f(x_i, y_j) + \min(D_{i,j-1}, D_{i-1,j}, D_{i-1,j-1})$

$$f(x_i, y_j) = \frac{1}{2}\left(\frac{|v_x(i) - \bar{v}|}{|\bar{v}|} + \frac{|v_y(j) - \bar{v}|}{|\bar{v}|}\right) \tag{15}$$

where $\bar{v}$ denotes average vector.
$D_{0,0} = 0$

3. $D_{M,N}$ is the distance between time series x and y
$D_{dte}(x, y) = D_{M,N}$
where M and N are length of time series x and y.

4 Simulation Study

The efficiency of the proposed modification as a new similarity measure DTE is evaluated by comparing it with other similarity measures by using the classification accuracy of 1NN classifier. Nearest neighbour classifier is supposed to achieve the best classification accuracy for time series classification suggested by many researchers [27].

4.1 Data Set Used

The benchmark data sets consisting of 43 different time series data from University of California, Riverside (UCR) time series repository [28] are used for the simulation experiments. The training data is used as labeled data for classifier and classification accuracy is calculated on the test set. The average classification accuracy for 20 trials on different partitions of training and test data are noted for all the data sets.

4.2 Simulation Results

The simulation results of classification accuracies with different similarity measures are shown in Table 1. CTE(1) and CTE(2) represent two implementations of the measure. In CTE(1), CTE is calculated with fixed m and τ where m is taken as 3 and $\tau = 2$. In CTE(2), CTE is calculated with different m and τ for different time series. m and τ for each time series is calculated using popular method of mutual information [24]. Similarly for Dynamic Translation error measure, DTE(1) corresponds to fixed m and τ with $m = 3$ and $\tau = 2$. While DTE(2) corresponds to different values of embedding parameters calculated for different time series. From Table 1 it is found that no similarity measure exists which performs the best for all the data sets. The bold figures in Table 1 represents the best value for each data set. Table 2 represents the average classification accuracy over all the data sets of the individual measures and the associated order of computational cost.

It is evident from the simulation results that our new measures DTE(1) and DTE(2) produce the best classification accuracy in 12 data sets and 9 data sets respectively. In combination, i,e if either of them is used, they cover 19 data sets. TWED and DTW are the closest competitor producing best classification accuracy for 11 and 10 data sets respectively. Regarding classification accuracy, CTE(2) is better than CTE(1), but computational cost of CTE(2) is greater than CTE(1). In case of the modified measure DTE, DTE(1) is better than DTE(2) in the number of best accuracy data set, though average classification accuracy over 43 data sets for DTE(2) is higher than DTE(1) as shown in Table 2. Thus the effect of setting individual embedding parameters has less pronounced in the modified measure DTE than earlier measure CTE in terms of achieving best classification results in individual data sets as well as in the average over all the data sets. Regarding average classification accuracy over 43 sets, TWED and DTW are better in average classification accuracy than our proposed measures as found in Table 2. But in most of the data sets our measure produces better classification accuracy. TWED also depends on parameter selection. Computational cost of DTW, TWED and DTE are of same order while CTE and Euclid are computationally light.

Table 1 Classification accuracy with different similarity measures

Data set name	No. of class	Classification accuracy with									
		Euclid	Fourier	AR	DTW	EDR	CTE(1)	DTE(1)	CTE(2)	TWED	DTE(2)
50 words	50	0.67	0.63	0.21	0.71	0.72	0.48	**0.78**	0.64	0.76	0.68
Adiac	37	0.6	**0.65**	0.29	0.59	0.54	0.51	**0.65**	0.60	0.60	**0.65**
Beef	5	0.5	0.6	0.53	0.5	0.57	0.7	**0.87**	0.60	0.50	0.83
CBF	3	0.89	0.75	0.53	**0.99**	0.98	0.32	0.47	0.72	**0.99**	0.42
Cl. conc.	3	0.62	0.67	0.68	0.62	0.66	0.64	0.72	0.64	0.63	**0.73**
CinCECGtorso	4	0.94	0.88	0.54	0.89	0.94	0.70	0.73	0.79	0.76	**0.97**
Coffee	2	0.75	0.93	0.86	0.82	0.5	0.61	0.90	0.79	0.75	**0.96**
CricketX	12	0.59	0.51	0.30	**0.78**	0.62	0.2	0.74	0.41	0.69	0.58
CricketY	12	0.67	0.49	0.22	0.76	0.64	0.18	0.72	0.39	**0.77**	0.57
CricketZ	12	0.64	0.53	0.28	**0.78**	0.53	0.18	0.71	0.41	0.75	0.58
DaiSizeRed	4	0.93	0.93	0.77	**0.96**	0.94	0.81	0.94	0.92	0.95	0.93
ECG200	2	0.89	**0.9**	0.79	0.79	0.89	0.85	0.86	0.78	0.85	0.84
ECG5days	2	0.78	0.74	0.74	0.79	0.94	0.78	0.86	0.92	0.80	**1.00**
FaceAll	14	0.72	**0.80**	0.36	0.77	**0.80**	0.57	0.74	0.58	0.78	0.75
Face4	4	0.84	0.73	0.42	0.84	**0.90**	0.5	0.16	0.49	0.88	0.16
FacesUCR	14	0.80	0.72	0.34	0.94	0.93	0.58	0.88	0.84	**0.95**	0.90
Fish	7	0.79	0.79	0.35	0.86	0.91	0.44	0.91	0.85	**0.94**	**0.94**
GunPt	2	0.95	0.92	0.79	0.88	0.97	0.86	**1.00**	0.97	0.98	0.96
Haptics	5	0.36	0.38	0.30	0.36	0.36	0.33	0.37	**0.40**	**0.40**	0.37
InlineSkate	7	0.35	0.30	0.35	0.37	0.3	0.23	**0.49**	0.28	0.43	0.32
ItalyPD	2	**0.96**	0.95	0.61	0.95	0.94	0.73	0.92	0.85	0.95	0.95

(continued)

Table 1 (continued)

Data set name	No. of class	Classification accuracy with									
		Euclid	Fourier	AR	DTW	EDR	CTE(1)	DTE(1)	CTE(2)	TWED	DTE(2)
Lighting2	2	0.82	0.66	0.67	0.80	0.80	0.64	0.46	0.52	**0.84**	0.58
Lighting7	7	0.71	0.53	0.37	**0.77**	0.66	0.33	0.14	0.42	**0.77**	0.53
MALLAT	8	**0.92**	0.90	0.49	0.91	0.87	0.60	0.80	0.86	0.90	0.91
MedImage	10	0.70	0.69	0.5	**0.76**	0.66	0.53	0.73	0.61	0.74	0.63
MoteStrain	2	0.86	0.86	0.57	0.89	0.87	0.81	**0.93**	0.85	0.88	0.86
Oliveoil	4	0.83	0.83	0.50	**0.87**	**0.87**	0.6	0.8	0.83	0.83	**0.87**
OSULeaf	6	0.55	0.53	0.44	0.64	0.61	0.42	**0.88**	0.60	0.81	0.65
SonyAiboRS	2	0.69	0.69	**0.93**	0.73	0.71	0.64	0.43	0.82	0.68	0.86
SonyAiboRS2	2	**0.87**	0.84	0.86	0.84	0.85	0.80	0.63	0.84	**0.87**	0.86
StarLightC	3	0.85	0.82	0.83	0.89	0.87	0.78	**0.92**	0.87	0.88	0.89
SwedishLeaf	15	0.79	0.75	0.60	0.80	0.87	0.55	**0.88**	0.78	0.87	0.86
Symbols	6	0.90	0.87	0.74	0.95	0.95	0.82	0.24	0.92	**0.97**	0.77
Syncontl	6	0.88	0.79	0.51	**0.99**	0.92	0.31	0.48	0.66	0.96	0.43
Trace	4	0.76	0.80	0.84	**0.99**	0.68	0.46	0.97	0.87	0.97	0.81
TwoPattern	4	0.96	0.78	0.23	**1.00**	0.98	0.29	0.25	0.64	0.99	0.39
TwoLeadECG	2	0.74	0.78	0.67	0.95	0.79	0.81	0.84	**0.98**	0.97	**0.98**
uWGLX	8	0.74	0.73	0.26	0.73	0.64	0.41	0.12	0.55	**0.76**	0.71
uWGLY	8	**0.67**	0.63	0.29	0.64	0.48	0.34	0.12	0.41	0.66	0.59
uWGLZ	8	0.65	0.64	0.27	0.66	0.45	0.39	0.12	0.51	**0.67**	0.63
wafer	2	0.99	0.99	0.99	0.98	0.99	0.92	**1.00**	0.99	0.99	**1.00**
WordsSyn	25	0.63	0.59	0.21	0.67	0.54	0.46	**0.75**	0.58	0.70	0.71
yoga	2	0.83	0.83	0.61	0.84	0.51	0.71	**0.88**	0.79	0.86	0.82

Table 2 Average classification accuracy and computational cost

Average classification accuracy with

Euclid	Fourier	AR	DTW	EDR	CTE(1)	DTE(1)	CTE(2)	TWED	DTE(2)
0.76	0.73	0.59	0.79	0.75	0.61	0.68	0.69	0.81	0.73

Computational cost with

Euclid	Fourier	AR	DTW	EDR	CTE(1)	DTE(1)	CTE(2)	TWED	DTE(2)
$O(M)$	$O(M^2)$	$O(M^2)$	$O(M^2)$	$O(M^2)$	$O(M)$	$O(M^2)$	$O(M)$	$O(M^2)$	$O(M^2)$

5 Conclusion

Time series classification is an important area for practical data processing and has applications in various fields. As time series data is generated in huge quantity, its processing needs to be quick. For summarization of time series data, classification or categorization of the data is needed. As traditional machine learning algorithms for static data become unsuitable for processing huge MTS data, new methodologies are needed to be developed. Time series data poses two types of challanges, one in their representation and the other in measuring their similarity for grouping. There are a lot of similarity measures developed so far, but their performances are not same in all types of data. It is necessary to know the best suitable similarity measure for a particular application. Though euclid distance is the most simple similarity measure, it cannot be used for unequal length time series. Dynamic time warping and its variants are popular measures but their computational costs are high.

In this work, new similarity measures are proposed from a different view point of representation of the time series based on multidimensional delay vector (MDV) for reconstruction in the phase space. The previously defined measure CTE is computationally very light though recognition accuracy is poor compared to DTW. But CTE has the merit of adjusting the measure to a particular time series by using the embedding parameters of the particular time series which also increases recognition accuracy. The new measure DTE proposed here (DTE(1) with fixed parameters and DTE(2) with embedding parameters set to the particular time series) is not much influenced by the setting of the embedding parameters compared to our earlier measure CTE. The proposed measures DTE(1) and DTE(2) seem to be efficient compared to the existing popular measures as is evident from the simulation results over 43 benchmark data sets. Though the computational cost of proposed measures DTE(1) and DTE(2) is higher than our previously proposed measure, it is comparable to DTW. Currently we are doing several experiments to reduce computational cost of our proposed measures while keeping the classification accuracy high by reducing the dimension of the time series. We hope to report our result in near future.

Acknowledgements We would like to thank all the donors as well as the maintainer of the benchmark data sets for providing us the access for downloading the data from The UCR Time Series Classification/Clustering Homepage: www.cs.ucr.edu/~eamonn/time_series_data/.

References

1. Kantz, H., Schreiber, T.: Nonlinear time series analysis. Cambridge University Press, Cambridge, UK (2004)
2. Buza, K., Nanopoulos, A., Schmidt-Thieme, L.: Time series classification based on individual error prediction. In: Proceedings of IEEE International Conference on Computational Science and Engineering, pp. 48–54 (2010)
3. Chakraborty, B.: Feature selection and classification techniques for multivariate time series. In: Proceedings of ICICIC 2007, (2007)
4. Zhang, X., Wu, J., et al.: A novel pattern extraction method for time series classification. Opt. Eng. **10**(2), 253–271 (2009)
5. Buza, K., Nanopoulos, A., Schmidt-Thieme, L.: Fusion of similarity measure for time series classification. LNCS **6679**, 253–261 (2011)
6. Ding, H., Trajcevski, G., Schuermann, P., Wang, X., Keogh, E.: Querying and mining of time series data: experimental comparison of representations and distance measures. In: Proceedings of 34th VLDB, pp. 1542–1552 (2008)
7. Yang, K., Sahabi, C.: An efficient knearest neighbor search for multivariate time series. Inf. Comput. **205**(1), 65–98 (2007)
8. Xu, J.W., Paiva, A.R., Park, I., Principe, J.C.: A reproducing kernel Hillbert space framework for information theoretic learning. IEEE Trans. Signal Process. **56**(12), 5891–5902 (2008)
9. Chakraborty, B.: A proposal for classification of multi sensor time series databased on time delay embedding. In: Proceedings of 8th International Conference on Sensing Technology ICST, pp. 31–35 (2014)
10. Chakraborty, B., Manabe, Y.: An efficient approach for person authentication using online signature verification. In: Proceedings of SKIMA **2008**, 12–17 (2008)
11. Giusti, R., Batista, G.E.A.P.A.: An empirical comparison of dissimilarity measures for time series classification. In: Proceedings of Brazilian Conference on Intelligent Systems (BRACIS'13), pp. 82–88 (2013)
12. Xing, Z., Pei, J., Keogh, E.: A brief survey on sequence classification. ACM SIGKDD Explor. Newslett. **12**(1), 40–48 (2010)
13. Lal, T.N., et al.: Support vector channel selection in BCI. IEEE Trans. Biomed. Eng. **51**(6), 1003–1010 (2004)
14. Ye, L., Keogh, E.: Time series shapelets: A new primitive for data mining. In: Proceedings 15th ACM SIGKDD International Conference on Knowledge Discovery and Data Mining, pp. 947–956 (2009)
15. Lesh, N., Zaki, M.J., Ogihara, M.: Mining features for sequence classification. In: Proceedings of 5th ACM SIGKDD International Conference on Knowledge Discovery and Data Mining, pp. 342–346 (1999)
16. Ji, X., Baily, J., Dong, G.: Mining minimal distinguishing subsequence patterns with gap constraints. Knowl. Inf. Syst. **11**(3), 259–286 (2007)
17. Yoon, H., Yang, K., Sahabi, C.: Feature subset selection and feature ranking for multivariate time series. IEEE Trans. Knowl. Data Eng. **17**(9), 1186–1198 (2005)
18. Kim, S.B., Han, K.S., et al.: Some effective techniques for naive bayes text classification. IEEE Trans. Knowl. Data Eng. **18**(11), 1457–1466 (2006)
19. Wang, X., Mueen, A., Ding. H., Trajcevski, G., Scheuermann, P., Keogh, E.: Experimental comparison of representation methods and distance measures for time series data. Data Min. Knowl. Discov. **26**, 275–309 (2013)
20. Ratanamahatana, C.A., Keogh, E.J.: Making time series classification more accurate using learned constraints. In: Proceedings of SIAM International Conference on Data Mining, pp. 11–22 (2004)
21. Levenshtein, I.: Binary codes capable of correcting deletions, inssertions and reversals. Sov. Phys. Doklady **10**, 707–710 (1966)
22. Marteau, R.F.: Time warp edit distance with stiffness adjustment for time series matching. IEEE Trans. PAMI **31**(2), 306–318 (2009)

23. Alligood, K., Sauer, T., Yorke, J.A.: Chaos : An Introduction to Dynamical Systems. Springer, New York (1997)
24. Aberbanel, H.D.I.: Analysis of Observed Chaotic Data. Springer, New York (1996)
25. Manabe, Y., Chakraborty, B.: Identity detection from online handwriting time series. In: Proceedings of SMCia08, pp. 365–370 (2008)
26. Yoshida, S., Chakraborty, B.: A comparative study of similarity measures for time series classification. In: The Proceedings of International Workshop on Time Series Data Analysis and its Applications (TSDAA 2015), (2015)
27. Serra, J., Arcios, J.: An empirical evaluation of similarity measures for time series classification. Knowl. Based Syst. **67**, 305–314 (2014)
28. Keogh, E., Zhu, Q., Hu, B., Hao, Y., Xi, X., Wei, L., Ratanamahatana, C.A.: The UCR Time Series Classification/Clustering Homepage: http://www.cs.ucr.edu/~eamonn/time_series_data/ (2011)

A Fuzzy Time Series Model with Customized Membership Functions

Tamás Jónás, Zsuzsanna Eszter Tóth and József Dombi

Abstract In this study, a fuzzy time series modeling method that utilizes a class of customized and flexible parametric membership functions in the fuzzy rule consequents is introduced. The novelty of the proposed methodology lies in the flexibility of this membership function, which we call the composite kappa membership function, and its curve may take various shapes, such as a symmetric or asymmetric bell, triangular, or quasi trapezoid. In our approach, the fuzzy c-means clustering algorithm is used for fuzzification and for the establishment of fuzzy rule antecedents and a heuristic is introduced for identifying the quasi optimal number of clusters to be formed. The proposed technique does not require any preliminary parameter setting, hence it is easy-to-use in practice. In a real-life example, the modeling capability of the proposed method was compared to those of Winters' method, the Autoregressive Integrated Moving Average technique and Adaptive Neuro-Fuzzy Inference System. Based on the empirical results, the proposed method may be viewed as a viable time series modeling technique.

Keywords Fuzzy inference · Fuzzy time series · Membership functions · Short-term forecasting

1 Introduction

Following the fuzzy time series model proposed by Song and Chissom [14], many researchers have developed various fuzzy time series modeling techniques to enhance the modeling capability and forecasting performance of previous approaches. These efforts have resulted in a great variety of time series modeling methods with

T. Jónás (✉) · Z. Eszter Tóth
Department of Management and Corporate Economics, Budapest University of Technology and Economics, Magyar Tudósok Körútja 2, Budapest 1117, Hungary
e-mail: jonas@mvt.bme.hu

J. Dombi
Institute of Informatics, University of Szeged, Árpád Tér 2, Szeged 6720, Hungary

I. Rojas et al. (eds.), *Advances in Time Series Analysis and Forecasting*, Contributions to Statistics, DOI 10.1007/978-3-319-55789-2_20

significant achievements that are founded on the application of fuzzy theory (e.g. [3, 8, 11, 13, 15]).

Fuzzy inference systems (FIS) are one of the well-known applications of fuzzy logic and fuzzy set theory. Their advantages are two-fold, namely they are able to handle linguistic concepts and they are universal approximators by performing non-linear mappings between inputs and outputs. Guillaume [7] gives comprehensive overviews on how fuzzy inference systems can be constructed from data. In general, fuzzy time series models are founded on fuzzy inference systems that are built up from data.

In our study, we introduce a fuzzy time series modeling method which after input data normalization uses the fuzzy c-means clustering algorithm for fuzzification and for the establishment of fuzzy rule antecedents. The membership functions of the fuzzy rule consequents are based on a parametric function that is derived from Dombi's Kappa function. We call this membership function the composite kappa membership function. The novelty of the proposed methodology lies in the flexibility of this membership function as its curve may take various shapes (e.g. symmetric or asymmetric bell, triangular or quasi trapezoid). This property of our membership function enhances its applicability.

It is worth mentioning that the proposed technique does not require any preliminary parameter setting, and so the method can be readily applied. The model parameters, which affect the modeling results, are identified by the method itself. Namely, as part of our method, we introduce a heuristic to identify the quasi optimal number of clusters to be formed by the fuzzy c-means clustering algorithm that is used to establish the fuzzy rule antecedents.

The application of our method, its advantages and modeling capability are demonstrated through a real-life example. In this example, the modeling capability of our method is compared to those of Winters' method, the Autoregressive Integrated Moving Average (ARIMA) technique and Adaptive Neuro-Fuzzy Inference System (ANFIS).

The paper is organized as follows. Section 2 describes the details of constructing a fuzzy inference system beginning with data preparation, followed by the fuzzy c-means clustering of input vectors and the forming of fuzzy rules. Then we introduce the composite kappa membership function and its main properties, the fuzzy inference and the optimization of system outputs. Section 3 outlines the utilization of our fuzzy inference system for time series modeling and forecasting including the impact of model parameters. The application of our method, its advantages and modeling capability will be demonstrated via a real-life example in Sect. 4. In Sect. 5, we give a short summary of our results and conclusions.

2 The Methodology

We will construct a fuzzy inference system that can model the time series $X_1, X_2, \ldots, X_n$. Let $X_j, X_{j+1}, \ldots, X_{j+r-1}$ be an r-period long sub-time series and X_{j+r} its one period long continuation in the time series $X_1, X_2, \ldots, X_n$ ($j = 1, 2, \ldots, n-r$). r is the number of historical periods that we use for generating a one period ahead forecast,

$r \geq 1$, $r+1 \leq n$. Let $d = n - r$ denote the number of r-period long sub-time series in $X_1, X_2, \ldots, X_n$.

2.1 Data Normalization

For each r-period long sub-time series $X_j, X_{j+1}, \ldots, X_{j+r-1}$ and its X_{j+r} continuation, we define vector $\mathbf{x}_j = (x_{j,1}, x_{j,2}, \ldots, x_{j,r})$ and the scalar y_j as follows.

Case 1. If $X_j, X_{j+1}, \ldots, X_{j+r-1}$ are not all equal, then

$$x_{j,p} = \frac{X_{j+p-1} - \min\limits_{q=1,\ldots,r} (X_{j+q-1})}{\max\limits_{q=1,\ldots,r} (X_{j+q-1}) - \min\limits_{q=1,\ldots,r} (X_{j+q-1})} \tag{1}$$

$$y_j = \frac{X_{j+r} - \min\limits_{q=1,\ldots,r} (X_{j+q-1})}{\max\limits_{q=1,\ldots,r} (X_{j+q-1}) - \min\limits_{q=1,\ldots,r} (X_{j+q-1})} \tag{2}$$

$p = 1, 2, \ldots, r$.

Case 2. If $X_j, X_{j+1}, \ldots, X_{j+r-1}$ are all equal and non zero, say have the value of a ($a > 0$), then $\mathbf{x}_j = (1, 1, \ldots, 1)$, and $y_j = X_{j+r}/a$.

Case 3. If $X_j, X_{j+1}, \ldots, X_{j+r-1}$ are all zeros, then $\mathbf{x}_j = (0, 0, \ldots, 0)$, and $y_j = X_{j+r}$.

Owing to this transformation, each component of vector $\mathbf{x}_j$ is normalized to the [0, 1] interval. The $(\mathbf{x}_j, y_j)$ pairs represent the time series $X_1, X_2, \ldots, X_n$ such that the normalized vector $\mathbf{x}_j$ is followed by y_j, and our goal is to build a FIS that can adequately map the relation between $\mathbf{x}_j$ and y_j ($j = 1, 2, \ldots, d$). In other words, we wish to build a fuzzy inference system which has an r-dimensional vector $\mathbf{x}$ as input, and a scalar output y. The system approximates the $y = f(\mathbf{x})$ relation based on the sample $(\mathbf{x}_j, y_j)$, where $\mathbf{x}_j$, y_j are corresponding observations on $\mathbf{x}$ and y, respectively. In order to identify typical patterns in the time series, we cluster the $\mathbf{x}_j$ vectors using the fuzzy c-means clustering algorithm [1].

2.2 Fuzzy C-Means Clustering of Input Vectors

Let α be the exponent used in the fuzzy c-means clustering algorithm ($\alpha \in \mathbb{R}$, $0 < \alpha < \infty$) that determines the fuzziness of clusters, and let $\mathbf{C}_1, \mathbf{C}_2, \ldots, \mathbf{C}_m$ be the clusters formed with the $\mathbf{c}_1, \mathbf{c}_2, \ldots, \mathbf{c}_m$ cluster centroids, respectively ($1 \leq m \leq d$). (The typical value of α used in practice is 2.) Furthermore, let $\mathbf{U}$ be the fuzzy partition matrix

$$\mathbf{U} = \begin{bmatrix} u_{1,1} & u_{1,2} & \cdots & u_{1,d} \\ u_{2,1} & u_{2,2} & \cdots & u_{2,d} \\ \vdots & \vdots & \ddots & \vdots \\ u_{m,1} & u_{m,2} & \cdots & u_{m,d} \end{bmatrix}, \tag{3}$$

where $u_{i,j} \in [0, 1]$, $\sum_{i=1}^{m} u_{i,j} = 1$ for each j, and

$$u_{i,j} = \mu_i(\mathbf{x}_j) = \left(\frac{1}{\|\mathbf{x}_j - \mathbf{c}_i\|_2^2} \right)^{\frac{1}{\alpha-1}} \frac{1}{\sum\limits_{v=1}^{m} \left(\frac{1}{\|\mathbf{x}_j - \mathbf{c}_v\|_2^2} \right)^{\frac{1}{\alpha-1}}} \tag{4}$$

is the membership of vector $\mathbf{x}_j$ to cluster $\mathbf{C}_i$ $(i = 1, 2, \ldots, m,\ j = 1, 2, \ldots, d)$. The centroid vector $\mathbf{c}_i$ of cluster $\mathbf{C}_i$ is calculated as

$$\mathbf{c}_i = \frac{\sum\limits_{j=1}^{d} u_{i,j}^{\alpha} \mathbf{x}_j}{\sum\limits_{j=1}^{d} u_{i,j}^{\alpha}}. \tag{5}$$

2.3 Forming Fuzzy Rules

The $\mathbf{C}_1, \mathbf{C}_2, \ldots, \mathbf{C}_m$ fuzzy clusters represent fuzzy partitions in the input space. From a time series perspective, the r-period long normalized sub-time series are classified into m clusters, and the cluster centroids represent r-period long typical normalized patterns. From it, the following fuzzy rules can be formed.

$$\begin{array}{l} \text{Rule 1: if } \mathbf{x} \text{ is in } \mathbf{C}_1 \text{ then } \mathbf{B} = \mathbf{B}_1 \\ \text{Rule 2: if } \mathbf{x} \text{ is in } \mathbf{C}_2 \text{ then } \mathbf{B} = \mathbf{B}_2 \\ \quad \vdots \qquad\qquad \vdots \qquad\quad \vdots \qquad \vdots \\ \text{Rule m: if } \mathbf{x} \text{ is in } \mathbf{C}_m \text{ then } \mathbf{B} = \mathbf{B}_m, \end{array} \tag{6}$$

where $\mathbf{x} \in \mathbb{R}^r$ is an input vector normalized according to the method described in Sect. 2.1, $\mathbf{B}$ is the fuzzy output of the FIS and $\mathbf{B}_1, \mathbf{B}_2, \ldots, \mathbf{B}_m$ are fuzzy sets defined over set $\mathbf{Y}$, which is the domain of crisp system outputs. Based on (4), for any r-dimensional normalized input vector $\mathbf{x}$, its membership value $\mu_i(\mathbf{x})$ for cluster $\mathbf{C}_i$ can be computed as

$$\mu_i(\mathbf{x}) = \left(\frac{1}{\|\mathbf{x} - \mathbf{c}_i\|_2^2} \right)^{\frac{1}{\alpha-1}} \frac{1}{\sum\limits_{v=1}^{m} \left(\frac{1}{\|\mathbf{x} - \mathbf{c}_v\|_2^2} \right)^{\frac{1}{\alpha-1}}}. \tag{7}$$

The value of $\mu_i(\mathbf{x})$ may be viewed as the activation level of Rule i for the input $\mathbf{x}$. As results of the fuzzy c-means clustering, the rule antecedents are identified. In order to make the fuzzy inference system complete, the consequent of each rule needs to be identified; that is, the membership function of each $\mathbf{B}_i$ fuzzy set needs to be given ($i = 1, 2, \dots, m$). Furthermore, as we wish to use the FIS for time series modeling, a defuzzification method, which turns the fuzzy output into a crisp one, needs to be identified as well.

2.4 Membership Functions of the Fuzzy Rule Consequents

Let the $\mathbf{Y}$ domain of the crisp outputs of our FIS be

$$\mathbf{Y} = \left[y_l - \varepsilon, y_h + \varepsilon\right], \tag{8}$$

where $y_l = \min_{j=1,\dots,d}(y_j)$, $y_h = \max_{j=1,\dots,d}(y_j)$, $\varepsilon = c(y_h - y_l)$, $c > 0$, $c \in \mathbb{R}$. In our implementation, $c = 0.1$. In our method, the membership function of the consequent of the ith fuzzy rule; that is, the membership function $\mu_{B_i}^{(\mathbf{p}_i)}(y)$ of $\mathbf{B}_i$ is composed based on Dombi's kappa function that is treated as an operator in fuzzy theory [5, 6] and it is given by

$$\kappa_{\mu,\nu}^{(\lambda)}(x) = \frac{1}{1 + \frac{1-\nu}{\nu}\left(\frac{\mu}{1-\mu}\frac{1-x}{x}\right)^{\lambda}}, \tag{9}$$

where $x, \mu, \nu \in (0, 1)$, $\lambda \neq 0$. $\mu_{B_i}^{(\mathbf{p}_i)}(y)$ is given by a left hand side $l_i(y)$ and a right hand side $r_i(y)$ function as follows.

$$\mu_{B_i}^{(\mathbf{p}_i)}(y) = \begin{cases} 0, & \text{if } y \leq \alpha_i - \delta_{l,i} \\ l_i(y), & \text{if } \alpha_i - \delta_{l,i} < y \leq \alpha_i \\ r_i(y), & \text{if } \alpha_i < y < \alpha_i + \delta_{r,i} \\ 0, & \text{if } y \geq \alpha_i + \delta_{r,i}, \end{cases} \tag{10}$$

where the parameter vector $\mathbf{p}_i = (\alpha_i, \delta_{l,i}, \lambda_{l,i}, \delta_{r,i}, \lambda_{r,i})$, $\alpha_i \in \mathbf{Y}$, $\delta_{l,i}, \delta_{r,i} \in \mathbb{R}^+$, $\lambda_{l,i}, \lambda_{r,i} \in [1, +\infty)$, $i = 1, 2, \dots, m$. Functions $l_i(y)$ and $r_i(y)$ are derived from the kappa function in (9) by applying $\mu = \nu = 0.5$ and projecting its domain from (0, 1) to the intervals $(\alpha_i - \delta_i, \alpha_i)$ and $(\alpha_i, \alpha_i + \delta_i)$, respectively:

$$l_i(y) = \frac{1}{1 + \left(\frac{\alpha_i - y}{y - \alpha_i + \delta_{l,i}}\right)^{\lambda_{l,i}}} \tag{11}$$

$$r_i(y) = \frac{1}{1 + \left(\frac{\alpha_i + \delta_{r,i} - y}{y - \alpha_i}\right)^{-\lambda_{r,i}}}. \tag{12}$$

Here, we call function $\mu_{B_i}^{(\mathbf{p}_i)}(y)$ the composite kappa membership function.

2.4.1 The Basic Properties of the Composite Kappa Membership Function

It can be seen that the $l_i(y)$ function is strongly monotonously increasing from 0 to 1, and

$$l_i\left(\alpha_i - \frac{\delta_{l,i}}{2}\right) = 0.5. \tag{13}$$

The derivative of $l_i(y)$ at $\alpha_i - \delta_i/2$ is

$$\left.\frac{\mathrm{d}l_i(y)}{\mathrm{d}y}\right|_{\alpha_i - \frac{\delta_{l,i}}{2}} = \frac{\lambda_{l,i}}{\delta_{l,i}}. \tag{14}$$

That is, if $\delta_{l,i}$ is fixed, then parameter $\lambda_{l,i}$ determines the slope of curve of $l_i(y)$ at the point $(\alpha_i - \delta_{l,i}/2, 0.5)$. It can also be seen that $r_i(y)$ is strongly monotonously decreasing from 1 to 0,

$$r_i\left(\alpha_i + \frac{\delta_{r,i}}{2}\right) = 0.5, \tag{15}$$

and

$$\left.\frac{\mathrm{d}r_i(y)}{\mathrm{d}y}\right|_{\alpha_i + \frac{\delta_{r,i}}{2}} = -\frac{\lambda_{r,i}}{\delta_{r,i}}. \tag{16}$$

It means that if $\delta_{r,i}$ is fixed, then parameter $\lambda_{r,i}$ determines how quickly the curve of $r_i(y)$ changes from 1 to 0.

The curve of membership function $\mu_{B_i}^{(\mathbf{p}_i)}(y)$ is bell-shaped, if $\lambda_{l,i} > 1$ and $\lambda_{r,i} > 1$, and it is triangular, if $\lambda_{l,i} = \lambda_{r,i} = 1$. If $\lambda_{r,i} = \lambda_{l,i}$ and $\delta_{l,i} = \delta_{r,i}$, then the curve of $\mu_{B_i}^{(\mathbf{p}_i)}(y)$ is symmetric, otherwise it is asymmetric. Based on the properties of $\mu_{B_i}^{(\mathbf{p}_i)}(y)$ discussed so far, we may state that $\mu_{B_i}^{(\mathbf{p}_i)}(y)$ is a flexible membership function and its curve can have various shapes. Figure 1 shows some examples of the curve of membership function $\mu_{B_i}^{(\mathbf{p}_i)}(y)$.

2.5 *The Fuzzy Inference and Optimization of System Outputs*

The membership function $\mu_B(y)$ of the fuzzy output of the FIS that we wish to construct is computed by using Mamadani's implication and max aggregation. That is,

$$\mu_B(y) = \max_{i=1,\ldots,m}\left(\min\left(\mu_i(\mathbf{x}_j), \mu_{B_i}^{(\mathbf{p}_i)}(y)\right)\right). \tag{17}$$

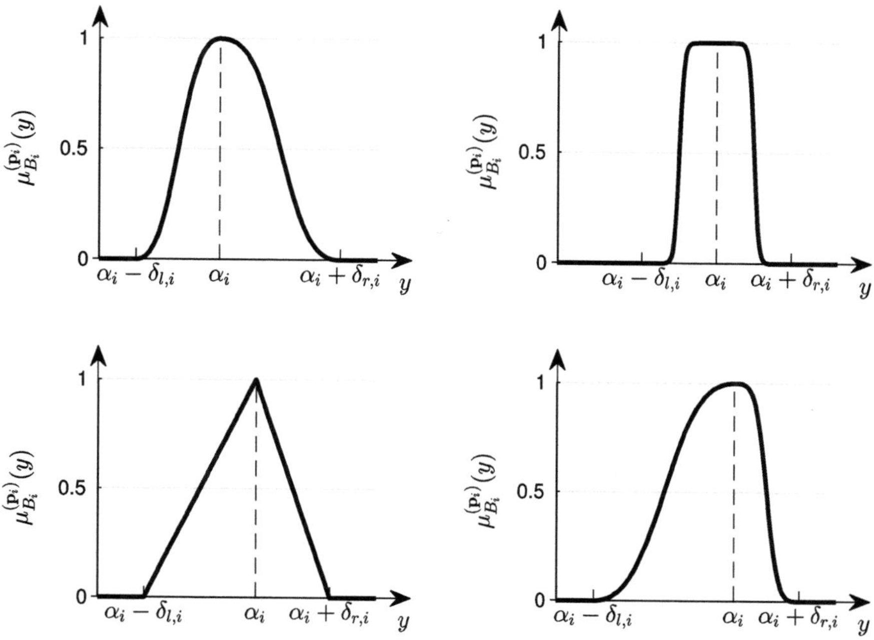

Fig. 1 Examples of membership functions of fuzzy set B_i

Let the matrix $\mathbf{P}$ consist of the parameter vectors $\mathbf{p}_1, \mathbf{p}_2, \dots, \mathbf{p}_m$ of the membership functions $\mu_{B_1}^{(\mathbf{p}_1)}(y), \mu_{B_2}^{(\mathbf{p}_2)}(y), \dots, \mu_{B_m}^{(\mathbf{p}_m)}(y)$, respectively. In other words

$$\mathbf{P} = \begin{bmatrix} \mathbf{p}_1 \\ \mathbf{p}_2 \\ \vdots \\ \mathbf{p}_m \end{bmatrix} = \begin{bmatrix} \alpha_1 & \delta_{l,1} & \lambda_{l,1} & \delta_{r,1} & \lambda_{r,1} \\ \alpha_2 & \delta_{l,2} & \lambda_{l,2} & \delta_{r,2} & \lambda_{r,2} \\ \vdots & \vdots & \vdots & \vdots & \vdots \\ \alpha_m & \delta_{l,m} & \lambda_{l,m} & \delta_{r,m} & \lambda_{r,m} \end{bmatrix}. \tag{18}$$

The optimal model parameter matrix $\mathbf{P}^*$ may be identified as

$$\mathbf{P}^* = \arg\min_{\mathbf{P}} \left(\sum_{j=1}^{d} \left(\hat{y}_j(\mathbf{P}) - y_j\right)^2 \right), \tag{19}$$

where $\hat{y}_j(\mathbf{P})$ is the crisp output of the fuzzy inference system with the parameter matrix $\mathbf{P}$ for the input vector $\mathbf{x}_j$. $\hat{y}_j(\mathbf{P})$ is computed via the center of gravity (COG) defuzzification:

$$\hat{y}_j(\mathbf{P}) = \frac{\int\limits_{y\in\mathbf{Y}} y\mu_B(y)\mathrm{d}y}{\int\limits_{y\in\mathbf{Y}} \mu_B(y)\mathrm{d}y} = \frac{\int\limits_{y\in\mathbf{Y}} y \max\limits_{i=1,\dots,m}\left(\min\left(\mu_i(\mathbf{x}_j), \mu_{B_i}^{(\mathbf{p}_i)}(y)\right)\right)\mathrm{d}y}{\int\limits_{y\in\mathbf{Y}} \max\limits_{i=1,\dots,m}\left(\min\left(\mu_i(\mathbf{x}_j), \mu_{B_i}^{(\mathbf{p}_i)}(y)\right)\right)\mathrm{d}y}. \tag{20}$$

The optimization problem given by (19) can be solved by using the Interior Point Algorithm [16]. In this algorithm, the parameters may be initialized as follows:

$$\alpha_i = \frac{\sum\limits_{j=1}^{d} \mu_i(\mathbf{x}_j)y_j}{\sum\limits_{j=1}^{d} \mu_i(\mathbf{x}_j)} \tag{21}$$

$\delta_{l,i} = (\alpha_i - y_l)/2$, $\delta_{r,i} = (y_h - \alpha_i)/2$, $\omega_{l,i} = \omega_{r,i} = 1$, $i = 1, \dots, m$. Let us assume that $y_l \neq y_h$ and let

$$\Delta = \frac{y_h - y_l}{k}, \tag{22}$$

where $k \in \mathbb{N}$, $k \gg 1$ (in our implementation $k = 1000$). Then the fraction of integrals in the center of gravity defuzzification in (20) can be numerically approximated as follows:

$$\hat{y}_j(\mathbf{P}) \approx \frac{\sum\limits_{t=0}^{k} (y_l + t\Delta) \max\limits_{i=1,\dots,m}\left(\min\left(\mu_i(\mathbf{x}_j), \mu_{B_i}^{(\mathbf{p}_i)}(y_l + t\Delta)\right)\right)}{\sum\limits_{t=0}^{k} \max\limits_{i=1,\dots,m}\left(\min\left(\mu_i(\mathbf{x}_j), \mu_{B_i}^{(\mathbf{p}_i)}(y_l + t\Delta)\right)\right)}.$$

3 Modeling and Forecasting

Let $X_1, X_2, \dots, X_r$ be an r-period long known time series and X_{r+1} its one period long unknown continuation, $r \geq 1$. Furthermore, let $\hat{X}_{r+1}$ denote the forecast for X_{r+1} generated based on the $X_1, X_2, \dots, X_r$ values using a fuzzy inference system constructed according to the method that was introduced in Sect. 2. $\hat{X}_{r+1}$ can be generated as follows. Depending on the $X_1, X_2, \dots, X_r$ values, the normalized input vector $\mathbf{x} = (x_1, x_2, \dots, x_r)$ for the FIS is applied according to one of the following cases.

Case 1. If $X_1, X_2, \dots, X_r$ are not all equal, then

$$x_p = \frac{X_p - \min\limits_{q=1,\dots,r}(X_q)}{\max\limits_{q=1,\dots,r}(X_q) - \min\limits_{q=1,\dots,r}(X_q)} \tag{23}$$

$p = 1, 2, \dots, r$.

Case 2. If $X_1, X_2, \ldots, X_r$ are all the same and non-zero, they have the value of a ($a > 0$), then $\mathbf{x}_j = (1, 1, \ldots, 1)$.
Case 3. If $X_1, X_2, \ldots, X_r$ are all zero, then $\mathbf{x} = (0, 0, \ldots, 0)$.

Let $\hat{y}$ be the system output for the input vector $\mathbf{x}$. Depending on how vector $\mathbf{x}$ was created from the time series $X_1, X_2, \ldots, X_r$ according to the above cases, the $\hat{X}_{r+1}$ forecast is computed according to one of the following denormalizations.

Case 1. If $x_1, x_2, \ldots, x_r$ are not all equal, then

$$\hat{X}_{r+1} = \hat{y}\left(\max_{q=1,\ldots,r}(X_q) - \min_{q=1,\ldots,r}(X_q)\right) + \min_{q=1,\ldots,r}(X_q) \tag{24}$$

Case 2. If $\mathbf{x} = (1, 1, \ldots, 1)$, then $\hat{X}_{r+1} = ay$.
Case 3. If $\mathbf{x} = (0, 0, \ldots, 0)$, then $\hat{X}_{r+1} = y$.

If we apply this forecasting method for X_{j+r} continuation of each r-period long sub-time series $X_j, X_{j+1}, \ldots, X_{j+r-1}$ in the time series $X_1, X_2, \ldots, X_n$, ($j = 1, 2, \ldots,$ $n - r$), then $\hat{X}_{r+1}, \hat{X}_{r+2}, \ldots, \hat{X}_n$ are the predicted (simulated) values of $X_{r+1}, X_{r+2}, \ldots,$ X_n, respectively. That is, the $\hat{X}_{r+1}, \hat{X}_{r+2}, \ldots, \hat{X}_n$ values model the $X_{r+1}, X_{r+2}, \ldots, X_n$ values.

3.1 The Impact of Model Parameters

The fuzzy inference system introduced in Sect. 2, which we use for time series modeling, has two important parameters that significantly influence the goodness of the model. These are the number of historical periods (r) taken into account to form the input vectors and the number of clusters (m) used as an input to the fuzzy c-means clustering algorithm.

Now suppose r be a fixed number. Our FIS has that many fuzzy rules as many clusters of the input vectors $\mathbf{x}_j$ are formed in the fuzzy c-means algorithm ($j = 1, 2, \ldots, d$). On the one hand, if m is close to the number of input vectors d, the FIS can yield good fitting results, but in such cases its generalization capability might be questionable. On the other hand, if the number of clusters is low, the FIS can grab some generic relations in the time series, but the model fitting might be poor. Figure 2 shows an example of how the number of clusters m influences the modeling capability of our FIS. In this example, the original time series contains data of 77 periods and the number of historical periods is $r = 13$.

In the literature, there are well known methods and metrics that can be used for clustering validation and finding the optimal number of clusters, like the silhouette coefficient [12], the Calinski-Harabasz index [2] and the Davies-Bouldin index [4]. Based on the experiences that we gained from practical applications of our method to forecast demand for electronic products in the industry, the above methods overstate the number of clusters. As an alternative method, we used the following simple

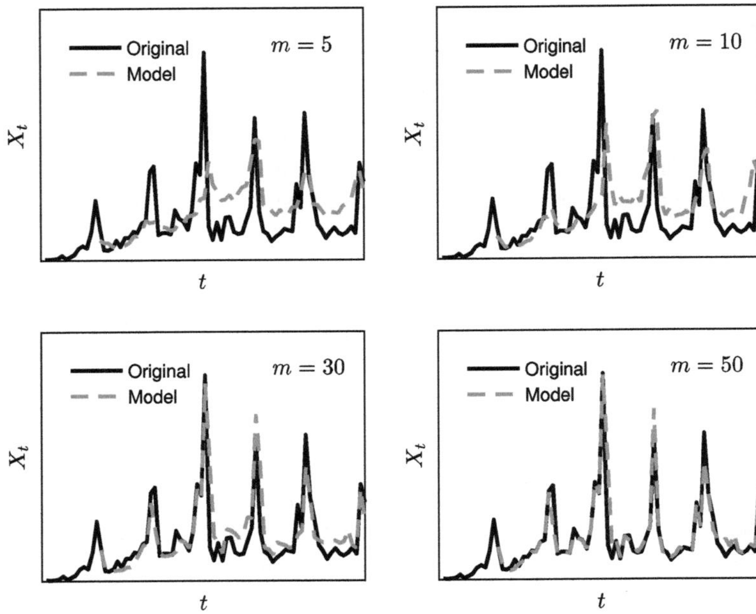

Fig. 2 The impact of number of clusters

heuristic, which we found more suitable in practice, to identify the number of clusters. As we used fuzzy c-means clustering, the $\mathbf{C}_v$ cluster is defined as set of those $\mathbf{x}_j$ vectors for which membership values to $\mathbf{C}_v$ are the highest among all the clusters; that is,

$$\mathbf{C}_v = \left\{ \mathbf{x}_j : \mu_v(\mathbf{x}_j) = \max_{s=1,\ldots,m} \mu_s(\mathbf{x}_j), j \in \{1, 2, \ldots, d\} \right\}, \tag{25}$$

and if a vector $\mathbf{x}_j$ has 0.5 membership vale in two different clusters, then let it be in the cluster with the lower index ($v = 1, 2, \ldots, m$). Let $\mathcal{I}_v$ be

$$\mathcal{I}_v = \left\{ j : \mathbf{x}_j \in \mathbf{C}_v; j \in \{1, 2, \ldots, d\} \right\}, \tag{26}$$

that is, $\mathcal{I}_v$ is the set of indexes of those $\mathbf{x}_j$ vectors that are in cluster $\mathbf{C}_v$, and furthermore, let n_v denote the number of vectors in cluster $\mathbf{C}_v$ ($v = 1, 2, \ldots, m$). Using these notations the sum of squared totals (SST), the sum of squares between (SSB) and the sum of squared errors (SSE) over all the $\mathbf{x}_j$ vectors taking into account the $\mathbf{C}_1, \mathbf{C}_2, \ldots, \mathbf{C}_m$ clusters can be written as

$$SSB(m) = \sum_{i=1}^{m} n_i \|\mathbf{c}_i - \bar{\mathbf{x}}\|_2^2; \quad SSE(m) = \sum_{i=1}^{m} \sum_{j \in \mathcal{I}_i} \|\mathbf{x}_j - \mathbf{c}_i\|_2^2;$$
$$SST = \sum_{i=1}^{m} \sum_{j \in \mathcal{I}_i} \|\mathbf{x}_j - \bar{\mathbf{x}}\|_2^2 = SSB(m) + SSE(m), \tag{27}$$

where $\bar{\mathbf{x}}$ is the centroid of all $\mathbf{x}_j$ vectors ($j = 1, 2, \ldots, d$). Here, we treat $SSB(m)$ and $SSE(m)$ as quantities that depend on the number of clusters m. It can be seen that $s(m) = SSB(m)/SST$ is a monotonously increasing function of m. If $m = 1$, then $s(m) = 0$, and if $m = d$, then $s(m) = 1$. $s(m)$ increases from 0 to 1 in $d - 1$ steps, that is, its average increase is $1/(d - 1)$. With a binary search approach we identify the smallest integer m^* for which

$$s(m^* + 1) - s(m^*) \leq \frac{1}{d - 1} \tag{28}$$

and we set the number of clusters m to m^*. The idea behind this heuristic is that if $m > m^*$, then $s(m)$ increases by less than the $1/(d - 1)$ average increment, and so the $m > m^*$ values do not yield much better clustering.

A quasi optimal value r^* of the number of historical periods r can be obtained by minimizing the mean square error (MSE) of the fitted values. For each iteration on r ($r = 1, 2, \ldots, n - 1$), the number of clusters can be identified in the way described above. Note that since r is the dimension of the input vectors, if r is large, then problems of clustering high-dimensional data might arise [10]. In our industrial applications, $r \leq 20$.

4 A Practical Application

A time series representing the real-life historical demand for an electronic product at an electronics manufacturer company over 77 consecutive weeks was modeled using a fuzzy inference system that was constructed according to the method introduced in Sect. 2. Here, we refer to this fuzzy inference system as the composite kappa membership function-based fuzzy inference system and use the CKM-FIS abbreviation for it. The number of historical periods r and the number of clusters m in CKM-FIS were identified based on the approach described in Sect. 3.1, namely $r = 13$ and $m = 27$. Iterations on r were carried out in the $1, \ldots, 20$ range to get the value that yielded the lowest MSE of the fitted values. Here, the parameter α of the fuzzy c-means clustering was set to 2.0. The MSE values of the CKM-FIS based fits were compared to those of the ARIMA, Winters' and ANFIS methods. In Winters' method multiplicative seasonality was assumed, and both the ARIMA and the Winters' methods were applied with seasonal length equal to parameter r of CKM-FIS; that is, a seasonality of 13 weeks was considered. The Hyndman and Khandakar algorithm [9] was used to identify the best fitting ARIMA model. The ANFIS method, implemented

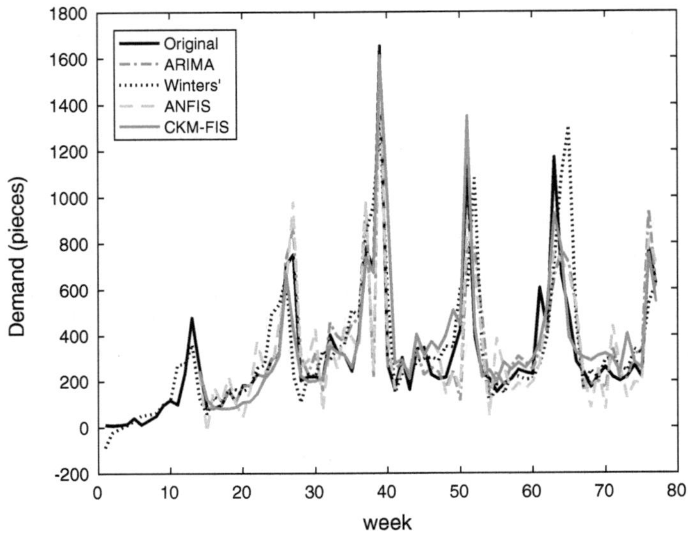

Fig. 3 The original demand time series and some models

Table 1 MSE values of fits

Method	Modeled periods	MSE
ARIMA	15, ..., 77	22092.52
Winters'	1, ..., 77	38826.28
ANFIS	14, ..., 77	21067.27
CKM-FIS	14, ..., 77	21302.54

in MATLAB R2016a by using its ANFIS tool, was applied with the same input and output pairs as the CKM-FIS method. The demand time series and the models investigated are shown in Fig. 3. The periods modeled by each method and the corresponding MSE values are summarized in Table 1. The ARIMA method gives forecasts from week 15 to week 77. Since $r = 13$, the CKM-FIS and ANFIS methods give forecasts from week 14 to week 77, while Winters' method provides forecasts to the entire time range of the time series. From a modeling accuracy viewpoint, we can say that the ARIMA method, the CKM-FIS and ANFIS methods give similar results, their MSE values are similar, but Winters' method gives a lower accuracy.

5 Conclusions

In this study, a novel methodology was presented for building a fuzzy inference system with customized membership functions. The main reason for building such a system was to introduce a technique that can be easily applied for short-term forecasting. The flexibility of the composite kappa membership functions of fuzzy sets in the rule consequents proves the novelty of our approach. The curve of the introduced membership function may have various shapes, and this feature enhances the practical applicability of the CKM-FIS method. An application and the advantages of the proposed methodology were demonstrated through a real-life industrial example. The MSE values of the CKM-FIS based fits were compared to those of the ARIMA, Winters' and ANFIS methods. Taking modeling accuracy into consideration, we may conclude that the ARIMA, ANFIS and CKM-FIS methods provide similar results, while the Winters' method results in a lower accuracy. It should be added here that our method does not require any preliminary parameter settings, as following the proposed methodology, the number of historical periods r and the number of clusters used in then fuzzy c-means clustering m are obtained during the execution of the algorithm itself. This is a practical advantage of the proposed method compared to the ARIMA and Winters' methods, which require certain input parameters, such as the model identification for the ARIMA, and the seasonal length both for the ARIMA and the Winters' methods. Besides the model free nature of the CKM-FIS method, which is a generic characteristic of fuzzy methods, another useful property of the proposed method is that the constructed fuzzy inference system embodies certain semantics. Namely, the rule antecedents are cluster centroids that represent typical normalized patterns in the time series, while a linguistic value, such as very-very low, very low, low, etc., may be associated with the consequent of each fuzzy rule. It allows the users to give a linguistic interpretation of the rules that "drive" the time series.

References

1. Bezdek, J.C.: Pattern Recognition with Fuzzy Objective Function Algorithms. Plenum Press, New York (1981)
2. Calinski, T., Harabasz, J.: A dendrite method for cluster analysis. Commun. Stat. **3**(1), 1–27 (1974)
3. Cheng, C.H., Cheng, G.W., Wang, J.W.: Multi-attribute fuzzy time series method based on fuzzy clustering. Expert Syst. Appl. **34**(2), 1235–1242 (2008)
4. Davies, D.L., Bouldin, D.W.: A cluster separation measure. IEEE Trans. Pattern Anal. Mach. Intell. **1**(2), 95–104 (1979)
5. Dombi, J.: Modalities. In: Eurofuse 2011. Advances in Intelligent and Soft Computing. Springer, London pp. 53–65 (2012)
6. Dombi, J.: On a certain type of unary operators. In: Proceedings of 2012 IEEE International Conference on Fuzzy Systems, Brisbane, QLD, pp. 1–7, 10–15 June 2012
7. Guillaume, S.: Designing fuzzy inference systems from data: an interpretability-oriented review. IEEE Trans. Fuzzy Syst. **9**, 426–443 (2001)

8. Huarng, K.: Heuristic models of fuzzy time series for forecasting. Fuzzy Sets Syst. **123**, 369–386 (2001)
9. Hyndman, R., Khandakar, Y.: Automatic time series forecasting: the forecast package for r. J. Stati. Softw. **27** (2008). doi:10.18637/jss.v027.i03
10. Kriegel, H., Kröger, P., Zimek, A.: Clustering high-dimensional data. ACM Trans. Knowl. Discov. Data **3**(1) (2009). doi:10.1145/1497577.1497578
11. Li, S.-T., Cheng, Y.-C., Lin, S.-Y.: A FCM-based deterministic forecasting model for fuzzy time series. Comput. Math. Appl. **56**(12), 3052–3063 (2008)
12. Rousseeuw, P.J.: Silhouettes: a graphical aid to the interpretation and validation of cluster analysis. J. Comput. Appl. Math. **20**, 53–65 (1987)
13. Singh, S.R.: A robust method of forecasting based on fuzzy time series. Appl. Math. Comput. **188**(1), 472–484 (2007)
14. Song, Q., Chissom, B.S.: Fuzzy time series and its models. Fuzzy Sets Syst. **54**, 269–277 (1993)
15. Štěpnička, M., Dvořák, A., Pavliska, V., Vavříčková, L.: A linguistic approach to time series modeling with the help of F-transform. Fuzzy Sets Syst. **180**(1), 164–184 (2011)
16. Waltz, R., Morales, J., Nocedal, J., Orban, D.: An interior algorithm for nonlinear optimization that combines line search and trust region steps. Math. Program. **9**, 391–408 (2006)

Model-Independent Analytic Nonlinear Blind Source Separation

David N. Levin

Abstract Consider a time series of signal measurements, $x(t)$, where x has two or more components. This paper shows how to perform nonlinear blind source separation; i.e., how to determine if these signals are equal to linear or nonlinear mixtures of the state variables of two or more statistically independent subsystems. First, the local distributions of measurement velocities are processed in order to derive vectors at each point in x-space. If the data are separable, each of these vectors must be directed along a subspace of x-space that is traversed by varying the state variable of one subsystem, while all other subsystems are kept constant. Because of this property, these vectors can be used to construct a small set of mappings, which must contain the "unmixing" function, if it exists. Therefore, nonlinear blind source separation can be performed by examining the separability of the data after they have been transformed by each of these mappings. The method is analytic, constructive, and model-independent. It is illustrated by blindly recovering the separate utterances of two speakers from nonlinear combinations of their audio waveforms.

Keywords Blind source separation · Nonlinear signal processing · Invariants · Sensor · Analytic · Model-independent

1 Introduction

Consider an evolving physical system that is being observed by making time-dependent measurements ($x_k(t)$ for $k = 1, \dots, N$ where $N \geq 2$), which are invertibly related to the system's state variables. The objective of blind source separation (BSS) is to determine if the measurements are mixtures of the state variables of statistically independent subsystems. Specifically, we want to know if there is an invertible, possibly nonlinear, N-component "unmixing" function, f, that transforms the measurement time series into a time series of separable states:

D.N. Levin (✉)
Department of Radiology, University of Chicago,
1310 N. Ritchie Ct., Unit 26 AD, Chicago, IL 60610, USA
e-mail: d-levin@uchicago.edu
URL: http://radiology.uchicago.edu/directory/david-n-levin

I. Rojas et al. (eds.), *Advances in Time Series Analysis and Forecasting*,
Contributions to Statistics, DOI 10.1007/978-3-319-55789-2_21

$$s(t) = f[x(t)]. \tag{1}$$

Here, $s(t)$ denotes a set of components, $s_k(t)$ for $k = 1, \dots, N$, which can be grouped to form the state variables of "statistically independent", possibly multidimensional subsystems. In other words, can the data be transformed from the measurement coordinate system, x, to another coordinate system, s, in which the data's components form statistically independent groups?

There is a variety of methods for solving this blind source separation (BSS) problem for the special case in which the signals are linearly related to the system states [1]. However, some observed signals (e.g., from biological or economic systems) may be nonlinear functions of the underlying system states. Computational methods of separating such nonlinear mixtures are limited [2], even though humans seem to do it in an effortless manner.

This paper utilizes a criterion for "statistical independence" [3] that differs from the conventional one. Specifically, let $\rho_S(s, \dot{s})$ be the probability density function (PDF) in $(s, \dot{s})$-space, where $\dot{s} = ds/dt$. Namely, let $\rho_S(s, \dot{s})dsd\dot{s}$ be the fraction of total time that the location and velocity of $s(t)$ are within the volume element $dsd\dot{s}$ at location $(s, \dot{s})$. In this paper, the data are defined to be separable if and only if there is an unmixing function that transforms the measurements so that $\rho_S(s, \dot{s})$ is the product of the density functions of individual components (or groups of components)

$$\rho_S(s, \dot{s}) = \prod_{a=1,2,\dots} \rho_{Sa}(s_{(a)}, \dot{s}_{(a)}). \tag{2}$$

where $s_{(a)}$ is a subsystem state variable, comprised of one or more of the components of s. This criterion for separability is consistent with our intuition that the statistical distribution of the state and velocity of one independent subsystem should not depend on the particular state and velocity of any other independent subsystem.

This criterion for statistical independence should be compared to the conventional criterion, which is formulated in s-space (i.e., state space) instead of $(s, \dot{s})$-space (the space of states and state velocities). The system is said to be separable if there is an unmixing function that transforms the measurements so that $\rho_S(s)$ is the product of the density functions of individual components (or groups of components)

$$\rho_S(s) = \prod_{a=1,2,\dots} \rho_{Sa}(s_{(a)}), \tag{3}$$

In *every* formulation of BSS, multiple solutions can be created by applying "subsystem-wise" transformations, which permute the subsystem state variables and/or transform their components among themselves. These solutions are the same as one another, except for differing choices of the coordinate systems used to describe each subsystem. However, the criterion in (3) is so weak that it suffers from a much worse non-uniqueness problem: namely, solutions can almost always be created by mixing the subsystem state variables of other solutions (see [2, 4]).

There are at least two reasons why (2) is the preferred way of defining "statistical independence":

1. If a physical system is comprised of two independent subsystems, we normally expect that there is a unique way of identifying the subsystems. As mentioned above, (3) is too weak to meet this expectation. On the other hand, (2) is a much stronger constraint than (3). Specifically, (3) can be recovered by integrating both sides of (2) with respect to velocity. This shows that the solutions of (2) are a subset of the solutions of (3). Therefore, it is certainly possible that (2) reformulates the BSS problem so that it has a unique solution (up to subsystem-wise transformations), although this is not proved in this paper.
2. For all systems that obey the laws of classical mechanics and are in thermal equilibrium, the PDF in $(s, \dot{s})$-space is the Maxwell-Boltzmann distribution, which is proportional to the exponential of the system's energy. If the system consists of two non-interacting subsystems, it's energy is the sum of the subsystem energies, and, therefore, this distribution factorizes exactly as in (2). Thus, for classical physical systems, non-interacting subsystems are statistically independent in the sense of (2).

There are several other ways in which the proposed method of nonlinear BSS differs from methods in the literature:

1. As stated above, in this paper the BSS problem is reformulated in the joint space of states and state velocities. Although there is some earlier work in which BSS is performed with the aid of velocity information [5, 6], these papers utilize the *global* distribution of measurement velocities (i.e., the distribution of velocities at all points in state space). In contrast, the method proposed here exploits additional information that is present in the *local* distributions of measurement velocities (i.e., the velocity distributions in each neighborhood of state space).
2. Many investigators have attempted to simplify the BSS problem by assuming prior knowledge of the nature of the mixing function; i.e., they have modelled the mixing function. For example, the mixing function has been assumed to have parametric forms that describe post-nonlinear mixtures, linear-quadratic mixtures, and other combinations [1]. In contrast, the present paper proposes a model-independent method that can be used in the presence of any invertible diffeomorphic mixing function.
3. In most other approaches, nonlinear BSS is reduced to the optimization problem of finding the unmixing function that maximizes the independence of the source signals corresponding to the observed mixtures. This usually requires the use of iterative algorithms with attendant issues of convergence and computational cost [1]. In contrast, the method proposed in this paper is analytic and constructive.

In an earlier paper [7], the criterion in (2) was used in order to perform nonlinear BSS. However, this method was quite different from the one proposed here. The current paper shows how the measurement time series endows state space with local vectors that contain crucial information about the separability of the data. In contrast, the presence of these vectors played no role whatsoever in [7]. Instead, BSS was performed by deriving a large number of local scalars that must lie in low dimensional subspaces, if the data were separable.

2 Method

This section describes the proposed method of performing nonlinear blind source separation, which was briefly presented in [8] and in [9]. The overall strategy is to determine if the system can be separated into two (possibly multidimensional) independent subsystems. If the data cannot be so separated, they are simply inseparable. If such a two-fold separation is possible, we can then examine the data describing the evolution of each independent subsystem in order to determine if it can be further separated into two smaller subsystems. This recursive process can be repeated until each independent subsystem cannot be further divided into smaller components. For example, for $N = 3$, we can first determine if the system can be separated into a subsystem with one degree of freedom and a subsystem having two degrees of freedom. If such a separation is possible, the data from the two-dimensional subsystem can then be examined to determine if it can be further subdivided into two one-dimensional subsystems.

At each stage, a five-step procedure is used to determine separability:

1. Use the local distributions of measurement velocities to construct N vectors at each point x: $V_{(i)}(x)$ for $i = 1, \ldots, N$.
The first step is to construct second-order and fourth-order local correlations of the data's velocity

$$C_{kl}(x) = \langle(\dot{x}_k - \bar{\dot{x}}_k)(\dot{x}_l - \bar{\dot{x}}_l)\rangle_x \tag{4}$$

$$C_{klmn}(x) = \langle(\dot{x}_k - \bar{\dot{x}}_k)(\dot{x}_l - \bar{\dot{x}}_l)(\dot{x}_m - \bar{\dot{x}}_m)(\dot{x}_n - \bar{\dot{x}}_n)\rangle_x \tag{5}$$

where $\bar{\dot{x}} = \langle\dot{x}\rangle_x$, where the bracket denotes the time average over the trajectory's segments in a small neighborhood of x, and where all subscripts are integers less than or equal to N. Because $\dot{x}$ is a contravariant vector, $C_{kl}(x)$ and $C_{klmn}(x)$ are local contravariant tensors of rank 2 and 4, respectively. The definition of the PDF implies that $C_{kl}(x)$ and $C_{klmn}(x)$ are two of its moments; e.g.,

$$C_{kl\ldots}(x) = \frac{\int \rho(x,\dot{x})(\dot{x}_k - \bar{\dot{x}}_k)(\dot{x}_l - \bar{\dot{x}}_l)\ldots d\dot{x}}{\int \rho(x,\dot{x})d\dot{x}}, \tag{6}$$

where $\rho(x,\dot{x})$ is the PDF in the x coordinate system, where "…" denotes possible additional subscripts on the left side and corresponding factors of $\dot{x} - \bar{\dot{x}}$ on the right side, and where all subscripts are integers less than or equal to N. Although (6) is useful in a formal sense, in practical applications all required correlation functions can be computed directly from local time averages of the data (e.g., (4) and (5)), without explicitly computing the data's PDF. Also, note that velocity "correlations" with a single subscript vanish identically

$$C_k(x) = 0. \tag{7}$$

Next, let $M(x)$ be any local $N \times N$ matrix, and use it to define M-transformed velocity correlations, I_{kl} and I_{klmn}

$$I_{kl}(x) = \sum_{1 \le k', l' \le N} M_{kk'}(x) M_{ll'}(x) C_{k'l'}(x), \tag{8}$$

$$I_{klmn}(x) = \sum_{1 \le k', l', m', n' \le N} M_{kk'}(x) M_{ll'}(x) M_{mm'}(x) M_{nn'}(x) C_{k'l'm'n'}(x). \tag{9}$$

Because $C_{kl}(x)$ is generically positive definite at any point x, it is possible to find a particular form of $M(x)$ that satisfies

$$I_{kl}(x) = \delta_{kl} \tag{10}$$

$$\sum_{1 \le m \le N} I_{klmm}(x) = D_{kl}(x), \tag{11}$$

where $D(x)$ is a diagonal $N \times N$ matrix. Such an $M(x)$ can be constructed from the product of three matrices: (1) a rotation that diagonalizes $C_{kl}(x)$, (2) a diagonal rescaling matrix that transforms this diagonalized correlation into the identity matrix, (3) another rotation that diagonalizes

$$\sum_{1 \le m \le N} \tilde{C}_{klmm}(x),$$

where $\tilde{C}_{klmn}(x)$ is the fourth-order velocity correlation ($C_{klmn}(x)$) after it has been transformed by the first rotation and the rescaling matrix. As long as D is not degenerate, $M(x)$ is unique, up to arbitrary *local* permutations and/or reflections. In almost all applications of interest, the velocity correlations will be continuous functions of x. Therefore, in any neighborhood of state space, there will always be a continuous solution for $M(x)$, and this solution is unique, up to arbitrary *global* permutations and/or reflections.

In any other coordinate system x', the most general solution for M' is given by

$$M'_{kl}(x') = \sum_{1 \le m, n \le N} P_{km} M_{mn}(x) \frac{\partial x_n}{\partial x'_l}, \tag{12}$$

where M is a matrix that satisfies (10) and (11) in the x coordinate system and where P is a product of permutation and reflection, matrices. This can be proven by substituting this equation into the definition of $I'_{kl}(x')$ and $I'_{klmn}(x')$ and by noting that these quantities satisfy (10) and (11) in the x' coordinate system because (8) and (9) satisfy them in the x coordinate system. By construction, M is not singular, and, therefore, it has a non-singular inverse.

Notice that (12) shows that the rows of M transform as local covariant vectors, up to global permutations and/or reflections. Likewise, the same equation implies that the columns of M^{-1} transform as local contravariant vectors (denoted as $V_{(i)}(x)$ for $i = 1, \ldots, N$), up to global permutations and/or reflections. As shown in the following, these particular vectors contain significant information about the separability of the data. In fact, they can be used to construct a small group of mappings that includes an unmixing function, if one exists.

2. Use the $V_{(i)}(x)$ to construct a small group of mappings, $\{u(x)\}$, each of which has N components, $u_k(x)$ for $k = 1, \ldots, N$.
One such mapping will be constructed for each way of partitioning the $V_{(i)}$ into two groups (groups 1 and 2), without distinguishing the order of the two groups or the order of vectors within each group. For example, for a three-dimensional system ($N = 3$), three $u(x)$ functions must be constructed, each one corresponding to one of the three distinct ways of partitioning three vectors into two groups: $\{\{V_1\}, \{V_2, V_3\}\}$, $\{\{V_2\}, \{V_1, V_3\}\}$, and $\{\{V_3\}, \{V_1, V_2\}\}$. On the other hand, for two-dimensional systems, there is only one way to divide the vectors into two groups, and, therefore, only one function, $u(x)$, has to be constructed in order to perform BSS.

For each grouping, let N_1 and N_2 denote the number of vectors in groups 1 and 2, respectively, and let G_1 and G_2 denote the collections of N_1 and N_2 values of i for the vectors $V_{(i)}$ in groups 1 and 2, respectively. Each mapping, $u(x)$, is comprised of the union of the components of an N_1-component function, $u_{(1)}(x)$, and the components of an N_2-component function, $u_{(2)}(x)$, which are constructed as described in the next paragraph. For example, for the above-mentioned three-dimensional system, the first mapping, $u(x)$, has three components, comprised of the single component of $u_{(1)}(x)$ and the two components of $u_{(2)}(x)$.

The construction of $u_{(1)}(x)$ is initiated by picking any point x_0 in the x coordinate system. We then find an N_1-dimensional curvilinear subspace, consisting of all points that can be reached by starting at x_0 and by moving along all linear combinations of the local vectors in group 1. This subspace can be described by a function $X(\sigma)$, where σ ($\sigma = \sigma_i$ for $i \in G_1$) parameterizes the subspace by labelling its points in an invertible fashion. Formally, $X(\sigma)$ satisfies the differential equations

$$\frac{\partial X}{\partial \sigma_i} = V_{(i)}(X) \tag{13}$$

for $i \in G_1$ with the boundary condition, $X(0) = x_0$. Then, for each value of σ, we define an N_2-dimensional curvilinear subspace, consisting of all points that can be reached by starting at $X(\sigma)$ and by moving along all linear combinations of the local vectors in group 2. This subspace can be described by a function $Y(\tau)$, where τ ($\tau = \tau_j$ for $j \in G_2$) parameterizes the subspace by labelling its points in an invertible fashion. $Y(\tau)$ satisfies the differential equation

$$\frac{\partial Y}{\partial \tau_j} = V_{(j)}(Y) \tag{14}$$

for $j \in G_2$ with the boundary condition, $Y(0) = X(\sigma)$. Finally, the function $u_{(1)}(x)$ is defined so that it is constant on each one of the Y subspaces. Specifically, $u_{(1)}(x) = \sigma$ whenever x is in the Y subspace containing $X(\sigma)$. The function $u_{(2)}(x)$ is defined by following an analogous procedure in which the roles of groups 1 and 2 are switched. Finally, the union of the N_1 components of $u_{(1)}(x)$ and the N_2 components of $u_{(2)}(x)$ is taken to define the mapping, $u(x)$, that corresponds to the chosen grouping of the vectors, $V_{(i)}$, into groups 1 and 2.

The foregoing procedure can be illustrated by considering the construction of $u_{(1)}(x)$, corresponding to the first grouping of vectors in the three-dimensional case mentioned in the previous paragraph. In that case: (a) $X(\sigma)$ describes a curved line that passes through x_0, that is parallel to $V_{(1)}(x)$ at each point, and that is parameterized by σ; (b) each function, $Y(\tau)$, describes a curved surface, which intersects that curved line at some value of the parameter σ and which is parallel to all linear combinations of $V_{(2)}(x)$ and $V_{(3)}(x)$ at each point; (c) along each of these curved surfaces, $u_{(1)}(x)$ is equal to the corresponding value of σ. Likewise, for the construction of $u_{(2)}(x)$ in the three-dimensional case: (a) $X(\sigma)$ describes a curved surface that passes through x_0, that is parameterized by the two components of σ, and that is parallel to all linear combinations of $V_{(2)}(x)$ and $V_{(3)}(x)$ at each point; (b) each function $Y(\tau)$ describes a curved line, which intersects that surface at a value of the parameter σ and which is parallel to $V_{(1)}(x)$ at each point; (c) along each of these curved lines, $u_2(x)$ is equal to the corresponding value of σ.

3. Use each constructed mapping, $u(x)$, to transform the measured time series, $x(t)$.
For each $u(x)$, the transformed time series, $u[x(t)]$, is the union of the N_1 components of $u_{(1)}[x(t)]$ and the N_2 components of $u_{(2)}[x(t)]$.

4. Determine if at least one mapping, $u(x)$, is comprised of two statistically independent variables, $u_{(1)}(x)$ and $u_{(2)}(x)$.
Specifically, determine if at least one transformed time series, $u[x(t)]$, has a PDF that factorizes as

$$\rho_U(u, \dot{u}) = \prod_{a=1,2} \rho_{Ua}(u_{(a)}, \dot{u}_{(a)}). \tag{15}$$

Here, u denotes $u[x(t)]$, and $\dot{u}$ is its time derivative. Alternatively, we can compute a large set of correlations of multiple components of each transformed time series and then determine if they are products of lower-order correlations of two subsystems, as required by (15).

5. Use the result of step 4 to determine if the data are separable and, if they are, to determine an unmixing function. Specifically, if at least one mapping, $u(x)$, describes two statistically independent variables ($u_{(1)}(x)$ and $u_{(2)}(x)$), it is obvious that the data are separable and $u(x)$ is an unmixing function. On the other hand, if none of the mappings in $\{u(x)\}$ is comprised of two statistically independent variables, $u_{(1)}(x)$ and $u_{(2)}(x)$, the data are inseparable in any coordinate system.

This last statement is a consequence of the following fact, which is proved in the next two paragraphs: namely, if the data are separable, at least one constructed mapping, $u(x)$, describes a pair of statistically independent variables, $u_{(1)}(x)$ and $u_{(2)}(x)$.

The only remaining task is to prove the result asserted above: if the data are separable, at least one of the ways of grouping local vectors leads to a mapping, $u(x)$, that is an unmixing function. The first step is to show that the matrix M and the local vectors have simple forms in the separable (s) coordinate system. In particular, we prove that the following block-diagonal matrix is the M matrix in the s coordinate system

$$M_S(s) = \begin{pmatrix} M_{S1}(s_{(1)}) & 0 \\ 0 & M_{S2}(s_{(2)}) \, . \end{pmatrix} \tag{16}$$

Here, each submatrix M_{Sa} is the M matrix derived from the correlations between components of the corresponding subsystem state variable, $s_{(a)}$. For example, in the case of a separable three-dimensional system, (16) asserts that M_S consists of 1×1 and 2×2 blocks, which are the M matrices of one-dimensional and two-dimensional subsystems, respectively. In order to prove (16) in the general case, it is necessary to show that M_S satisfies (10) and (11) in the s coordinate system. To do this, first note that (2), (6), and (7) imply that velocity correlations vanish in the s coordinate system if their indices contain a solitary index from any one block. It follows that $C_{Skl}(s)$ consists of two blocks, each of which contains the second-order velocity correlations of an independent subsystem. This implies that (16) satisfies the constraint (10), because, by definition, each block of M_S transforms the corresponding block of C_{Skl} into an identity matrix. In order to prove that (16) satisfies (11), substitute it into the definition of

$$\sum_{1 \leq m \leq N} I_{Sklmm} . \tag{17}$$

Then, note that: (1) when k and l belong to different blocks, each term in this sum vanishes because it factorizes into a product of correlations, one of which has a single index and, therefore, must vanish because of (7); (2) when k and l belong to the same block and are unequal, each term with m in any other block contains a factor equal to I_{Skl}, which vanishes for $k \neq l$, as proved above; (3) when k and l belong to the same block and are unequal, the sum over m in the same block vanishes, because each block of M_S is defined to satisfy (11) for the corresponding subsystem. This completes the proof that M_S satisfies (10) and (11). It follows that M_S is the M matrix in the s coordinate system, as asserted above.

Recall that the local vectors in the s coordinate system are columns of the matrix, M_S^{-1}. Because of the block diagonality of M_S^{-1}, the local vectors can be sorted into two groups (groups 1 and 2) that consist of the columns passing through blocks 1 and 2, respectively, and that contain N_1 and N_2 vectors, respectively. Therefore, at each point s, the local vectors in the first group are linear combinations of the unit vectors parallel to the first N_1 axes of the s coordinate system, and the local vectors in the second group are linear combinations of the unit vectors parallel to the last N_2 axes of the s coordinate system. Hence, in the s coordinate system, the function, $X(\sigma)$, which was used to define $u_{(1)}(x)$, describes the linear subspace that contains the point $s[x_0]$ and that is spanned by the first group of unit vectors. Likewise, each $Y(\tau)$, which was used to define $u_{(1)}(x)$, describes a linear subspace that contains $s[X(\sigma)]$ for some value of σ and that is spanned by the second group of unit vectors. This implies that the state variable of the first subsystem, $s_{(1)}(x)$, is constant within each Y subspace, being equal to the value of $s_{(1)}(x)$ at the intersection of that Y subspace with the X subspace. But, recall that $u_{(1)}(x)$ is also constant within each Y subspace, being equal to the value of σ at its intersection with the X subspace. Therefore, because σ is defined to be invertibly related to the points in the X subspace and because the values of $s_{(1)}$ are also invertibly related to the points in the X subspace, these paired values must be invertibly related to one another; i.e., $\sigma = h_1(s_{(1)})$ where h_1 is an invertible function. It follows that $u_{(1)}(x)$ and $s_{(1)}(x)$ must also be invertibly related at each point; i.e., $u_{(1)}(x) = h_1[s_{(1)}(x)]$. In a similar manner, it can be shown that $u_{(2)}(x)$ and $s_{(2)}(x)$ are also related by some invertible function. Because $s_{(1)}$ and $s_{(2)}$ are the state variables of independent subsystems and because $u_{(1)}$ and $u_{(2)}$, respectively, are invertibly related to them, $u_{(1)}$ and $u_{(2)}$ must be subsystem state variables in some other subsystem coordinate systems. This completes the proof of the assertion at the beginning of the previous paragraph: namely, if the data are separable, at least one way of grouping the local vectors (e.g., the grouping corresponding to the above-mentioned blocks) leads to a mapping, $u(x)$, that describes a pair of statistically independent state variables ($u_{(1)}$ and $u_{(2)}$).

3 Experiments

In this section, the new BSS technique is illustrated by using it to blindly disentangle nonlinear mixtures of the audio waveforms of two male speakers. Each speaker's waveform was 30 s long and consisted of an excerpt from an audio book recording. The waveform of each speaker, denoted $s_k(t)$ for $k = 1$ or 2, was sampled 16,000 times per second with two bytes of depth. The thick gray lines in Fig. 1 show the two speakers' waveforms during a short (30 ms) interval. These waveforms were then mixed by the nonlinear functions

$$\mu_1(s) = 0.763s_1 + (958 - 0.0225s_2)^{1.5}$$
$$\mu_2(s) = 0.153s_2 + (3.75 * 10^7 - 763s_1 - 229s_2)^{0.5}, \tag{18}$$

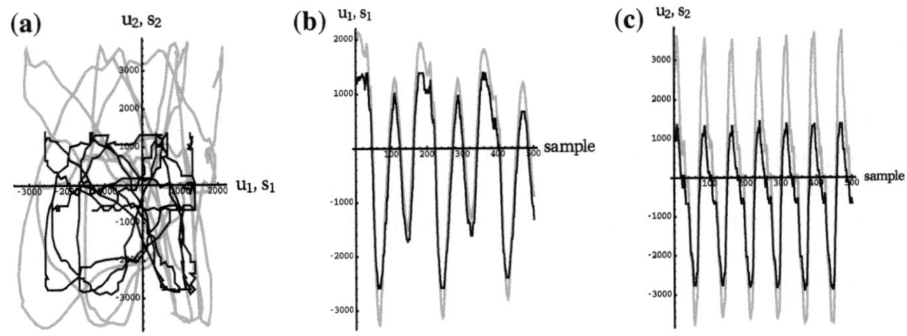

Fig. 1 **a** The thick gray line depicts the trajectory of 30 ms of the two speakers' unmixed speech in the s coordinate system, in which each component is equal to one speaker's speech amplitude. The thin black line depicts the waveforms (u) of the two speakers during the same time interval, recovered by blindly processing their nonlinearly mixed speech. Panels **b** and **c** show the time courses of s_1 and u_1 and of s_2 and u_2, respectively

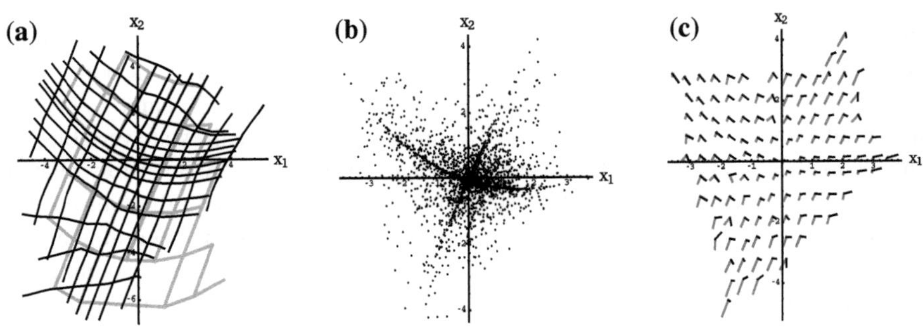

Fig. 2 **a** The thick gray curves comprise a regular Cartesian grid of lines in the s coordinate system, after they were nonlinearly mapped into the x coordinate system by the mixing in (18). The thin black lines depict lines of constant u_1 or of constant u_2, where u denotes a possibly separable coordinate system derived from the measurements. **b** A random subset of the measurements along the trajectory of the mixed waveforms, $x(t)$. **c** The thick gray and thin black lines show the local vectors, $V_{(1)}$ and $V_{(2)}$, respectively, after they have been uniformly scaled for the purpose of display

where $-2^{15} \leq s_1, s_2 \leq 2^{15}$. This is one of a variety of nonlinear transformations that were tried with similar results. The measurements, $x_k(t)$, were taken to be the variance-normalized, principal components of the sampled waveform mixtures, $\mu_k[s(t)]$. Figure 2a shows how this nonlinear mixing mapped an evenly-spaced Cartesian grid in the s coordinate system onto a warped grid in the x coordinate system. Figure 2b shows the distribution of measurements created by randomly sampling $x(t)$, and Fig. 3 shows the time course of $x(t)$ during the same short time interval depicted in Fig. 1. When either waveform mixture ($x_1(t)$ or $x_2(t)$) was played as an audio file, it sounded like a confusing superposition of two voices, which were quite difficult to understand.

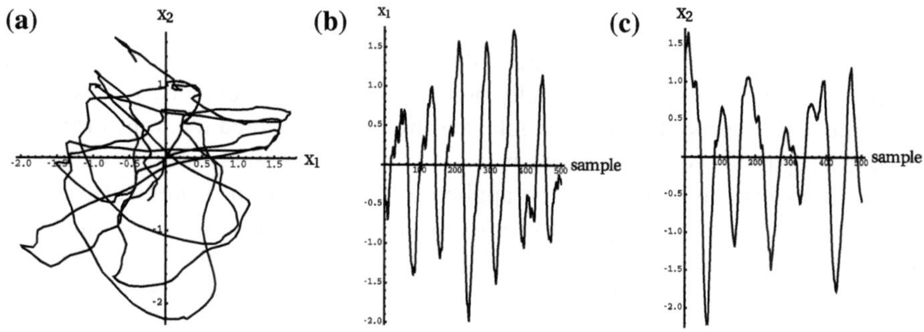

Fig. 3 **a** The trajectory of measurements, $x(t)$, during the 30 ms time interval depicted in Fig. 1. Panels **b** and **c** show the time courses of x_1 and x_2, respectively

The proposed BSS technique was then applied to these measurements as follows:

1. The entire set of 500,000 measurements, consisting of x and $\dot{x}$ at each sampled time, was sorted into a 16×16 array of bins in x-space. Then, the $\dot{x}$ distribution in each bin was used to compute local velocity correlations (see (4) and (5)), and these were used to derive M and $V_{(i)}$ for each bin. Figure 2c shows these local vectors at each point.
2. These vectors were used to construct the mapping, $u(x)$. As described in Method, the first step was to choose some point x_0 and then use the vectors $V_{(1)}(x)$ to construct the curvilinear line, $X(\sigma)$. Then, for each point σ on this curve, the local vectors $V_{(2)}(x)$ were used to construct a curvilinear line, $Y(\tau)$. Along each of these Y curves, $u_1(x)$ was defined to be a constant equal to the value of σ at the curve's point of intersection with $X(\sigma)$. The mapping, $u_2(x)$, was defined by an analogous procedure. In this way, each point x was assigned values of both u_1 and u_2, thereby defining the mapping, $u(x)$. A group of the thin black lines in Fig. 2a depict a family of curves having constant values of u_1, which are evenly-spaced and increase as one moves from curve to curve in the family. Figure 2a also shows a family of curves having constant values of u_2, which are evenly-spaced and increase as one moves from curve to curve in the family.
3. As proved in Method, if the data are separable, $u(x)$ must an unmixing function. Therefore, the separability of the data could be determined by seeing if $u[x(t)]$ has a factorizable density function (or factorizable correlation functions). If the density function does factorize, the data are patently separable, and the components of $u[x(t)]$ describe the evolution of the independent subsystems. On the other hand, if the density function does not factorize, the data must be inseparable.

In this illustrative example, the separability of the u coordinate system was verified by a more direct method. Specifically, Fig. 2a shows that the isoclines for increasing values of u_1 (or u_2) nearly coincide with the isoclines for increasing values of s_1 (or s_2). This demonstrates that the u and s coordinate systems differ by component-wise transformations of the form: $(u_1, u_2) = (h_1(s_1), h_2(s_2))$ where h_1 and h_2 are monotonic functions. Because the data are separable in the s coordinate system and because component-wise transformations do not affect separability, the data must also be separable in the u coordinate system. Therefore, we have accomplished the objectives of BSS: namely, by blindly processing the measurements, $x(t)$, we have determined that the system is separable, and we have computed the transformation, $u(x)$, to a separable coordinate system.

The transformation, $u(x)$, can be applied to the mixture measurements, $x(t)$, to recover the original unmixed waveforms, up to component-wise transformations. The resulting waveforms, $u_1[x(t)]$ and $u_2[x(t)]$, are depicted by the thin black lines in Fig. 1, which also shows the trajectory of the unmixed waveforms in the s coordinate system. Notice that the two trajectories, $u[x(t)]$ and $s(t)$, are similar except for component-wise transformations along the two axes. The component-wise transformation is especially noticeable as a stretching of $s_2(t)$ with respect to $u_2[x(t)]$ along the positive s_2 axis. When each of the recovered waveforms, $u_1[x(t)]$ and $u_2[x(t)]$, was played as an audio file, it sounded like a completely intelligible recording of one of the speakers. In each case, the other speaker was not heard, except for a faint "buzzing" sound in the background. Therefore, the component-wise transformations, which related the recovered waveforms to the original unmixed waveforms, did not noticeably reduce intelligibility.

4 Conclusion

This paper describes how to determine if time-dependent signal measurements are comprised of linear or nonlinear mixtures of the state variables of statistically independent subsystems. Specifically, the measurement time series is used to derive a small number of mappings, which must include a transformation to a separable coordinate system, if one exists. Therefore, separability can be determined by testing the separability of the data, after they have been transformed by each of these mappings.

Some comments on this result:

1. Most other approaches to nonlinear BSS are model-dependent because they assume that the mixing function has a specific parametric form [1]. In contrast, the BSS method described in this paper is model-independent in the sense that it can be used to separate data that were mixed by any invertible diffeomorphic mixing function.
2. Notice that the proposed method is analytic and constructive, in contrast to the iterative techniques that are commonly used in the literature [1].

References

1. Comon, P., Jutten, C. (eds.): Handbook of Blind Source Separation, Independent Component Analysis and Applications. Academic Press, Oxford (2010)
2. Jutten, C., Karhunen, J.: Advances in blind source separation (BSS) and independent component analysis (ICA) for nonlinear mixtures. Int. J. Neural Syst. **14**, 267–292 (2004)

3. Levin, D.N.: Using state space differential geometry for nonlinear blind source separation. J. Appl. Phys. **103**, art. no. 044906 (2008)
4. Hyvärinen, A., Pajunen, P.: Nonlinear independent component analysis: existence and uniqueness results. Neural Netw. **12**, 429–439 (1999)
5. Ehsandoust, B., Babaie-Zadeh, M., Jutten, C.: Blind source separation in nonlinear mixture for colored sources using signal derivatives. In: Vincent, E., et al. (eds.) Latent Variable Analysis and Signal Separation, LNCS 9237, Springer, pp. 193–200 (2015)
6. Lagrange, S., Jaulin, L., Vigneron, V., Jutten, C.: Analytic solution of the blind source separation problem using derivatives. In: Puntonet, C.G., Prieto, A.G. (eds.) Independent Component Analysis and Blind Signal Separation, LNCS, vol. 3195, pp. 81–88. Springer, Heidelberg (2004)
7. Levin, D.N.: Performing nonlinear blind source separation with signal invariants. IEEE Trans. Signal Process. **58**, 2131–2140 (2010)
8. Levin, D.N.: Model-independent analytic nonlinear blind source separation (2017). http://arxiv.org/abs/1703.01518
9. Levin, D.N.: Nonlinear blind source separation using sensor-independent signal representations. In: Proceedings of ITISE 2016: International Work-Conference on Time Series Analysis, Granada, Spain, pp. 84–95, 27–29 June 2016

Dantzig-Selector Radial Basis Function Learning with Nonconvex Refinement

Tomojit Ghosh, Michael Kirby and Xiaofeng Ma

Abstract This paper addresses the problem of constructing nonlinear relationships in complex time-dependent data. We present an approach for learning nonlinear mappings that combines convex optimization for the model order selection problem followed by non-convex optimization for model refinement. This approach exploits the linear system that arises with radial basis function approximations. The first phase of the learning employs the Dantzig-Selector convex optimization problem to determine the number and candidate locations of the RBFs. At this preliminary stage maintaining the supervised learning relationships is not part of the objective function but acts as a constraint in the optimization problem. The model refinement phase is a non-convex optimization problem the goal of which is to optimize the shape and location parameters of the skew RBFs. We demonstrate the algorithm on on the Mackey-Glass chaotic time-series where we explore time-delay embedding models in both three and four dimensions. We observe that the initial centers obtained by the Dantzig-Selector provide favorable initial conditions for the non-convex refinement problem.

Keywords Dantzig-Selector · Chaotic time-series prediction · Sparse radial basis functions · Model order selection · Mackey-Glass equation

1 Introduction

We are interested in learning patterns and nonlinear structures exhibited by high-dimensional trajectories for time-series prediction. In general, the temporally ordered data can be sampled in high-dimensions by observing the evolution of a dynamical system. Alternatively, scalar values of a high-dimensional process can be collected, and subsequently used to reconstruct the geometric structure, if it exists, of the full system which can't be observed.

T. Ghosh · M. Kirby (✉) · X. Ma
Colorado State University, Fort Collins, CO 80523, USA
e-mail: Michael.Kirby@Colostate.Edu

I. Rojas et al. (eds.), *Advances in Time Series Analysis and Forecasting*,
Contributions to Statistics, DOI 10.1007/978-3-319-55789-2_22

The motivation for this approach is provided by a theoretical result in dynamical systems known as Takens' theorem [1] as well as evidence that this is a practical tool for reconstructing the geometry of time series [2]. A tutorial introduction to Takens' theorem is provided in [3]. It has been characterized as a tool for linking the analysis of experimental time-series and the theory of dynamical systems. Loosely speaking, Takens theorem states that if the data corresponds to the evolution of a trajectory on a smooth manifold of dimension m, then a $2m + 1$-tuple of time ordered samples can be used to reconstruct a *copy* of the geometric structure that can't be otherwise observed. Takens' theorem rigorously establishes this surprising and inherently useful idea.

Takens' theorem may be viewed as a geometric "bridge" that allows us to map scalar observations to structures in high-dimensions using the "method of delays" [3]. It provides a formalism for taking ordered samples $\{x_1, x_2, x_3, \dots\}$ and constructing delay vectors, e.g., (x_n, x_{n-T}, x_{n-2T}). For example, given an appropriate delay time T, a copy of topological circle ($m = 1$) could be reconstructed in three dimensions using this delay embedding.

This embedding strategy has natural consequences for the time-series prediction problem, the goal of which is to determine the future value x_{n+T} from previous values $\{x_n, x_{n-T}, x_{n-2T}, \dots\}$. Having additional mathematical rigor and insight supporting this framework led to a flurry of activity related to constructing nonlinear models for prediction using a variety of techniques including artificial neural networks, see, e.g., [4]. In this setting time-series prediction is a data fitting problem, i.e., we are concerned with the construction of nonlinear mappings of the form

$$x_{n+T} = f(x_n, x_{n-T}, x_{n-2T}, \dots)$$

from observed data. The issue now is that this data fitting problem is very challenging. Determining f from data alone results immediately in several difficult questions, not the least of which is how many parameters are required for the model?

An attractive approach for modeling such time-series data sets, proposed in [5], is provided by skew Radial Basis Function (RBF) expansions of the form

$$f(x) = w_0 + \sum_{k=1}^{m} w_k z(\Lambda_k^T (x - c_k))\phi(\|x - c_k\|_{W_k}). \tag{1}$$

The addition of the skew term z provides additional shape to the RBFs making them capable of fitting more complex data with fewer terms. See Fig. 1 for a contrast between the radially symmetric Gaussian RBF and its skew counterpart. As a compelling example of this fact, it was demonstrated that a standard RBF expansion requires $N_c = 13$ non-skew RBFs using

$$f(x) = w_0 + \sum_{k=1}^{N_c} w_k \phi_k(\|x - c_k\|) \tag{2}$$

to approximate a single skew Gaussian function [5].

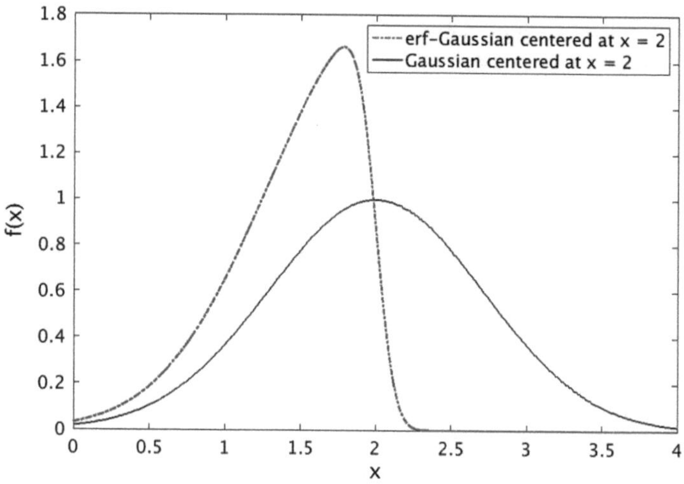

Fig. 1 The regular Gaussian function centered at x = 2 compared to the erf-Gaussian skew function centered at x = 2 with $\lambda = -7$. See [5]

In these definitions above we assume that f is of the form

$$f : U \subset \mathbb{R}^n \to V \subset \mathbb{R}$$

The number of basis functions is denoted N_c, and their locations $\{c_k\}$, are generally unknown and must be determined empirically from the data. The radial basis functions ϕ can be selected from a family of functions, either with or without shape parameters. For example, thin-plate spline RBFs have the form $\phi(r) = r^2 \ln r$ and no parameters, while Gaussian RBFs have the form $\phi(r) = \exp(-r^2/\alpha^2)$ where the parameter α determines the width of the Gaussian. When different RBFs are used at each center we use the notation ϕ_k to emphasize the presence of, e.g., potentially different parameter values.

In general, shape parameters provide RBFs with more flexibility to fit the data at the expense of having additional computation cost to estimate them. RBF expansions as described by Eq. (1) allow for significant additional parametric complexity [5, 6]. This can be very effective for improving data fitting given breaking the symmetry of the RBFS greatly enhances the representational power. This approach has been seen to have significant impact when data has edges, or even discontinuities [5–7]. The main issue that remains today with using data driven techniques for determining nonlinear functions f from data is the fact that the model order, N_c in the case of radial basis functions, is unknown and hard to discover empirically from the data.

In this paper we demonstrate how the Dantzig-Selector RBF model with nonconvex refinement is a very attractive framework for addressing the model order determination problem. We will show by example that this method is capable of modeling the chaotic dynamics of the Mackey-Glass Equation with high accuracy.

The Dantzig-Selector optimization problem that was originally suggested in the context of compressive sensing [8, 9]. Here we leverage these ideas to create an automated nonlinear function fitting tool. Our methodology requires two basic steps. The first phase involves solving the Dantzig-Selector optimization problem for model order determination and an initialization of RBF center location. The second phase implements a nonlinear optimization refinement of the phase one output tuning the RBF shape and location parameters to better fit the data. While we focus on the time-delayed embedding paradigm of the time-series prediction problem, the proposed algorithm does not require a one-dimensional parameterization of the data. It is applicable to the general setting of supervised learning, i.e., mapping given input-output pairs for the discovery of nonlinear relationships [12]. The approach is also conducive to representing data as the graph of a function in the spirit of Whitney's theorem as outlined in [10, 11].

This paper extends the preliminary results presented in [12]. There the focus was on illustrating the behavior of the algorithm on simple synthetic examples and exploring different norm constraints. Here, we focus exclusively on the Dantzig-Selector with nonconvex refinement and benchmark this algorithm on the Mackey-Glass Equation that has served as a tool for RBF algorithm development and evaluation in the setting of time-series prediction.

2 The Radial Basis Function Optimization Problem

One of the attractive features of radial basis function expansions for nonlinear function approximation is the versatility with which one can approach the optimization problem to determine the function parameters. Centers can be randomly selected from the data and clustering algorithms applied to refine the model. Or, more computational effort can be made to optimize center locations using gradient descent on the error function. Weights can also be solved in isolation using least-squares techniques including the computation of the pseudo-inverse, or they may be integrated with the other variables in a gradient based minimization problem.

Despite this flexibility, one of the main complications of RBF modeling is determining how complex the representation should be for a given set of data. *How many radial basis functions are required?* Substantial progress has been made in this direction with a growing algorithm that terminates when the residuals are deemed to be noise [5, 6, 13]. Here we take a different route using recent ideas from sparsity promoting optimization [14–16].

At the heart of this problem is the fact that learning the center location and shape parameters in a Radial Basis Function expansion is an inherently non-convex optimization problem. Applying the interpolation condition $y^{(\mu)} = f(x^{(\mu)})$ at each data point $x^{(\mu)}$, $\mu = 1, \dots, P$, gives rise to the generally under-determined system

$$\Phi w = b$$

where the μ'th row is $\Phi_i = (1, \phi(\|x^{(\mu)} - c_1\|_2), \dots, \phi(\|(x^{(\mu)} - c_{N_c}\|_2))$ and $b = (y^{(1)}, \dots, y^{(P)})^T$. If the centers are determined and fixed, as well as the shape parameters, then one solves the convex problem

$$w^* = \arg \underset{w}{\text{minimize}} \, \|\Phi w - y\|_2$$

for the weights. This convex optimization problem for w has the well-known solution $w^* = \Phi^{\dagger} x$ where $\Phi^{\dagger}$ is the pseudo-inverse, see, e.g., [17]. The expansion weights w_k play a special role in this optimization problem in the sense that they can be determined either as the solution to a linear system, or can be determined incrementally using gradient descent.

However, the identification of the number of centers in the model N_c, known as the *model order determination problem*, is a non-convex optimization problem. Given a number of centers N_c the determination of the optimal location of these centers is also a non-convex optimization problem which can be done simultaneously with the weights via

$$(w^*, \{c_1^*, \dots, c_{N_c}^*\}) = \arg \underset{w, \{c_k\}}{\text{minimize}} \, \|\Phi w - y\|_2$$

In practice, the RBF training algorithms generally go between two extremes: picking centers randomly leaving only the least squares computation, or solving the fully nonlinear optimization problem. We propose a mixed algorithm that involves a preliminary convex step for model order determination and center location using sparsity promoting optimization, followed by a non-convex phase of model refinement. This algorithm exploits the full flexibility of the RBF structure, where parameter dependencies are both linear and nonlinear.

2.1 Dantzig-Selector Optimization for Model Order Selection

In particular, we are concerned with determining sparse representations for time-series data using functions of the form given by Eqs. (1) and (2) by solving

$$\text{minimize} \, \|w\|_1 \tag{3}$$

subject to

$$\|\Phi^T \Phi w - \Phi^T b\|_\infty \le \epsilon. \tag{4}$$

This convex optimization problem appears in a different context in compressed sensing and has been referred to as the *Dantzig-Selector* [8, 9].

The use of the one-norm for the expansion coefficients w_i is seen to be a sparsity promoting feature, i.e., the geometry of the solution favors zero entries in w [14, 15]. The ℓ_1-norm serves as a proxy for the ℓ_0-norm. This is a class of problems that can be solved using ideas from convex optimization [18]. Indeed, this problem can be readily converted to the linear program of the form

$$\text{minimize } c^T x$$

subject to

$$Ax \leq b.$$

Let $R = \Phi^T \Phi$ and $r = \Phi^T b$, with $f = r + \epsilon e_n, g = -r + \epsilon e_n$ and where $e_n^T = (1 \ \dots, 1)$ is a vector of n ones. Then we have

$$A = \begin{pmatrix} I_{n \times n} & -I_{n \times n} \\ -I_{n \times n} & -I_{n \times n} \\ R & 0_{n \times n} \\ -R & 0_{n \times n} \end{pmatrix}$$

and

$$b = \begin{pmatrix} 0_n \\ 0_n \\ f \\ g \end{pmatrix}, \quad c = \begin{pmatrix} 0_n \\ e_n \end{pmatrix}.$$

In prior work we found by empirical exploration that the ℓ_∞ constraint outperformed other p-norms for our problem [12]. Although we maintain our focus on the ℓ_∞ norm in this paper, we certainly consider the other norms still of potential interest.

Setting a constraint bound on ϵ is required to implement this linear program. We note that if ϵ is taken too small, then the LP will be infeasible. If it is taken too large, then we expect the resulting weights to be less relevant to the final objective of fitting the data. We view the value

$$\epsilon_c = \|\Phi w^* - b\|_\infty$$

where w^* is the solution to the least squares problem to be a good candidate value for ϵ in the linear program.

3 The Prediction of Chaotic Time-Series

The Mackey-Glass Equation proposed in [19] models a physiological control system in terms of first-order nonlinear differential-delay equations. The resulting time-series has been shown to be *chaotic*, i.e., in particular it is extremely sensitive to initial conditions in the sense that two nearby trajectories separate exponentially

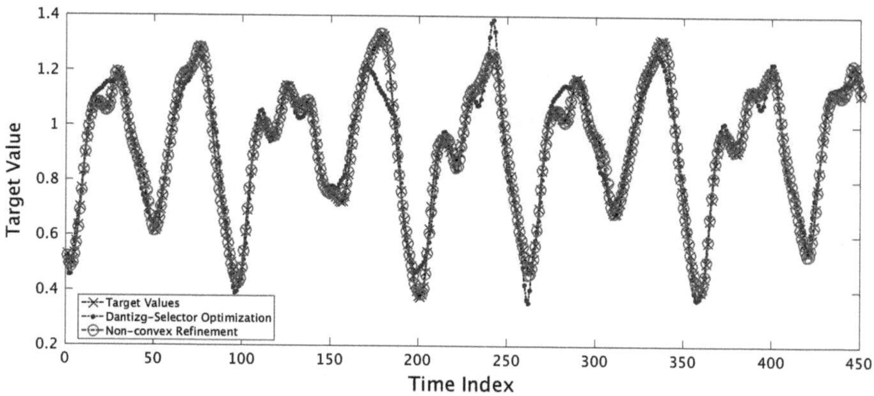

Fig. 2 The output of Dantzig-Selector skew RBF model with non-convex refinement for the testing data compared to the target values and output of Dantzig-Selector skew RBF model without non-convex refinement

fast. This time series has been widely used in benchmarking studies as described in [6, 13, 20], and references therein.

The Mackey-Glass time-delay equation

$$\frac{dx}{dt} = bx(t) + \frac{ax(t-\tau)}{1 + x(t-\tau)^{10}}$$

generates a chaotic time series that exhibits sensitive dependence on initial conditions. The time series is generated by integrating the equation with model parameters $a = 0.2, b = 0.1$, and $\tau = 17$ using the trapezoidal rule with $\Delta t = 1$, with initial conditions $x(t-\tau) = 0.3$ for $0 \leq t \leq \tau$.

Using the Mackey-Glass Equation, we generated 1000 scalar values for training the RBFs and 450 points for testing, as shown in Fig. 2. Following [6], and references therein, a four dimensional time delayed embedding is used where

$$x_{n+50} = f(x_n, x_{n-6}, x_{n-12}, x_{n-18}),$$

i.e., four points separate by 6 samples are predicting 50 steps ahead in time. An autocorrelation analysis was used to determine the delay time $T = 6$ and numerical embedding tools determined that the appropriate embedding dimension should be four. For the purposes of exploring the behavior of the proposed algorithm, we consider three and four dimensional embeddings in this paper.

The Dantzig-Selector selected 34 Gaussian RBFs each with initial Gaussian width 0.5. For the non-convex refinement, a skewing term $z(x; \Lambda_k)$ modulates each of the Gaussian RBFs, i.e., the model Equation (1) employs

$$z(\rho) = \frac{1}{\pi}\arctan(\rho) + \frac{1}{2}$$

Table 1 Accuracy on the test data set for the 4 dimensional embedding using skew RBFs

Training phase	Dantzig-Selector	Non-convex(iter = 75)	iter = 200	iter = 500	iter = 1000
RMSE	5.1e-2	3.95e-2	2.39e-2	1.39e-2	7.2e-3

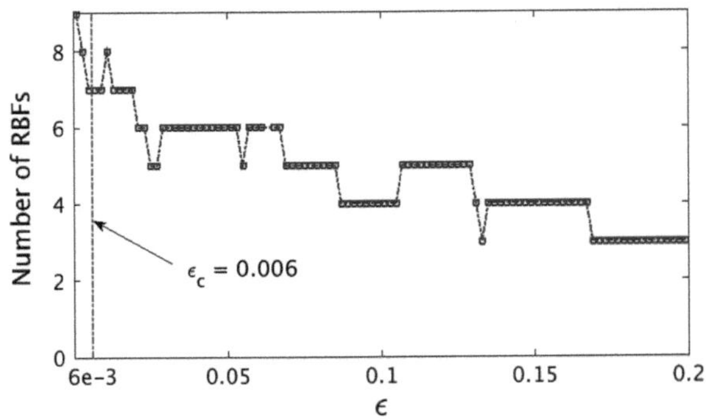

Fig. 3 The number of selected centers using Dantzig-Selector optimization as a function of tolerance ϵ on constraints while the center candidates are fixed. The *red vertical line* is the ℓ_∞-norm of the pseudo-inverse error, i.e., ϵ_c. In this experiment the domain is three dimensional with delay $T = 1$ and predicting one-step ahead

in addition to

$$\phi(\mu) = \exp(\frac{-\mu^2}{\alpha^2}).$$

See [6] and references therein for additional details and options for skew functions. After 1000 iterations of non-convex refinement using scaled conjugate gradient descent, an RMSE of 0.0072 is obtained. Table 1 shows how the RMSE on the test data is decreasing over the training phases. The non-convex training improves the overall accuracy by nearly an order of magnitude.

In Figs. 3 and 4 we explore the impact of varying ϵ on the solution to the Dantzig-Selector algorithm. We use 1000 data points for training. The RBFs are picked to be Gaussian with unit width. We randomly sample 400 initial centers from the domain of the training data and keep these fixed across different values of ϵ. The tolerance for the constraint ϵ is varying from 0.001 to 0.2. The domain is three dimensional. In Fig. 3 we see the impact of the value of ϵ on the number of centers selected, i.e., fewer centers are selected as the constraint is relaxed. As can be seen from Fig. 4, the minimum RMSE on test data is obtained when the tolerance ϵ is near the pseudo-inverse established ϵ_c, which might suggest that we should start with the pseudo-inverse predicted epsilon. With fewer centers we see the expected result of the RMSE growing as a function of ϵ.

In Fig. 5, we explore the sensitivity of the Dantzig-Selector algorithm on the choice of the the Gaussian width α. For simplicity, we take each α to be the same in

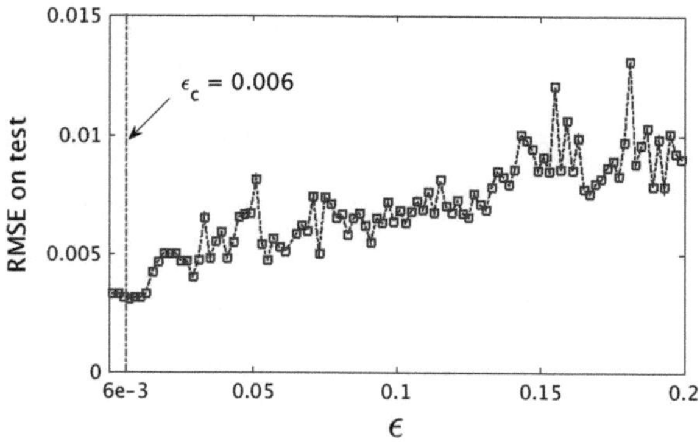

Fig. 4 The RMSE of the Dantzig-Selector RBF model with non-convex refinement applied on the test data as a function of tolerance ϵ on constraints to the Dantzig-Selector while the initial center candidates are fixed. The *red vertical line* is the ℓ_∞-norm pseudo-inverse error ϵ_c. The domain is three dimensional with a time delay $T = 1$ and predicting one-step ahead

Fig. 5 RMSE of the Dantzig-Selector RBF model applied on the test data as a function of Gaussian width

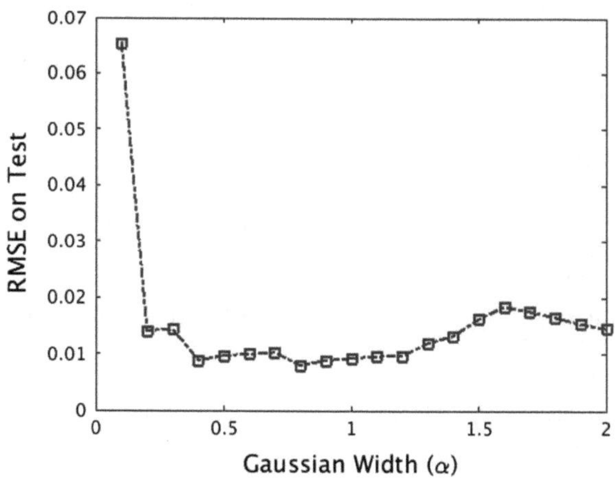

the convex phase of the algorithm. To establish that this is a reasonable approach, we build models for a range of $\alpha \in [0.1, 2]$. The Dantzig-Selector is solved with 1000 data points each embedded into three dimensions. A total of 400 initial centers are randomly sampled from the domain of the training data; they are not selected to be training points. Figure 5 shows the RMSE of the model produced by the convex optimization problem (before refinement) on testing data as a function with respect to Gaussian width. From this experiment, we can see that RMSE varies with respect to the Gaussian width but but produces relatively robust results over a range of α, i.e., the RMSE is more or less constant. The RMSE on the test data after convex refinement is similarly robust, but does increase (not unexpectedly) in the vicinity of $\alpha > 1.2$.

Lastly, we discuss the numerical experiments using standard thin-plate spline RBFs, i.e., without any skewing shape parameters. In Table 2 we present results on

Table 2 Results on the 3D data using delay $T = 1$ and predicting one-step ahead with thin-plate spline RBFs. The model is trained on 1020 points and tested on the following 300 data points in the time-series

Initial no. of centers	250	500	750	1000	1250	1500
No. of selected centers	17	18	17	18	17	17
Dantzig Selector RMSE	4.264e-03	3.689e-03	3.670e-03	3.457e-03	3.476e-03	3.694e-03
Refined RMSE 75 iters	3.552e-03	3.280e-03	2.989e-03	3.033e-03	2.947e-03	3.348e-03

Table 3 Results on the 4D data using delay $T = 6$ and predicting fifty steps ahead with thin-plate spline RBFs. The model is trained on 1020 points and tested on the following 300 data points in the time-series

Initial no. of centers	250	500	750	1000	1250	1500
No. of selected centers	33	37	38	35	35	38
Dantzig Selector RMSE	3.9590e-02	3.6736e-02	3.5396e-02	3.2442e-02	3.4522e-02	3.8012e-02
Refined RMSE 75 iters	3.3293e-02	3.2699e-02	3.1988e-02	3.1079e-02	3.1909e-02	3.2941e-02
Refined RMSE 250 iters	2.9881e-02	2.8668e-02	2.9044e-02	2.9701e-02	3.0114e-02	3.0013e-02
Refined RMSE 500 iters	2.7277e-02	2.7423e-02	2.6548e-02	2.6988e-02	2.7699e-02	2.7469e-02
Refined RMSE 750 iters	2.6414e-02	2.4848e-02	2.4703e-02	2.4971e-02	2.6312e-02	2.4972e-02
Refined RMSE 1000 iters	2.5657e-02	2.4071e-02	2.3482e-02	2.3993e-02	2.5507e-02	2.4161e-02

the 3D embedding of the Mackey-Glass time-series. The experiment here has been changed to look only one step ahead, not 50. Also, we use $T = 1$ for the delay. We note that the accuracy for the RBF is already very high after the Dantzig-Selector.

This fact, in addition to the fact that there are no shape parameters to tune, only center locations, results in only modest accuracy improvements during the non-convex learning phase of the algorithm. We also see that the number of centers selected by the Dantzig-Selector is steady at approximately 17 as we vary the initial number of centers used. We see similar behavior using the 4 dimensional embedding in Table 3. In this case the errors after the Dantzig-Selector are even lower, at the expense of a larger model, i.e., approximately 35–40 RBFs.

4 Gradient Computation of Skew RBF

For the convenience of the reader, in this section we present some of the details required to implement the non-convex refinement phase of the model. The involves computing the gradients of the error functions in terms of the parameters. The RBF model consists of m skew RBF functions, i.e.,

$$f(x) = w_0 + \sum_{k=1}^{m} w_k z(\Lambda_k^T(x - c_k))\phi(\|x - c_k\|_{W_k})$$

where z is the shape function and ϕ is the radial basis function. Suppose the training data consists of p input output pairs $\{\mathbf{x}^{(\mu)}, y^{(\mu)}\}_{\mu=1}^{p}$, where $\mathbf{x}^{(\mu)} \in \mathbb{R}^n$ and $y^{(\mu)} \in \mathbb{R}$. Correspondingly we have c_k, $\Lambda_k \in \mathbb{R}^n$ and $W_k \in \mathbb{R}^{n\times n}$. We define $\|x - c_k\|_{W_k} = \sqrt{(x - c_k)^T W_k (x - c_k)}$ where W_k is a positive semi-definite matrix. The non-convex optimization problem is then

$$\underset{w_k,\Lambda_k,c_k,W_k}{\text{minimize}} \quad E(w_k, \Lambda_k, c_k, W_k) = \sum_{i=1}^{p} (y^{(\mu)} - f(x^{(\mu)}))^2.$$

The gradient of this objective function can be computed in what follows. Note, in order to ensure that W_k still remains a positive semi-definite matrix during the gradient descent method, we define $W_k = A_k A_k^T$ and compute the gradient with respect to A_k. (Initially, A_k is always picked to be identity matrix.) Differentiating gives

$$\frac{\partial E}{\partial w_k} = 2\sum_{\mu=1}^{p} (f(x^{(\mu)}) - y^{(\mu)}) \frac{\partial f^{(\mu)}}{\partial w_k},$$

$$\frac{\partial E}{\partial \Lambda_k} = 2\sum_{\mu=1}^{p} (f(x^{(\mu)}) - y^{(\mu)}) \frac{\partial f^{(\mu)}}{\partial \Lambda_k},$$

$$\frac{\partial E}{\partial c_k} = 2\sum_{\mu=1}^{p} (f(^{(\mu)}) - y^{(\mu)}) \frac{\partial f^{(\mu)}}{\partial c_k},$$

and

$$\frac{\partial E}{\partial A_k} = 2\sum_{\mu=1}^{p}(f(x^{(\mu)}) - y^{(\mu)})\frac{\partial f^{(\mu)}}{\partial A_k}.$$

Then the derivatives $\frac{\partial f^{(\mu)}}{\partial w_k}, \frac{\partial f^{(\mu)}}{\partial \Lambda_k}, \frac{\partial f^{(\mu)}}{\partial c_k}, \frac{\partial f^{(\mu)}}{\partial A_k}$ are computed, respectively. If we let $\rho = \Lambda_k^T(x^{(\mu)} - c_k)$ and $\gamma = \|x^{(\mu)} - c_k\|_{A_k A_k^T}$, then

$$\frac{\partial f^{(\mu)}}{\partial w_0} = 1,$$

$$\frac{\partial f^{(\mu)}}{\partial w_k} = z_k\,\phi_k,$$

$$\frac{\partial f^{(\mu)}}{\partial \Lambda_k} = w_k\,\phi_k\,z_k'\frac{\partial \rho}{\partial \Lambda_k},$$

$$\frac{\partial f^{(\mu)}}{\partial A_k} = w_k\,z_k\,\phi_k'\frac{\partial \gamma}{\partial A_k},$$

and

$$\frac{\partial f^{(\mu)}}{\partial c_k} = w_k\,z_k'\,\phi_k\,\frac{\partial \rho}{\partial c} + w_k\,z_k\,\phi_k'\frac{\partial \gamma}{\partial c_k}.$$

Furthermore,

$$\frac{\partial \rho}{\partial \Lambda_k} = x^{(\mu)} - c_k,$$

$$\frac{\partial \gamma}{\partial A_k} = A_k(x^{(\mu)} - c_k)(x^{(\mu)} - c_k)^T/\gamma,$$

$$\frac{\partial \rho}{\partial c_k} = -\Lambda_k,$$

and

$$\frac{\partial \gamma}{\partial c_k} = -A_k A_k^T(x^{(\mu)} - c_k)/\gamma.$$

In this paper we employ the scaled conjugate-gradient descent algorithm to minimize the objective function over w_k, Λ_k, c_k, W_k for $k \in \{1, 2, \ldots, m\}$.

5 Background and Related Work

Radial basis functions now have a long history. They were proposed in [21] as an alternative to feed-forward Artificial Neural Networks for arbitrary function approximation. Theoretical justification for using RBFs as a tool for universal function approximation was provided in [22]. A fast learning algorithm was already proposed in [23], but this still required the tuning of a variety of *ad hoc* parameters and the model order determination problem remained open.

We proposed the first steps towards a black box algorithm for construction such functions where the only required user parameter is the confidence level of a statistical test [13, 20]. This work proposed the use of skew-radial basis functions expansions to approximate time-series data. Also, it has been suggested that the RBF be compactly supported [24]. The skew-radial basis functions are capable of capturing data with sharp gradients [7] and even discontinuities [5]. An accelerated version of this algorithm was presented in [6] with detailed convergence proofs of the algorithm.

More recently RBF networks have been proposed for sparse signal recovery [25]. LASSO and LARS techniques have also been incorporated with RBF networks [26]. Sparse RBF kernel with ℓ_1-norm penalty has shown promising results [27]. The paper [12] proposed the ℓ_1-norm minimization of the weight vector only in the objective functions while imposing the RBF fitting problem as a constraint on the feasible set.

6 Conclusions and Future Work

We have proposed an algorithm consisting of two stages for creating models of time series data. The first stage is a Dantzig-Selector convex optimization problem. The solution to this problem provides the number of basis functions required for the model, i.e., addresses the challenging model order determination problem. It also provides initial estimates for the locations of the model functions, i.e., an initialization of the centers of the RBF. This preliminary set of center locations provides surprisingly good estimate for the initial conditions of the RBF centers as evidenced by the surprisingly low prediction errors on the test data before the non-convex refinement step. The refinement stage of the algorithm solves a non-convex optimization problem to improve the location of the RBFs and associated shape parameters. In the case of the non-convex problem with skew RBFs on four dimensional domains we observed that the non-convex refinement substantially improved results over the preliminary Dantzig-Selector. This suggests that the initial conditions for the centers were well chosen in the sense that they were in the basin of attraction of the optimal solution produced by the non-convex problem. Often non-convex optimization problems for data fitting problems must be resolved many times, at great expense, to overcome the presence of poor local minima. Indeed, we have found that that the Dantzig-Selector produces uniformly good initial conditions for the non-convex problem. In other words, we have not observed that the initial conditions lead to inferior non-convex solutions after refinement. The consistency of the quality of the Dantzig-Selector centers, both in terms of number and location, is surprisingly robust.

Acknowledgements This paper is based on research partially supported by the National Science Foundation under Grants Nos. DMS-1322508 and IIS-1633830. Any opinions, findings, and conclusions or recommendations expressed in this material are those of the authors and do not necessarily reflect the views of the National Science Foundation.

References

1. Takens, F.: Detecting strange attractors in turbulence. In: Dynamical Systems and Turbulence, Warwick, pp. 366–381. Springer, Berlin (1980)
2. Casdagli, M., Eubank, S., Doyne Farmer, J., Gibson, J.: State space reconstruction in the presence of noise. Phys. D: Nonlinear Phenom. **51**(1), 52–98 (1991)
3. Huke, J.P.: Embedding nonlinear dynamical systems: a guide to Takens' theorem. The Manchester Institute for Mathematical Sciences Eprint 2006.26 (2006)
4. Weigend, A.S., Gershenfeld, N.A. (eds.): Time Series Prediction: Forecasting the Future and Understanding the Past, pp. 105–129. Addison-Wesley, Reading, MA (1993)
5. Jamshidi, A., Kirby, M.: Skew-radial basis function expansions for empirical modeling. SIAM J. Sci. Comput. **31**(6), 4715–4743 (2010)
6. Jamshidi, A.A., Kirby, M.J.: A radial basis function algorithm with automatic model order determination. SIAM J. Sci. Comput. **37**(3), A1319–A1341 (2015)
7. Jamshidi, A., Kirby, M.: Skew-radial basis functions for modeling edges and jumps. In: Mathematics in Signal Processing Conference Digest, Royal Agricultural College, Cirencester, U.K., The Institute for Mathematics and its Applications, Dec 2008
8. Candes, E., Romberg, J.: ℓ_1-magic: recovery of sparse signals via convex programming, vol. 4, p. 46. http://www.acm.caltech.edu/l1magic/downloads/l1magic.pdf (2005)
9. Candes, E., Tao, T.: The Dantzig selector: statistical estimation when p is much larger than n. Ann. Stat. 2313–2351 (2007)
10. Broomhead, D.S., Kirby, M.: A new approach for dimensionality reduction: theory and algorithms. SIAM J. Appl. Math. **60**(6), 2114–2142 (2000)
11. Broomhead, D.S., Kirby, M.: The Whitney reduction network: a method for computing autoassociative graphs. Neural Comput. **13**, 2595–2616 (2001)
12. Ghosh, T., Kirby, M., Ma, X.: Sparse skew radial basis functions for time-series prediction. In: Proceedings International Work Conference on Time Series Analysis, pp. 296–307. Granada, Spain, June 2016
13. Jamshidi, A.A., Kirby, M.: Towards a black box algorithm for nonlinear function approximation over high-dimensional domains. SIAM J. Sci. Comput. **29**, 941 (2007)
14. Donoho, D.L.: Compressed sensing. IEEE Trans. Inf. Theory **52**(4), 1289–1306 (2006)
15. Candes, E.J., Tao, T.: Decoding by linear programming. IEEE Trans. Inf. Theory **51**(12), 4203–4215 (2005)
16. Daubechies, I., Defrise, M., De Mol, C.: An iterative thresholding algorithm for linear inverse problems with a sparsity constraint. Commun. Pure Appl. Math. **57**(11), 1413–1457 (2004)
17. Kirby, M.: Geometric Data Analysis: An Empirical Approach to Dimensionality Reduction and the Study of Patterns. Wiley (2001)
18. Boyd, S., Vandenberghe, L.: Convex Optimization. Cambridge University Press (2004)

19. Mackey, M.C., Glass, L.: Oscillation and chaos in physiological control systems. Science **197**(4300), 287–289 (1977)
20. Jamshidi, A.: Modeling Spatio-Temporal Systems with Skew Radial Basis Functions: Theory, Algorithms and Applications. Ph.D. dissertation, Colorado State University, Department of Mathematics (2008)
21. Broomhead, D.S., Lowe, D.: Radial basis functions, multi-variable functional interpolation and adaptive networks. Technical report, DTIC Document (1988)
22. Park, J., Sandberg, I.W.: Universal approximation using radial-basis-function networks. Neural Comput. **3**(2), 246–257 (1991)
23. Moody, J., Darken, C.J.: Fast learning in networks of locally-tuned processing units. Neural Comput. **1**(2), 281–294 (1989)
24. Jamshidi, A.A., Kirby, M.J.: Examples of compactly supported functions for radial basis approximations. In: Kozerenko, E., Arabnia, H.R., Shaumyan, S. (eds.) Proceedings of The 2006 International Conference on Machine learning; Models. Technologies and Applications, pp. 155–160. CSREA Press, Las Vegas (2006)
25. Vidya, L., Vivekanand, V., Shyamkumar, U., Mishra, D.: RBF-network based sparse signal recovery algorithm for compressed sensing reconstruction. Neural Netw. **63**, 66–78 (2015)
26. Zhou, Q., Song, S., Cheng, W., Huang, G.: Kernelized lars-lasso for constructing radial basis function neural networks. Neural Comput. Appl. **23**(7), 1969–1976 (2013)
27. Lihua, F., Zhang, M., Li, H.: Sparse RBF networks with multi-kernels. Neural Process. Lett. **32**(3), 235–247 (2010)

A Soft Computational Approach to Long Term Forecasting of Failure Rate Curves

Gábor Árva and Tamás Jónás

Abstract In this study, a soft computational method for modeling and forecasting bathtub-shaped failure rate curves of consumer electronic goods is introduced. The empirical failure rate time series are modeled by a flexible function the parameters of which have geometric interpretations, and so the model parameters grab the characteristics of bathtub-shaped failure rate curves. The so-called typical standardized failure rate curve models, which are derived from the model functions through standardization and fuzzy clustering processes, are applied to predict failure rate curves of consumer electronics in a method that combines analytic curve fitting and soft computing techniques. The forecasting capability of the introduced method was tested on real-life data. Based on the empirical results from practical applications, the introduced method may be viewed as a novel, alternative reliability prediction technique.

Keywords Reliability · Failure rate curve model · Fuzzy clustering · Forecasting empirical failure rates · Consumer electronics

1 Introduction

In reliability theory, the quantity $h(t)\Delta t$ is known as the conditional probability that a component or a product will fail in the time interval $(t, t + \Delta t]$, given that it has survived until time t. Function $h(t)$ is called the failure rate function or hazard function that can be estimated as

$$h(t) \approx \frac{N(t) - N(t + \Delta t)}{N(t)\Delta t}, \tag{1}$$

where $N(t)$ is the number of components or products that have survived until time t from the number of products or components that were put into operation. If $\Delta t = 1$

G. Árva · T. Jónás (✉)
Department of Management and Corporate Economics, Budapest University of Technology and Economics, Magyar Tudósok Körútja 2, Budapest 1117, Hungary
e-mail: jonas@mvt.bme.hu

I. Rojas et al. (eds.), *Advances in Time Series Analysis and Forecasting*, Contributions to Statistics, DOI 10.1007/978-3-319-55789-2_23

and t is taken in discrete times then the estimated failure rate λ_i for period i may be given by

$$\lambda_i = \frac{N(i\Delta t) - N((i+1)\Delta t)}{N(i\Delta t)} = \frac{N(i) - N(i+1)}{N(i)}, \quad (2)$$

$i = 0, 1, \ldots, n$, and so $\lambda_0, \lambda_1, \ldots, \lambda_n$ may be viewed as an empirical failure rate time series.

In the electronic industry, consumer electronic goods are typically tested functionally, while application of specific reliability tests to these products is not a common practice. In addition to that, considering the shortening product life-cycles, the curves of failure rate functions of these products are bathtub-shaped containing all the three characteristic phases of the traditional bathtub curve: the decreasing first phase, called infancy, the quasi constant second phase, which is also called the period of normal operation or useful life, and the third, increasing one, called the wear-out period. The failure rate functions represent the characteristics of product lives [4, 10]. Many papers that discuss models of failure rate functions deal with extensions of analytic models that are based on Weibull-, exponential, Marshall-Olkin extended uniform, and log-normal distributions (e.g. [1, 11]). In some models, new parameters are introduced that refer to the circumstances of usage, production, etc.

During the past few years, besides adding new parameters or factors to existing models, new approaches that are founded on soft computing methods, such as fuzzy logic or artificial neural networks have been discovered (e.g. [2, 5, 15]). Recently, artificial neural networks have been widely used not only in the electronic industry, but also in several other industries to predict failure rates. Son et al. [12] presented a soft computing based technique for acquiring a proper maintenance plan for individual parts in a complex system. They used a combination of neural network and evolutionary algorithm to discover the relationship between individual parts of a complex system to optimize its reliability.

The complete empirical failure rate time series of end-of-life consumer electronic products of the same commodity may be viewed as an empirical knowledge base of product reliability. This knowledge can be built up from field data, and can be used to predict the unknown failure rate curves of newly marketed products of the same commodity. Our approach is founded on an analytic decomposition of empirical failure rate time series by using a flexible model function that can describe well all the three parts of the traditional bathtub-shaped hazard function curve. The most important property of the introduced model function is that each of its parameters has a geometric interpretation that is related to shape of the failure rate curve. After applying appropriate transformations to the model functions, those can be standardized and the standardized models can be clustered based on their parameters. This process results in the typical standardized failure rate curve models that can be used to predict failure rates of active products. The standardized failure rate curve model, which we introduce here, can be considered as an alternative of the standardized line segments failure rate curve model [14], and its modified version presented by Dombi et al. [7]. The developed forecasting method is founded on measuring fuzzy similarity between the known fraction of failure rate time series of the studied active product and each fitted typical standardized failure rate curve model. In this sense, our method is a hybrid one that combines times series, analytic curve fitting and soft computing techniques.

The introduced method was tested on real life data and its goodness was compared to widely used forecasting techniques, such as the moving average, the exponential smoothing, the linear regression and the autoregressive integrated moving average (ARIMA) methods as well as to a forecasting method that utilizes a feedforward artificial neural network (FNN). Results of the practical application are discussed in a case study. Based on the empirical results, it may be concluded that our method and the FNN-based forecasting are able to indicate the turning points of the bathtub-shaped failure rate curve in advance, while the traditional, statistical forecasting techniques do not have such a capability.

2 The Modeling and Forecasting Methodology

Let us assume, that we have the $\lambda_{i,t_0}, \lambda_{i,t_1}, \ldots, \lambda_{i,t_{n_i}}$ time series ($i = 1, 2, \ldots, m$), each of them represents the complete empirical failure rate curve of a product, and the studied products are all from the same, well defined product category. The $\lambda_{i,t_0}, \lambda_{i,t_1}, \ldots, \lambda_{i,t_n}$ values denote the failure rates of the ith product week by week, and from this point, the simplified $\lambda_{i,0}, \lambda_{i,1}, \ldots, \lambda_{i,n_i}$ notation is used for time series $\lambda_{i,t_0}, \lambda_{i,t_1}, \ldots, \lambda_{i,t_{n_i}}$. The approach introduced here is based on the phenomenon that failure rate curves of the studied consumer electronic products are bathtub-shaped with three characteristic parts: the first, decreasing, the second, quasi constant, and the third increasing part as depicted in Fig. 1.

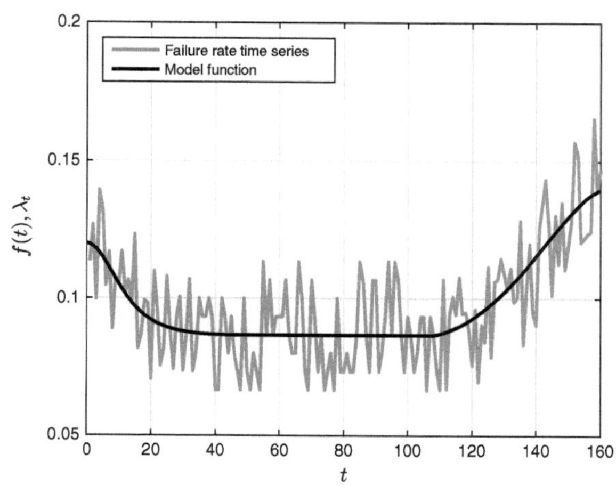

Fig. 1 An empirical failure rate time series and its $f(t)$ model

2.1 The Model Function

The parametric function that we will use as a model of each $\lambda_{i,0}, \lambda_{i,1}, \dots, \lambda_{i,n_i}$ historical failure rate time series of end-of-life products ($i = 1, \dots, m$) is based on Dombi's kappa function that is known as an operator in fuzzy theory [8, 9] and is given by

$$\kappa_{\mu,\nu}^{(\omega)}(x) = \frac{1}{1 + \frac{1-\nu}{\nu} \left(\frac{\mu}{1-\mu} \frac{1-x}{x} \right)^{\omega}}, \tag{3}$$

where $x, \mu, \nu \in (0, 1), \omega \neq 0$. In our implementation, $\nu = 0.5$ and the model function is founded on the following $g_{\mu,\omega} : [0, 1] \to [0, 1], x \mapsto g_{\mu,\omega}(x)$ function:

$$g_{\mu,\omega}(x) = \begin{cases} 0, & \text{if } (x = 0 \text{ and } \omega > 0) \text{ or } (x = 1 \text{ and } \omega < 0) \\ \kappa_{\mu,\nu}^{(\omega)}(x), & \text{if } 0 < x < 1, \omega \neq 0 \\ 1, & \text{if } (x = 0 \text{ and } \omega < 0) \text{ or } (x = 1 \text{ and } \omega > 0) \end{cases}, \tag{4}$$

where $0 < \mu < 1$. It can be seen that function $g_{\mu,\omega}(x)$ is monotonously increasing from 0 to 1, if the parameter ω is positive, and it is monotonously decreasing from 1 to 0, if ω is negative. Parameter ω determines the slope of the function curve in the $(\mu, 0.5)$ point. The function has value of 0.5 at μ. If $|\omega| \neq 1$, then the curve has an inflection point in the $(0, 1)$ interval. If $|\omega| = 1$, then $g_{\mu,\omega}(x)$ is either convex, or concave, or a line in the $(0, 1)$ interval, depending on the value of μ. If $\omega = 0$, then $g_{\mu,\omega}(x)$ is constant with value of 0.5. Main properties of function $g_{\mu,\omega}(x)$ are summarized in Table 1.

Table 1 Main properties of function $g_{\mu,\omega}(x)$

ω	μ	Monotony	Shape in $(0, 1)$
$0 < \omega < 1$	$0 < \mu < 1$	Increasing	Turns from concave to convex
$\omega = 1$	$0 < \mu < 0.5$	Increasing	Concave
$\omega = 1$	$\mu = 0.5$	Increasing	Line
$\omega = 1$	$0.5 < \mu < 1$	Increasing	Convex
$\omega > 1$	$0 < \mu < 1$	Increasing	Turns from convex to concave
$-1 < \omega < 0$	$0 < \mu < 1$	Decreasing	Turns from convex to concave
$\omega = -1$	$0 < \mu < 0.5$	Decreasing	Convex
$\omega = -1$	$\mu = 0.5$	Decreasing	Line
$\omega = -1$	$0.5 < \mu < 1$	Decreasing	Concave
$\omega < -1$	$0 < \mu < 1$	Decreasing	Turns from concave to convex

Since the curve of function $g_{\mu,\omega}(x)$ may have various shapes, its appropriate linearly transformed variants are suitable to model the decreasing first and increasing third phases of bathtub-shaped failure rate curves of electronic products. It is also worth mentioning that parameters ω and μ are responsible for the shape of function curve; that is, these have geometric interpretations, and so modeling based on function $g_{\mu,\omega}(x)$ has certain semantics.

Let $\lambda_0, \lambda_1, \dots, \lambda_n$ be the complete historical failure rate time series of a product. As a model for time series $\lambda_0, \lambda_1, ..., \lambda_n$, we use the following $f(t)$ function that is built upon appropriate linear transformations of function $g_{\mu,\omega}(x)$. Function $f(t)$ consist of three main parts representing the three sections of a traditional bathtub-shaped failure rate curve: the declining left phase $l(t)$, the constant mid phase λ_c, and the increasing right phase $r(t)$, and λ_l is the left most value of the model function; that is, $f(0) = \lambda_l$.

$$f(t) = \begin{cases} \lambda_l, & \text{if } t = 0 \\ l(t), & \text{if } 0 < t < t_{e,l} \\ \lambda_c, & \text{if } t_{e,l} \le t \le t_{s,r} \\ r(t), & \text{if } t_{s,r} < t \le t_{e,r} \end{cases}, \tag{5}$$

where

$$l(t) = \lambda_c + \frac{\lambda_l - \lambda_c}{1 + \left(\frac{t_{a,l}}{t_{e,l} - t_{a,l}} \frac{t_{e,l} - t}{t}\right)^{-\omega_l}} \tag{6}$$

$$r(t) = \lambda_c + \frac{\lambda_r - \lambda_c}{1 + \left(\frac{t_{a,r} - t_{s,r}}{t_{e,r} - t_{a,r}} \frac{t_{e,r} - t}{t - t_{s,r}}\right)^{\omega_r}}. \tag{7}$$

The model parameters need to meet the following criteria:

$$0 < t_{a,l} < t_{e,l} < t_{s,r} < t_{a,r} < t_{e,r}; \lambda_c < \lambda_l, \lambda_r; \omega_l, \omega r > 0. \tag{8}$$

$l(t)$ is defined in the $\left(0, t_{e,l}\right)$ domain and has the λ_l, λ_c, $t_{a,l}$, $t_{e,l}$ and ω_l parameters with the following roles: λ_c is the lowest value of $l(t)$ as well as the value of constant part of $f(t)$; $t_{a,l}$ is the place where $l(t) = (\lambda_l + \lambda_c)/2$; $t_{e,l}$ is the place of the end of the left side curve; the slope of $l(t)$ at t_{a_l} is proportional to ω_l.

$r(t)$ is defined in the $\left(t_{s,r}, t_{e,r}\right]$ domain and has the λ_r, λ_c, $t_{s,r}$, $t_{a,r}$, $t_{e,r}$ and ω_r parameters with the following roles: λ_r is the last value of $r(t)$; that is, it is the end value of the third segment of the life-cycle curve; λ_c is the lowest value of $r(t)$ as well as the value of constant part of $f(t)$; $t_{s,r}$ is the place of the start of the right side curve, same as the end locus of the constant middle segment of $f(t)$; $t_{a,r}$: is the place where $r(t) = (\lambda_r + \lambda_c)/2$; $t_{e,r}$: is the place of the end of the right side curve, $t_{e,r} = n$; the slope of $r(t)$ at t_{a_r} is proportional to ω_r.

The unknown model parameters can be determined by minimizing the quantity

$$\sum_{i=0}^{n}(f(i) - \lambda_i)^2. \tag{9}$$

It can be done by using the Interior Point Algorithm [3]. Function $f(t)$ is the *failure rate curve model (FCM)* of the empirical failure rate time series $\lambda_0, \lambda_1, \dots, \lambda_n$. Figure 1 shows how well function $f(t)$ can be used to model an empirical failure rate time series.

2.2 Standardizing the Failure Rate Curve Models

Once the parameters of $f(t)$ for a particular failure rate times series $\lambda_0, \lambda_1, \dots, \lambda_n$ have been identified, the $f(t)$ model can be standardized to the $s : [0, 1] \to [0, 1]$, $x \mapsto s(x)$ function by applying the following transformation:

$$x = \frac{t}{n}; s(x) = \frac{f(nx) - \lambda_c}{\max\{\lambda_l, \lambda_r\} - \lambda_c}. \tag{10}$$

Applying the transformation given by (10) to the model function $f(t)$ results in the following parameters of the $s(x)$ *standardized failure rate curve model (SFCM)* function.

$$\begin{aligned} y_l &= \frac{\lambda_l - \lambda_c}{\max\{\lambda_l, \lambda_r\} - \lambda_c}, y_c = 0, x_{s,l} = \frac{t_{s,l}}{n} = 0, x_{e,l} = \frac{t_{e,l}}{n}, x_{a,l} = \frac{t_{a,l}}{n} \\ y_r &= \frac{\lambda_r - \lambda_c}{\max\{\lambda_l, \lambda_r\} - \lambda_c}, x_{s,r} = \frac{t_{s,r}}{n}, x_{e,r} = \frac{t_{e,r}}{n} = 1, x_{a,r} = \frac{t_{a,r}}{n} \end{aligned} \tag{11}$$

The transformation given by (10) does not modify ω_l and ω_r, so function $s(x)$ is as follows.

$$s(x) = \begin{cases} y_l, & \text{if } x = 0 \\ s_l(x), & \text{if } 0 < x < x_{e,l} \\ 0, & \text{if } x_{e,l} \le x \le x_{s,r} \\ s_r(x), & \text{if } x_{s,r} < x \le 1, \end{cases} \tag{12}$$

where

$$s_l(x) = \frac{y_l}{1 + \left(\frac{x_{a,l}}{x_{e,l} - x_{a,l}} \frac{x_{e,l} - x}{x}\right)^{-\omega_l}}; s_r(x) = \frac{y_r}{1 + \left(\frac{x_{a,r} - x_{s,r}}{1 - x_{a,r}} \frac{1 - x}{x - x_{s,r}}\right)^{\omega_r}} \tag{13}$$

It is worth noting that due to the min-max standardization applied to $f(nx)$, one of the y_l, y_r values is 1. Each standardized failure rate curve model has eight parameters: $y_l, x_{a,l}, x_{e,l}, \omega_l, y_r, x_{a,r}, x_{s,r}, \omega_r$, and each parameter has a geometric interpretation related to the shape of the model curve. This semantics of the model parameters is an important property of the standardized failure rate curve functions, namely, it renders clustering them based on their parameters possible.

2.3 Identifying Typical Standardized Failure Rate Curve Models

Let $s_{\mathbf{p}_i}(x)$ denote the standardized failure rate curve model for the empirical failure rate time series $\lambda_{i,0}, \lambda_{i,1}, \dots, \lambda_{i,n_i}$, $(i = 1, 2, \dots m)$, where the parameter vector $\mathbf{p}_i$ is $\mathbf{p}_i = (y_{l,i}, x_{a,l,i}, x_{e,l,i}, \omega_{l,i}, y_{r,i}, x_{a,r,i}, x_{s,r,i}, \omega_{r,i})$. In order to identify typical standardized failure rate curve models, we cluster the $s_{\mathbf{p}_i}(x)$ models based on their parameter vectors $\mathbf{p}_i$ by applying the fuzzy C-means clustering method [6]. Let us assume that the $\mathbf{C}_1, \mathbf{C}_2, \dots, \mathbf{C}_N$ clusters $(N \le m)$ of the standardized failure rate curve models are formed. Let $\mathcal{I}_r$ be the index set of standardized failure rate curve models $s_{\mathbf{p}_i}(x)$ that belong to cluster $\mathbf{C}_r$ $(r \in 1, 2, \dots, N)$, that is,

$$\mathcal{I}_r = \left\{ i : \mathbf{p}_i \in \mathbf{C}_r, i \in \{1, 2, \dots, m\} \right\} \tag{14}$$

and let $\mathbf{c}_r$ be the centroid of $\mathbf{p}_i$ vectors in cluster $\mathbf{C}_r$. Vector $\mathbf{c}_r$ contains the parameters of the *cluster characteristic standardized failure rate curve model* $s_{\mathbf{c}_r}(x)$. The $s_{\mathbf{c}_1}(x), s_{\mathbf{c}_2}(x), \dots, s_{\mathbf{c}_N}(x)$ functions represent the *typical standardized failure rate curve models*, and as such can be taken as representative models of the empirical failure rate time series $\lambda_{i,0}, \lambda_{i,1}, \dots, \lambda_{i,n_i}$ $(i = 1, 2, \dots m)$. The typical SFCMs are generated from complete historical failure rate time series of a consumer electronic commodity, that is, they represent historical knowledge on failure rate curves of the studied product category. The knowledge represented by the typical standardized failure rate curve models can be used to predict the unknown failure rates of active products of which empirical failure rate time series are not complete yet.

2.4 Predicting Failure Rate Curves of Active Products

In our approach, active products of the studied product category are defined as ones with empirical failure rate time series that are not complete; that is, only a fraction of their failure rate time series is known. We may assume that products in the same product category have similar reliability properties. This assumption, which is empirically justified, lays the foundation of using the identified typical standardized failure rate models to predict the unknown continuations of failure rate curves of active products.

Let $\lambda_{F,0}, \dots, \lambda_{F,M}$ be a fractional failure rate time series of an active product $(M \ge 1)$. For each typical SFCM $s_{\mathbf{c}_r}(x)$ the $\alpha_r \ge M$, $\beta_r \ge 0$ and $\gamma_r \ge 0$ parameters are identified so that

$$g_r : [0, \alpha_r] \to \mathbb{R}^+ \cup \{0\}; g_r(t) = \gamma_r s_{\mathbf{c}_r}\left(\frac{t}{\alpha_r}\right) + \beta_r$$
$$d_r(\alpha_r, \beta_r, \gamma_r) = \sum_{i=0}^{M} \left(g_r(i) - \lambda_{F,i}\right)^2 \to \min . \tag{15}$$

α_r, β_r and γ_r in the optimization problems given by (15) can be found by applying the same Interior Point Algorithm referenced in Sect. 2.1. The d_r distance measures the level of dissimilarity between $g_r(t)$ and the fractional failure rate time series $\lambda_{F,0}, \dots, \lambda_{F,M}$ $(r = 1, 2, \dots, N)$. The normalized dissimilarity d_r^* is derived from d_r by applying the following transformation:

$$d_r^* = \begin{cases} \dfrac{d_r - \min(d_r)}{\max(d_r) - \min(d_r)}, & \text{if } \max(d_r) > \min(d_r) \\ 0, & \text{if } \max(d_r) = \min(d_r). \end{cases} \tag{16}$$

Each normalized dissimilarity d_r^* is turned to similarity w_r as follows [13]

$$w_r = \frac{e^{-d_r^*}}{\sum_{u=1}^{N} e^{-d_u^*}}. \tag{17}$$

Similarity w_r can be taken as a weight that expresses how well $s_{\mathbf{c}_r}(x)$ can be used as a model of fractional failure rate time series $\lambda_{F,0}, \dots, \lambda_{F,M}$. Based on this, we define the *prediction model* $p(x)$ for $\lambda_{F,0}, \dots, \lambda_{F,M}$ as

$$p(x) = \sum_{r=1}^{N} w_r s_{\mathbf{c}_r}(x), \tag{18}$$

and identify the $\alpha \geq M$, $\beta \geq 0$ and $\gamma \geq 0$ parameters to denormalize $p(x)$:

$$F : [0, \alpha] \to \mathbb{R}^+ \cup \{0\}; F(t) = \gamma p\left(\frac{t}{\alpha}\right) + \beta$$
$$\sum_{i=0}^{M} \left(F(i) - \lambda_{F,i}\right)^2 \to \min . \tag{19}$$

Solution for the fitting problem given by (19) can be found by applying the same Interior Point Algorithm referenced in Sect. 2.1. As $\alpha \geq M$, the $F(i)$ values for $\lfloor \alpha \rfloor \geq i > M$ may be viewed as predictions of the unknown future values $\lambda_{F,M+1}, \dots, \lambda_{F,\lfloor \alpha \rfloor}$; that is, the series $\lambda_{F,M+1}, \dots, \lambda_{F,\lfloor \alpha \rfloor}$ may be considered as a possible continuation of the fractional failure rate time series $\lambda_{F,0}, \dots, \lambda_{F,M}$.

3 A Case Study

3.1 *The Typical Standardized Failure Rate Curve Models of a Commodity*

The method discussed so far was applied to real-life empirical failure rate curves. 43 complete empirical failure rate curves of consumer electronic goods of the same commodity were used to generate typical standardized failure rate curve models. Each complete empirical failure rate time series represents weekly failure rates of a product. The standardized failure rate curve models of the studied 43 failure rate times series were clustered into 12 clusters; that is, 12 typical SFCMs were generated. Figure 2 shows graphs of the individual standardized failure rate curve models in each cluster (gray colored curves) and the cluster characteristic (typical) standardized failure rate curve models (black colored curves).

An empirical failure rate time series containing 176 weekly failure rates of a product, which had not been involved into establishing the typical standardized failure rate curve models, was selected to demonstrate how the typical SFCMs can be used to forecast future failure rates.

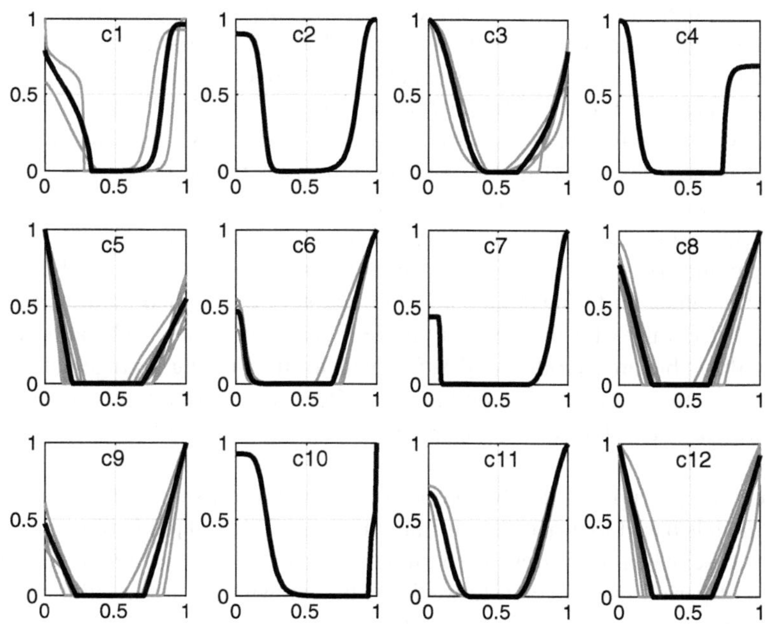

Fig. 2 Clustered standardized failure rate curve models

3.2 Results and Comparisons

The cluster characteristic SFCM based prediction was compared to five widely applied forecasting methods: the moving average, the exponential smoothing, the linear regression and the ARIMA methods as well as to a forecasting method that utilizes a feedforward artificial neural network. In the FNN-based method, each failure rate time series was modeled by a series of slope and intersection pairs of consecutive linear trends in the time series. Each trend line was generated based on data of 10 periods, and the slope and intersection pairs were used as inputs and outputs for training the neural network. In the FNN-based forecast computations, the output of each of the 43 FNNs was taken into account with a weight that represents the goodness of fitting of FNN output to the known fraction of the studied empirical failure rate time series.

The moving average was applied with span of 5, default weight for the exponential smoothing was computed by fitting an ARIMA (0,1,1) model to the data, and back-casting was used to calculate the initial smoothed value. The number of autoregressive terms (p), nonseasonal differences needed for stationarity (d), and lagged forecast errors (q) were indicated for the best fitting ARIMA model for each time series.

In order to evaluate goodness of the methods, we created forecasts based on data of the first 30, 80 and 130 weeks for the next 50 weeks; that is, we carried out almost year ahead forecasts. When we used the first 130 data as a known fraction of the studied failure rate time series, we were able to generate forecast only for the next 31 periods as parameter α of function $F(t)$ in this case was 160.4382. Note that the starting index of periods (weeks) is zero, and so if $\alpha = 160.4382$, then the last forecast period has the index of 160, that is, the latest forecast is for the 161st week. The mean square error (MSE) of the fitted and predicted values were calculated for each forecast to characterize goodness of the applied forecasting methods. The values of α, β and γ parameters of function $F(t)$ for the three forecasts are summarized in Table 2. The MSE values for fittings to the known fraction of the studied failure time series and the MSE values for the forecast failure rates are in Table 3. The abbreviations used in this table are as follows. MA(5) stands for the moving average with span of 5, Exp. S. is the exponential smoothing, Lin. Reg. is the linear regression and FNN is the neural network-based. Figure 3 shows the forecast results of the six studied methods.

There are a couple of notable properties of the typical standardized failure rate curve models based forecasts; that is, the function $F(t)$ based predictions. In the first forecast case, when predictions for period from week 30 to week 79 are given based on data of period from week 0 to week 29, the known part of the failure rate time series is in the first, declining phase of the bathtub curve. In this case, the linear regression gives the best fitting for the known fraction of the failure rate time series, and function $F(t)$ brings the forecast with the least MSE. The moving average and exponential smoothing methods show similar fitting results for the known fraction of the failure rate time series as the linear regression. As the failure rate time series

Table 2 Parameters of function $F(t)$

Parameters of $F(t)$	A	B	C
α	161.8701	176.2402	160.4382
β	0.0278	0.0221	0.0227
γ	0.0374	0.0449	0.0442

A: Forecast based on data of the first 30 weeks for the next 50 weeks
B: Forecast based on data of the first 80 weeks for the next 50 weeks
C: Forecast based on data of the first 130 weeks for the next 31 weeks

Table 3 MSE values for fits and forecast

	MA(5)	Exp. S.	Lin. Reg.	ARIMA	FNN	$F(t)$
Fits[a]	1.936E-05	1.823E-05	1.174E-05	1.217E-05	1.034E-05	1.175E-05
Forecasts[a]	2.434E-04	2.739E-04	1.074E-04	1.052E-04	5.002E-05	4.165E-05
Fits[b]	2.334E-05	2.290E-05	3.339E-05	2.062E-05	1.742E-05	1.657E-05
Forecasts[b]	4.486E-05	3.767E-05	7.382E-04	3.135E-04	3.491E-05	3.398E-05
Fits[c]	2.314E-05	2.226E-05	1.579E-05	2.716E-05	1.575E-05	1.626E-05
Forecasts[c]	2.608E-04	2.213E-04	2.803E-05	8.087E-05	1.415E-05	1.260E-05

[a]First 30 weeks → next 50 weeks, best fitting ARIMA is ARIMA(1,1,1)
[b]First 80 weeks → next 50 weeks, best fitting ARIMA is ARIMA(1,1,1)
[c]First 130 weeks → next 31 weeks, best fitting ARIMA is ARIMA(0,2,3)

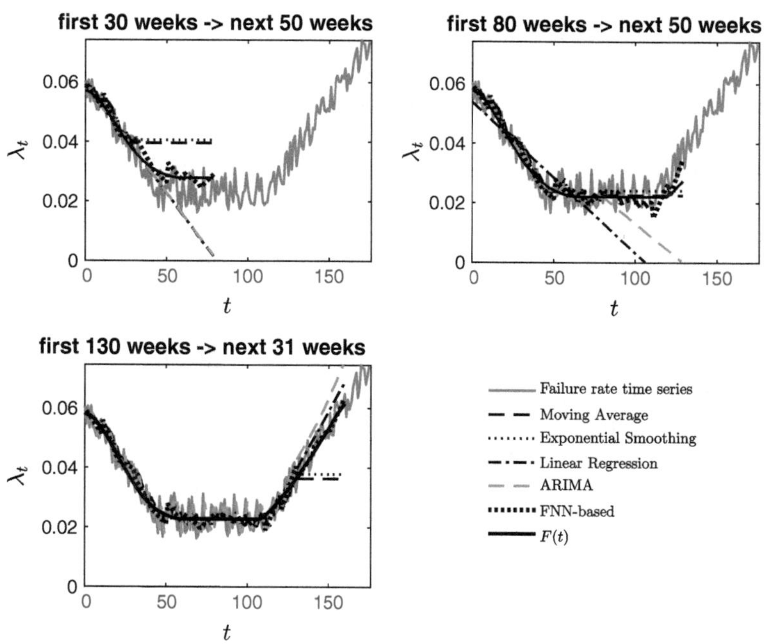

Fig. 3 Results from different forecast methods

is decreasing in the first 30 weeks and the moving average and exponential smoothing methods give constant forecasts, the latter two methods result in relatively weak forecast accuracy. The linear regression gives a decreasing forecast while the actual failure rates are getting quasi constant from week 50. It is important to highlight that even though the typical standardized failure rate curve models based forecast overestimates the actual values from week 40, it indicates that the failure rate curve turns from its decreasing phase to its quasi constant phase at around week 50. It is worth emphasizing that the FFN-based method also has this capability, but the other four methods are not able to predict the turning points of the bathtub curve.

In the second forecast case, when predictions for period from week 80 to week 129 are given based on data of period from week 0 to week 79, approximately first half of the known part of the failure rate time series is in the first, declining, while the other half is in the second, almost constant phase of the bathtub curve. The actual failure rate values start to increase at around week 115, and so the moving average, the exponential smoothing, function $F(t)$ and the FNN-based methods give similarly good forecast results for the period from week 80 to week 129. For this period, the linear regression and ARIMA result one order of magnitude worse forecast accuracy in terms of MSE than those of the moving average, exponential smoothing, function $F(t)$ or FNN-based methods. It is important mentioning that the function $F(t)$ and FNN-based forecasts are the ones among the studied six methods which are able to indicate that the failure rate curve will turn from its quasi constant, second phase to its increasing, third phase. Function $F(t)$ and the FNN-based method both suggest that the failure rate will take an increasing trend from approximately week 120, the actual figures show that in reality it happened a bit earlier, from approximately week 115.

In the third forecast case, predictions for period from week 130 to week 160 are given based on the first 130 weekly failure rates of the studied product. We know that from approximately week 115 to week 130 the failure rate curve is in its increasing, third phase, thus, the linear regression was applied for the period from week 115 to week 130 instead of using the first 130 data. The moving average and exponential smoothing methods are not able to predict well the failure rates in this phase of the failure rate curve. The ARIMA, the linear regression, function $F(t)$ and the FNN-based forecasts follow well the increasing trend of failure rate time series. A shortcoming of the function $F(t)$ based forecast is that the end of forecast period is determined by parameter α of function $F(t)$. In our case this is $\lfloor \alpha \rfloor = 160$. On the other hand, this property of our method is beneficial as value of α gives and indication on the time-wise length of the failure rate curve.

4 Conclusions

In this paper, we presented a hybrid technique for modeling and forecasting bathtub-shaped failure rate curves of consumer electronics. Complete empirical failure rate time series of products were considered as inputs for typifying SFCMs. Similarities among SFCMs are characterized by eight model parameters having geometric interpretations and cognitive aspect relating to the shape of the model curve and so clustering the SFCMs results in typical SFCMs. These typical SFCMs are applied for predicting unknown continuations of failure rate curves of active products of which complete empirical failure rate time series are unknown, that is, only a fraction of their failure rate time series is known. From a managerial perspective, discovering similarities among empirical failure rate curves generates added information both for the repair service providers and for the original equipment manufacturers. Electronic repair service provider companies can use it to predict resource needs for particular repair services, the latter ones can conclude on typical reliability characteristics of their products. In the case study, the accuracy of our methodology was compared to moving average, exponential smoothing, linear regression and ARIMA methods as well as to a FNN-based technique. In contrary to the traditional statistical forecasting methods, the $F(t)$ function-based forecasting methodology can indicate the turning points of the traditional bathtub shaped failure rate curve in advance. On the one hand, from this perspective, our method is similar to those which are based on soft computational techniques such as fuzzy systems or artificial neural networks. On the other hand, compared to the mentioned soft computational methods, the introduced method has a distinctive property that enhances its practical applicability. Namely, the parameters of the typical standardized failure rate curves, which the introduced method is based on, carry certain semantics that are related to the shapes of the modeled failure rate curves. A further practical advantage of the introduced method is that it does not require any preliminary knowledge of the failure rate probability distribution. Based on the encouraging results, our method may be viewed as a suitable alternative failure rate predicting technique.

References

1. Abid, S.H., Hassan, H.A.: Some additive failure rate models related with MOEU distribution. Am. J. Syst. Sci. **4**(1), 1–10 (2015)
2. Al-Garni, A.Z., Jamal, A.: Artificial neural network application of modeling failure rate for Boeing 737 tires. Qual. Reliab. Eng. Int. **27**, 209–219 (2011)
3. Bazaraa, M.S., Sherali, H.D., Shetty, C.M.: Nonlinear Programming: Theory and Algorithms, 3rd edn., pp. 315–519. Wiley, New Jersey (2006)
4. Campbell, D.S., Hayes, J.A., Jones, J.A., Schwarzenberger, A.P.: Reliability behaviour of electronic components as a function of time. Qual. Reliab. Eng. Int. **8**, 161–166 (1992)
5. Chen, K.Y.: Forecasting systems reliability based on support vector regression with genetic algorithms. Reliab. Eng. Syst. Saf. **92**, 423–432 (2007)
6. Chiu, S.L.: Fuzzy model identification based on cluster estimation. J. Intell. Fuzzy Syst. **2**(3), 267–278 (1994)
7. Dombi, J., Jónás, T., Tóth, Zs. E.: Clustering empirical failure rate curves for reliability prediction purposes in case of consumer electronic products. Qual. Reliab. Eng. Int. **32**(3), 1071–1083 (2016)
8. Dombi, J.: Modalities. In: Eurofuse 2011. Advances in Intelligent and Soft Computing, vol. 107, pp. 53–65. Springer, London (2012)

9. Dombi, J.: On a certain type of unary operators, In: Proceedings of 2012 IEEE International Conference on Fuzzy Systems, pp. 1–7, Brisbane, QLD, 10–15 June 2012
10. Goel, A., Graves, R.J.: Electronic system reliability: collating prediction models. IEEE Trans. Device Mater. Reliab. **6**, 258–265 (2006)
11. Marshall, A.W., Olkin, I.: A new method for adding a parameter to a family of distributions with application to the exponential and Weibull families. Biometrika **92**, 500–505 (1996)
12. Son, Y.T., Kim, B.Y., Park, K.J., Lee, H.Y., Kim, H.J., Suh, M.W.: Study of RCM-based maintenance planning for complex structures using soft computing-technique. Int. J. Automot. Technol. **10**, 635–644 (2009)
13. Tan, P.N., Steinbach M., Kumar, V.: Introduction to Data Mining, 1st edn., pp. 65–71. Addison-Wesley (2006)
14. Tóth, Zs. E., Jónás T.: Typifying empirical failure rate time series: a case study on consumer electronic products. In: Proceedings of International Work-Conference on Time Series (ITISE), pp. 396–407 (2014)
15. Xue, K., Xie, M., Tang, L.C., Ho, S.L.: Application of neural networks in forecasting engine systems reliability. Appl. Soft Comput. **2**, 255–268 (2003)

A Software Architecture for Enabling Statistical Learning on Big Data

Ali Behnaz, Fethi Rabhi and Maurice Peat

Abstract Most big data analytics research is scattered across multiple disciplines such as applied statistics, machine learning, language technology or databases. Little attention has been paid to aligning big data solutions with end-user's mental models for conducting exploratory and predictive data analysis. We are particularly interested in the way domain experts perform big data analysis by applying statistics to big data with a focus on statistical learning. In this paper we compare and contrast the different views about data between the fields of statistics and computer science. We review popular analysis techniques and tools within a defined analytics stack. We then propose a model-driven architecture that uses semantic and event processing technologies to achieve a separation of concerns between expressing the mathematical model and the computational requirements. The paper also describes an implementation of the proposed architecture with a case study in funds management.

1 Introduction

Data has long been utilized by scholars to run empirical tests or descriptive analysis. Increasing development in computer technologies in both software and hardware has enabled the storing, processing and exploring huge data (aka. Big Data) in the last decade. Analysis of these new datasets opens up a new breed of big data applications that are of significant interest to researchers and businesses in multiple areas. Peter Drucker, famous management guru, used to mention "you cannot manage what you don't measure." His wisdom is more highlighted in the era of big data; when we are collecting more data that we know what to do with. Should this

A. Behnaz (✉) · F. Rabhi
School of Computer Science, University of New South Wales, Sydney, NSW 2052, Australia
e-mail: ali.behnaz@unsw.edu.au

M. Peat
The University of Sydney Business School, Sydney, NSW 2006, Australia

I. Rojas et al. (eds.), *Advances in Time Series Analysis and Forecasting*, Contributions to Statistics, DOI 10.1007/978-3-319-55789-2_24

data be utilised properly, it will improve our knowledge of the world, and derived knowledge from data will improve performance and decision making. In this section, we provide background on big data and how it has redefined analytics.

1.1 What is Big Data?

However ubiquitous it appears now, "big data" is a nascent concept [1]. Diebold believes "big data … probably originated in lunch-table conversations at Silicon Graphics Inc. (SGI) in the mid-1990s, in which John Mashey figured prominently" [2]. The term big data became widespread just as recently as 2011, and its current hype is largely attributed to promotional motivations by IBM and other technology companies who have assembled multiple solutions for big data analytics problems [3]. Despite apparent association of big data to size, big data constitutes multiple characteristics. Namely, Lanley [1] proposed Volume, Variety, and Velocity (or the Three V's) as three dimensions in big data management. McKinsey Global Institute (MGI), a consultancy firm, defines big data in similar terms: "Big data refers to datasets whose size is beyond the ability of typical database software tools to capture, store, manage, and analyze. This definition is intentionally subjective and incorporates a moving definition of how big a dataset needs to be in order to be considered big data." As technology progresses over time, size of datasets that qualify as big data will also increase. In addition, the latter definition can be industry-specific or sector-specific [4].

Variety refers to different forms of data or structural heterogeneity in datasets. Advances in storage technologies have enabled businesses to collect various types of structured, semi-structured or unstructured data. Velocity refers to the speed at which data is generated and the required agility to be analysed and acted upon [3].

1.2 Big Data Analytics Process Model

Big data in a vacuum is worthless, its value is unravelled when it empowers the decision making process. Organisations need well-established processes to acquire meaningful insights from high volume, variety and velocity data. We split the overall process of data-driven insights into five stages, as shown in Fig. 1 [5]. These stages constitute two main sub processes: data management and analytics. Data management includes technologies and processes to collect and store data, and to prepare and restore it for analysis. On the other hand, analytics refers to techniques used for analyzing and acquiring insights from big data. According to [3], "big data analytics can be viewed as a sub-process in the overall process of 'insight extraction' from big data."

We provide a high-level overview of the analytics process model [6]. The steps are outlined below, and visualised in Fig. 2.

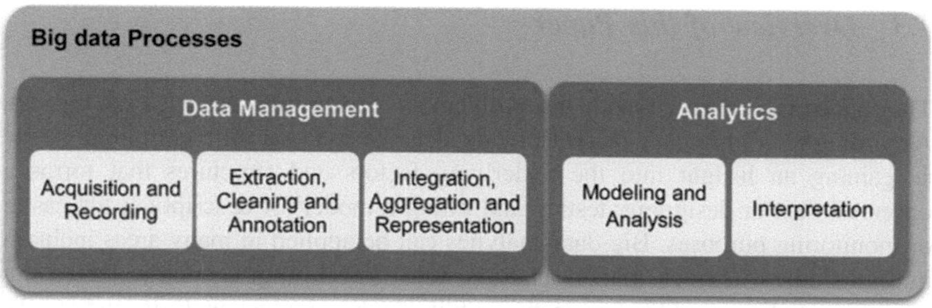

Fig. 1 Processes for extracting insights from big data

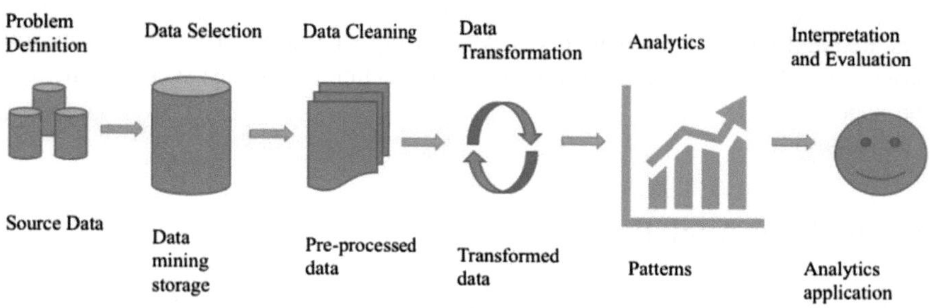

Fig. 2 Analytics process model

Analytics Steps

1. **Problem Definition**: Definition of the business problem to be solved with analytics.
2. **Data Selection**: All source data needs to be identified that could be of interest is identified and stored in data warehouse. Some preliminary data explanatory analysis can be performed at this stage.
3. **Data Cleaning**: This step focuses on handling inconsistencies in the data such as outliers, missing values and duplicates.
4. **Data Transformation**: Some aggregation, sorting or other pre-processing techniques can be applied to data at this stage.
5. **Analytics**: This step includes the algorithm or technique used to extract pattern in the data, such as churn prediction, fraud detection, customer segmentation, etc.
6. **Interpretation and Evaluation**: Analytics model usually generates many trivial patterns that needs to be assessed and interpreted by domain expert.

It is noteworthy to mention that the Analytics Process Model depicted in Fig. 2 is iterative. A data scientist may decide to revisit previous steps to generate different outcome. In that sense, an analytics process is a closed loop analysis system, which can get feedback loops from the output to input in any stage.

1.3 Overview of this Paper

This paper is concerned with the problem of performing analytics on big data. According to [7] big data analytics models can be used for different purposes such as gaining an insight into the underlying factors and structures that forms an observed data or designing, testing and fitting a model for descriptive, forecasting or monitoring purposes. Big data analytics can be applied in many areas including but not limited to stock market analysis, sales forecasting, economic forecasting, inventory studies, workload projection, utility studies, budget analysis, and etc. The main question addressed in this paper is how to design a software architecture that allows such analysis to be performed on big data possibly scattered across multiple computing platforms. The next section gives a review of existing software tools in this area and the challenges involved when using such tools. Then we discuss our proposed architecture and its prototype implementation. The last section concludes the paper.

2 Big Data Analytics

In this section we present an architectural view of big data analytics that could address the requirements outlined earlier and review popular techniques and tools that are utilised to perform analytics projects. We then define the research challenges involved when dealing with big data analytics.

2.1 Analysis Techniques and Tools

A hierarchy of abstractions for creating analytics solution is shown in the Fig. 3 based on the work of Milosevic et al. [8]. These abstractions facilitate demonstration of different capabilities and features in analytics solutions. Since we intend to study the complexities of drawing insights from big data for domain expert users the focus of this paper is on the 4th layer of the stack which is "Analytics tools." Such techniques and tools belong to different categories including:

- **Statistical Analysis**: As mentioned, the main objective of data processing/analysis is to suggest conclusion, support decision making and answer questions. Through application of statistical techniques raw data is converted to insightful information [9]. The goal of statistical analysis is to explore datasets and find insightful trends that can lead to actions [10]. Statistical analysis is comprised of multiple algorithms and different steps. "Data nature description", "data relation exploration", "data modelling on how the data relates to the underlying population", "validation of the model", and "predictive analytics"

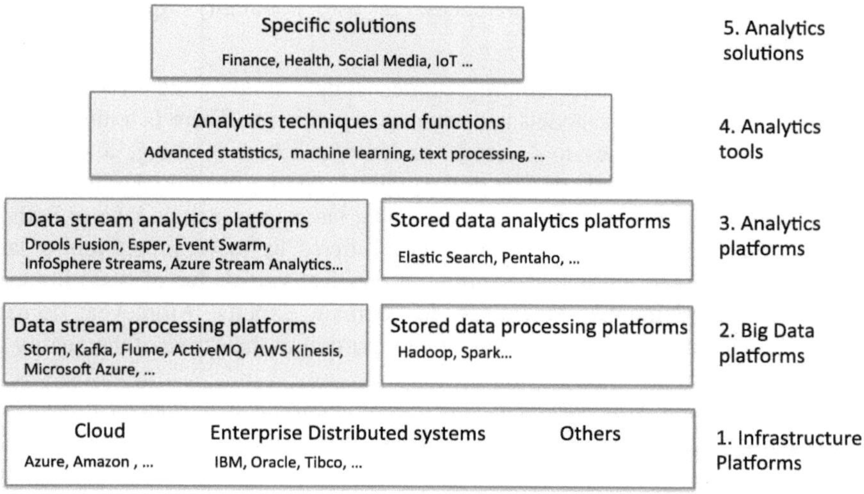

Fig. 3 Analytics stack, an architectural view

are five possible steps in statistical analysis. This kind of analysis is often followed by a data visualization to represent and interpret the outcomes [11].

- **Sentiment Analysis**: "Sentiment analysis is the process of algorithmically assigning a score to textual context that describes whether the author favoured positive or negative keywords and/or phrases within the text [12]."
- **Text Mining**: "Text mining, also known as text data mining or knowledge discovery from textual databases, refers generally to the process of extracting interesting and non-trivial patterns or knowledge from unstructured text documents [13]." For instance, [14] uses a method to predict stock markets movements by text mining articles from wall Street Journal [14].
- **Machine Learning/Deep Learning**: Machine learning/Deep Learning: Machine learning is a technique in data mining which constructs algorithms that can learn and make prediction from data [15]. Deep learning is a subset of machine learning that "attempts to model high-level abstractions in data by using multiple processing layers with complex structures [16]." For instance, [17] propose an algorithm that exploits the temporal correlation among global stock markets and multiple financial products to predict the next-day stock trend with the help of support vector machines.
- **Other Techniques**: There are multiple other areas of research that have been applied to big data for drawing inferences. For instance, neural network and genetic algorithms have been popular among traders in the last 20 years to predict stock market movement by defining best value for each parameter in an asset price.

There are many ways of using these techniques, each of which has its own advantages and shortcomings [18]. We classify all existing tools into three categories based on their features, ease of use, popularity, applicability and abstraction level:

- **Software Libraries**: Provide programmers simple building blocks for building sophisticated analysis models and running experiments. There is a multitude of packages and libraries to leverage statistics, machine learning, text-mining, sentiment analysis, and etc. The user can build a program tailored to their needs and utilize libraries built using programming languages such as Java or Python or "pre-packaged modules" such as those offered by the R programming language. The downside is that acquiring a practical knowledge of these libraries requires programming experience for domain experts. Moreover, domain experts would be willing to spend a larger portion of their time of the quality of their models rather than the implementation.
- **Statistical Data Analysis Tools**: These tool mainly include Statistical Analysis Systems (SAS), Stata and Statistical Package for the Social Sciences (SPSS). These tools are mostly user-friendly and their basic functions are easy to learn. They also offer handy data visualizations for each analysis. However, performing more complicated functions can be a hassle and learning curve becomes steep. They also trade-off vast benefits of programming languages for ease of use and user friendly interface.
- **Spreadsheets** such as Microsoft Excel, Google Spreadsheets, Apple Numbers, Gnumeric, etc. It has a simple, user-friendly interface which can be used to solve elementary to intermediate level analysis. However, there are many shortcomings in excel such as limited functions, rounding errors, miscalculations, etc.

For the purpose of this paper, we scope our analytics model to statistical learning. Statistical learning includes a set of tools for understanding and modelling complex datasets. This recently developed area of statistics couples with parallel progress in computer science, namely, machine learning and deep learning. Lasso and sparse regression, boosting and support vector machines, classification and regression tree are some of the methods in statistical learning.

We illustrate what statistical learning is using a simple example described in [19]. Suppose a researcher is interested to determine if the % change in inflation and the increase in population have an effect on beef consumption. In this context, the % inflation change and population increase are *independent variables* or *predictors* while beef consumption is the *dependent variable* or *response*. If the predictors are denoted as X_1, X_2 … X_n and the response is denoted as Y and Y is affected by the predictors, then we can define Y = f(X), where *f* is the function that connects the predictors X_1, X_2 … X_n to the response Y. This function, *f*, is generally unknown and one must estimate it based on observed data points. Statistical learning is a set of methods for estimating this function *f*. The two primary reasons for estimating *f* are *prediction* and *inference*. *Prediction* is about the using the estimated function *f* on a set of predictors, *X*, to calculate a predicted value for *Y*. *Inference* is concerned with how the response *Y* is affected as the predictors $\{X_1, X_2 \ldots Xn\}$ change. There are many linear and non-linear methods for estimating *f* and these methods can be broadly categorised as *parametric* and *non-parametric* methods [19]. We briefly provide an overview of parametric methods.

Given a set of data points or observations, these observations are referred to as *training data* as these observations will be used to train the method selected to estimate *f*. A parametric approach involves a two-step model based approach [19].

1. First, an assumption is made about the functional form or model of *f*, if *f* is linear in X it can be defined as[1]:

$$f(X) = \beta_0 + \beta_1 X_1 + \beta_2 X_2 + \beta_3 X_3 + \cdots + \beta_p X_p.$$

2. Once a model has been selected, the next step is to *fit* or *train* the model. In the previous step, if a linear model has been chosen the model estimator simply needs to estimate the parameters β_0, β_1, β_2 … β_p, once values of these parameters have been estimated the function *f* is defined as,

$$Y = f(X) = \beta_0 + \beta_1 X_1 + \beta_2 X_2 + \beta_3 X_3 + \cdots + \beta_p X_p.$$

One possible and quite commonly used approach to fitting the linear model is referred to as *ordinary least squares.* For our example, once the parameters have been estimated, we have a fitted linear model of the form:

$$\textit{beef consumption} = \beta_0 + \beta_1^* (\%\text{inflation change}) + \beta_2^* (\%\text{population change})$$

2.2 Challenge

Big data analytics solutions have considerably evolved from simple software tools to sophisticated problem solving environments for expert analysts. They are user-friendly and fit very well with end-user ways of conducting analytics. For example, a bank is interested to forecast the price of a commodity such as beef. A predictive model for beef price can use a combination of statistical learning techniques on multiple datasets. Depending on the performance and the accuracy of the model predictions, the user has several options: adjusting the mathematical model, changing some of the underlying variables or changing the way data is mapped into the variables. This is a highly ad hoc process which involves a mixture of both computational and statistical skills. What is needed is a way to allow

[1]This is a multiple linear regression, a widely used form in statistical learning.

complex statistical analysis to be conducted on big data, leveraging both big data analytics infrastructures and statistical platforms.

3 Proposed Model-Driven Architecture (MDE)

3.1 General Principles

Models are used in many applications of science. These models are usually created with the intention of demonstrating characteristics and features of other concepts or objects. The outcome of these models can be utilized in multiple disciplines [20]. Most domain-experts are educated to build models at high abstract levels. For example, the final outcome of an investment strategy or risk management approach is in the quality of the abstract models and availability of powerful tools for implementing these models.

Programming languages have merged as a major abstraction in dealing with the complexity of hardware and software. Over the last few decades, software engineering has been central in translating abstract ideas to machine language to harness the computational power of computers. The raising of abstraction levels for programming has since been improving. "Today's object-oriented languages enable programmers to tackle problems of a complexity undreamt of in the early days of programming [21]." Model Driven Development (MDD) is a natural improvement in this trend. It allows developers to use models to specify what system functionality is needed and what architecture must be used rather than defining how a system must be implemented [22]. Automation of model transformations is the key element in the vision for MDD. Developments tools must be able to offer the choice of applying pre-defined on-demand model transformations in addition to a language which helps advanced users defining their own model transformations and executing them on demand [23]. Software engineers require mature MDD tools and techniques to precisely implement a model [24]. This approach has been adopted and popularized in the industry and research community in the last few years, and therefore, a number of MDD tools have been proposed [25–27].

3.2 Overview of Proposed Architecture

The key idea of this architecture is use MDD principles to enable expert analysts to effectively integrate multiple data sources with a multitude of analysis techniques. As Fig. 4 shows, its central component is **a semantic reference model** that aligns big data characteristics with standard analysis models (statistical learning, etc.) Such a reference model leverages semantic Web technologies like RDF specifications [28, 29] which are increasingly used for defining formal descriptions of the

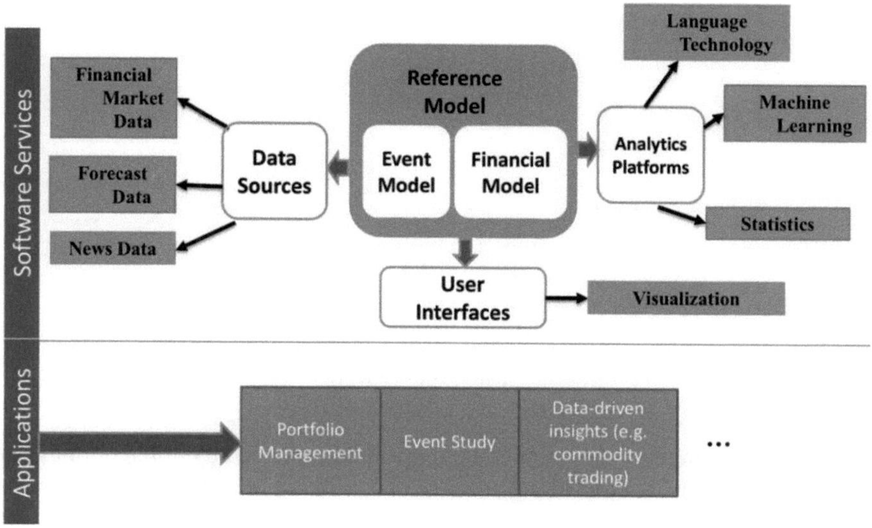

Fig. 4 Architectural vision

concepts, terms, and relationships within a given knowledge domain, so called *ontologies*. The main advantage of such models is that concepts and relationships to underlying services can be stored in a machine-readable form.

3.3 Current Implementation

Although Semantic Web technologies are ideal for defining domain-specific knowledge, individuals engaged in creating ontologies have little guidance beyond the technical standards documents [30]. In this implementation, we are using the CAPSICUM semantic modelling framework [12] to create our semantic reference model. The main feature of the CAPSICUM framework is a categorization of concepts across 9 cells which can be viewed across three layers: Business (i.e. concepts related to business analysis), Technical (i.e. design elements) and Platform (i.e. implementation-specific). Our current work focuses on the Business and Technical view models with concepts related to the area of big data analytics. The key concepts being modelled are illustrated in Fig. 5.

Figure 5 provides an overview of the key modelling concepts used in our semantic reference model. The left of the figure illustrates data represented at different levels of abstraction, each level obtained by applying a Data Processing Service. As an example, low level data on the left of Fig. 5 represent commodity trades and quotes (i.e. this is raw data) and high level data (data patterns) will result from the application of some sophisticated rules such as aggregating prices (e.g. weekly or monthly) taking into account various adjustment factors. Note that

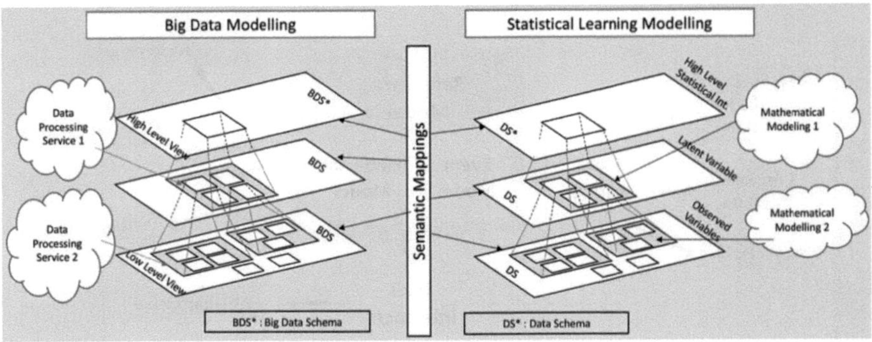

Fig. 5 Overview of the key modelling concepts to be represented

there could be different views on the same raw dataset created by different rules that can reflect different interpretations of the data. On the right of the figure, statistical learning variables are also represented as hierarchies. At the lowest level are observed variables (also called manifest variables in statistics) which represent variables that can be directly observed and directly measured [31]. At the higher levels are latent variables [32] which are inferred through a mathematical model from other variables that are observed.

Again, different interpretations can be built over observed variables via different mathematical models. The key innovation in this proposal is to represent big data analysis requirements separately from statistical analysis requirements (i.e. separate models) and represent the connections between data attributes and statistical learning variables using semantic relations. This approach has huge benefits. *A variable's association with a data attribute can be changed, or even moved at a higher level of abstraction in the big data hierarchy without affecting the mathematical models that rely on this variable.* In addition, these models will align statistical learning abstractions to Business View models of desired user outcomes. We specify an outcome as either the optimization of a model comprising variables or determining forecasted values for particular variables. The specification of such user outcomes will constitute the main query-interface through which the user will interact with the system.

3.4 Commodity Pricing Case Study

In this section we describe the case study for commodity pricing. We then describe the prototype implementation following MDD principles. Our implementation is being developed in conjunction with a real-life case study from banking industry. The case study was inspired by a Hackathon organized at University of New South Wales in partnership with ANZ Bank in Australia. Many of the banks' customers are interested in questions like "which countries and consumers will buy our

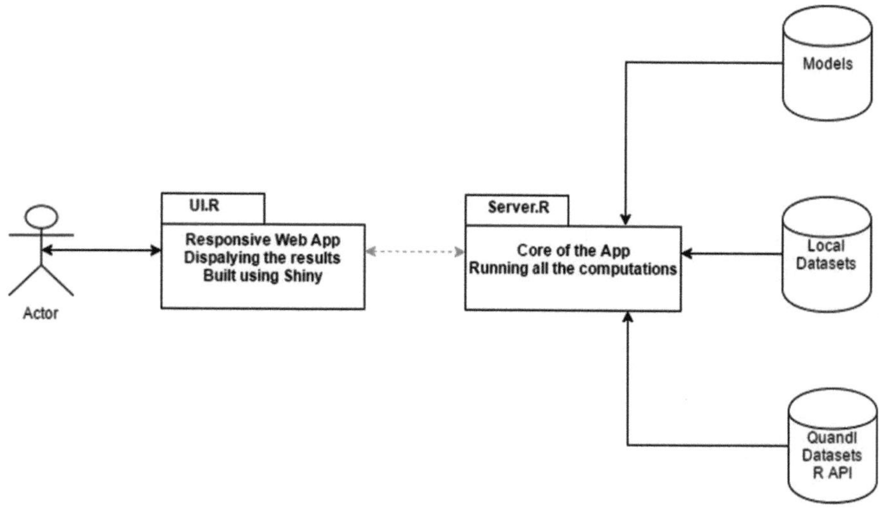

Fig. 6 Structure of the prototype

products? what prices and economic value is likely to be generated from this? what primary or processed food products should Australia seek to produce in future?". The idea of the competition was to use public and private data on this sector—macro-economic indicators, production volumes, weather patterns, prices, etc. to investigate what will drive this industry going forward [33]. Based on the available data, an instance of the analytics reference model was created to allow heterogeneous datasets to be analyzed.

For example, applying a functional form "Multiple linear regression" to the measure *Beef and Veal export* (dependent variable) and the measures *Export of goods* and *Employment in agriculture* (independent variables) would produce a linear function of the form:

$$\begin{aligned} \textit{Beef and Veal export} &= F(\textit{Export of goods}, \textit{Employment in agriculture}) \\ &= \beta_0 + \beta_1 \textit{Export of goods} + \beta_2 \textit{Employment in agriculture} \end{aligned}$$

We have restricted ourselves to regression model estimators, so the models produced are regression functions (characterised by regression factors). Using MDD principles, we have built an analytics tool for identifying price indicators of commodities. The structure of the prototype is illustrated in Fig. 6. The tool has been developed in R and built using libraries such as Shiny and ShinyBS for the User Interface, MASS library to perform stepwise regression, Quandl to get data from quandl.com and XLConnect to read local dataset saved as excel files.

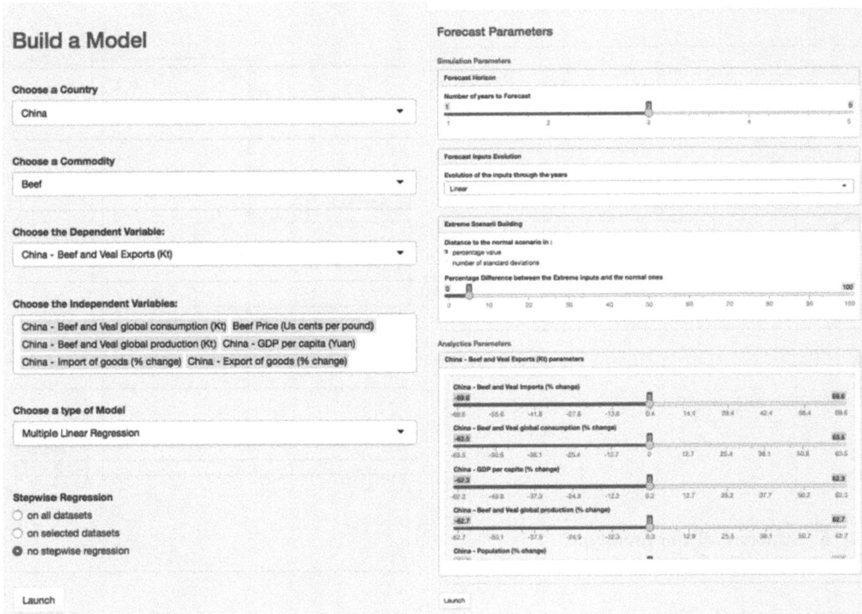

Fig. 7 Snap shots of commodity analytics tool

The tool works in two steps:

- A modelling step: where the user selects a variable to predict (dependent variable) and several possible explanatory variables (independent variables) in order to find the regression equation (model's equation) that describes the variations of the dependent variable.
- A forecasting step: using the equation of the first step and her own views on the independent variables, the user can forecast the dependent variable by inputting values for each of the independent variables which are fed into the model's equation to find a predicted value of the dependant variable.

In Fig. 7 we have provided a snapshot of the tool. The user interface is designed to enhance user interaction. We have grouped the measures by country and commodity. We have also provided an option for selecting models to deploy an analytics model. The tool is equipped with a predictive section which uses the outcome of the analytics model to generate different scenarios. To analyse scenarios, the user can tweak the tolerance of the measures (Forecast Parameters) and select the type of forecasting model.

The structure of the user interface in the model leverages our Semantic Reference Model. All measures shown to users are the result of querying the reference model. In addition, R code is automatically generated from a user query. For example, the snippet below shows generic code in R that implements multiple linear regression once the user has selected the dependent and independent variables.

```
regressionModel <- function(data_set, dep_var, indep_var){
    fit <- lm(dep_var~ indep_var)
return (fit)
}
```

Scalability is a property of this tool. Additional datasets can be added by modelling the appropriate measures in the reference model and such measures will be immediately available to the user via the User Interface. This architecture allows the user to create and add more analytics models or model estimators.

4 Conclusion

This paper proposes an innovative software architecture that improves the way domain experts apply statistical techniques involving big data by abstracting the underlying mathematical models away from the computational requirements. The main original features of the architecture are: (1) It has a semantic model that explicitly defines relationships between two types of abstractions: big data abstractions (used to represent various levels of understanding about big data) and statistical learning abstractions (used to represent statistical variables). In addition, this model will be linking these abstractions to desired user outcomes such as optimizing a model or forecasting the future value of a variable; (2) It leverages statistical platforms and big data processing technologies in a novel way to identify, extract and derive optimization and predictive models from event data sources and analytics services, based on a user-supplied expression of a desired outcome.

References

1. Laney, D.: 3-D data management: controlling data volume, velocity and variety. Application Delivery Strategies by META Group Inc. (2001)
2. Diebold, F.X.: A personal perspective on the origin(s) and development of "big data": the phenomenon, the term, and the discipline (Scholarly Paper No. ID 2202843). Social Science Research Network (2012)
3. Gandomi, A., Haider, M.: Beyond the hype: big data concepts, methods, and analytics. Int. J. Inf. Manag. **35**(2), 137–144. http://doi.org/10.1016/j.ijinfomgt.2014.10.007 (2015)
4. McKinsey & Company, Big data: The next frontier for innovation, competition, and productivity, p. 156. McKinsey Global Institute (2011)
5. Labrinidis, A., Jagadish, H.V.: Challenges and opportunities with big data. In: Proc. VLDB Endow. **5**(12), 2032–2033 (2012)

6. Baesens, B.: Analytics in a big data world: the essential guide to data science and its applications. Wiley and SAS Business Series (2014)
7. NIST/SEMATECH e-Handbook of statistical methods. http://www.itl.nist.gov/div898/handbook/. Accessed 15 Feb 2016
8. Milosevic, Z., Chen, W., Berry A., Rabhi, F.A.: An open architecture for event-based analytics, submitted to Computing (2015)
9. Lee, A.S., Hubona, G.S.: A scientific basis for rigor in information systems research. MIS Q. **33**(2), 237–262 (2009)
10. Schutt, R., O'Neil, C.: Doing Data Science: Straight Talk from the Frontline. O'Reilly Media Inc (2013)
11. Landau, S., Everitt, B.S.: A handbook of statistical analysis using SPSS, pp. 8–11. CRC Press (2004)
12. Robertson, C.S., Rabhi, F.A., Peat, M.: A service-oriented approach towards real time financial news analysis. In: Consumer Information Systems and Relationship Management: Design, Implementation, and Use: Design, Implementation, and Use (2013)
13. Tan, A.: Text mining: the state of the art and the challenges. In: Proceedings of the PAKDD 1999 Workshop on Knowledge Discovery from Advanced Databases, vol. 8. (1999)
14. Ming, F.: Stock market prediction from WSJ: text mining via sparse matrix factorization. In: 2014 IEEE International Conference on Data Mining (ICDM). IEEE (2014)
15. Kohavi, R., Provost, F.: Glossary of terms. Mach. Learn. **30**, 271–274 (1998)
16. Deng, L., Yu. D.: Deep learning: methods and applications. Found. Tr. Signal Process. **7**(3–4), 197–387 (2014)
17. Shen, S., Jiang, H., Zhang, T.: Stock market forecasting using machine learning algorithms (2012)
18. Zaidi, S., Nasir, M.: Teaching and Learning Methods in Medicine. Springer (2015)
19. James, G., Witten, D., Hastie, T., Tibshirani, R.: An introduction to statistical learning with applications in R. Springer, New York (2013)
20. Frankel, D.: Model Driven Architecture: Applying MDA to Enterprise Computing. OMG Press (2007)
21. Atkinson, C., Kühne, T.: Model-driven development: a metamodeling foundation. IEEE Softw. **20**(5), 36–41 (2003)
22. Soley, R.: OMG staff strategy group, model driven architecture. OMG White Paper, pp. 1–12. (April 2000)
23. Sendall, S., Kozaczynski, W.: Model transformation: the heart and soul of model-driven software development. IEEE Softw. **20**(5), 42–45 (2003)
24. Jouault, F., Allilaire, F., Bézivin, J., Kurtev, I.: ATL: a model transformation tool. Sci. Comput. Program. **72**(1–2), 31–39 (2008)
25. Agrawal, G., Karsai, Z., Kalmar, S., Neema, F., Vizhanyo, A.: The Design of a simple language for graph transformations. J. Softw. Syst. Model. (submitted for publication) (2005)
26. Gardner, T., Griffin, C.: A review of OMG MOF 2.0 Query/Views/Transformations Submissions and Recommendations Towards the Final Standard. IBM Hurley Development Lab., e-Business Integration Technologies (2003)
27. Varró, D., Varró, G., Pataricza, A.: Designing the automatic transformation of visual languages. J. Sci. Comput. Program. **44**, 205–227 (2002)
28. W3C Consortium, Semantic Web. https://www.w3.org/standards/semanticweb/. Accessed 18 Feb 2016
29. W3C Consortium: Resource Description Framework (RDF). http://www.w3.org/RDF/. Accessed 7 Nov 2014
30. Allemang, D., Hendler, J.: Semantic Web For The Working Ontologist: Effective Modeling in RDFS and OWL. Morgan Kaufmann (2008)
31. Dodge, Y.: The Oxford Dictionary of Statistical Terms. OUP (2003)

32. Wikipedia: Latent Variable (Definition). https://en.wikipedia.org/wiki/Latent_variable. Accessed 3 Feb 2016
33. Info Package for UNSW Data Science Hackathon. http://www.cse.unsw.edu.au/~fethir/HackathonInfo/HackathonStudentPack_v7.pdf. Accessed 10 Sep 2016

Part V
Applications in Time Series Analysis and Forecasting

Wind Speed Forecasting for a Large-Scale Measurement Network and Numerical Weather Modeling

Marek Brabec, Pavel Krc, Krystof Eben and Emil Pelikan

Abstract We investigate various problems encountered when forecasting wind speeds for a network of measurements stations using outputs of numerical weather prediction (NWP) model as one of the predictors in a statistical forecasting model. First, it is interesting to analyze prediction error properties for different station types (professional and amateur). Secondly, the statistical model can be viewed as a calibration of the original NWP model. Hence, careful semi-parametric smoothing of NWP input can discover various weak points of the NWP, and at the same time, it improves forecasting performance. It turns out that useful information is contained not only in the latest prediction available. It is beneficial to combine different horizon NWP predictions to one target time. GARCH sub-model for the residuals then shows complicated structure usable for short-term forecasts.

Keywords Semiparametric modeling · GAM · Wind speed forecasting · Numerical weather prediction model · Measurement network

1 Introduction

Wind forecasting is important for many practical purposes including safety, renewable energy generation, civil engineering, recreational activities like windsurfing or yachting, and others. Hence, it comes as no surprise that a substantial effort has

M. Brabec (✉) · P. Krc · K. Eben · E. Pelikan
Institute of Computer Science, Pod Vodarenskou vezi 2, 182 07 Prague 8, Czech Republic
e-mail: mbrabec@cs.cas.cz

M. Brabec · P. Krc · K. Eben · E. Pelikan
Institute of Informatics, Robotics, and Cybernetics, Czech Technical University in Prague, Zikova street 1903/4, 166 36 Prague 6, Czech Republic
e-mail: krc@cs.cas.cz

K. Eben
e-mail: eben@cs.cas.cz

E. Pelikan
e-mail: pelikan@cs.cas.cz

I. Rojas et al. (eds.), *Advances in Time Series Analysis and Forecasting*, Contributions to Statistics, DOI 10.1007/978-3-319-55789-2_25

been spent on designing and improving various wind speed prediction models and algorithms based on a wide variety of principles, reaching from linear methods [12], quantile regression [7] to neural networks [10]. Comprehensive reviews can be found e.g. in [4, 9]. Many windspeed forecasts for horizons of medium lengths are nowadays based on the numerical weather prediction models (NWP), either in a raw or somehow postprocessed form [1]. The problem of the windspeed forecasting based on the NWP output is far from being solved completely [4], however. In fact, a typical use in practice is governed by various ad hoc approaches amenable to potentially large improvements by formalized statistical analysis. This is true especially when speaking about less common purposes and forecasting schemes.

In this context, it is certainly interesting to investigate details of the prediction errors of the NWP forecasts for various reasons. First, a statistical regression model with measured windspeed as a dependent variable, using NWP output as inputs can be perceived as a calibration of the NWP. One of the challenges arises when formulating the model since it should be flexible enough to capture potentially nonlinear systematic patterns as well as complicated error behavior imposed by both weather complexity and imperfection of the NWP predictor. For practical purposes, it is important that the statistical model can serve as a pragmatic tool for improving the forecasts. But the analysis of random errors and systematic biases can also serve as a valuable feedback for the numerical modelers. By exploring deficiencies, it can focus various specific improvements.

2 Data

In this study, we will investigate the NWP predictions for speed of wind. The NWP predictions and output of statistical models using NWP as input will be compared to (ground) windspeed measurements at many (point) locations. Since the NWP model is computed and outputted on a grid of order of several kilometers, it is interesting to compare its prediction abilities for both measurement stations with relatively large spatial representativeness and for measurement points whose surroundings are complex, leading to smaller spatial representativity. In order to compare the performance of various models under these two circumstances, we use data from two sources: (i) from professional meteorological stations (run by the Czech Hydrometeorological Institute in the Czech Republic) and (ii) from the WindGURU project, [13] which collects data from many European wind stations installed mostly by yachting, windsurfing and other amateur enthusiasts mostly at locations near to the water bodies and sea. For this study, we analyze data from 35 type (i) stations and 58 type (ii) stations. While the professional stations are a priori located to have large spatial representativity, the WindGURU stations are amenable to local landscape properties and their representativity varies from one measurement point to another. The measurement locations are spread irregularly and the distance among them is typically so large that it does not pay off to model the data spatially or spatio-temporarily. We focus on the temporal component, instead. Figure 1 shows the placement of measurement

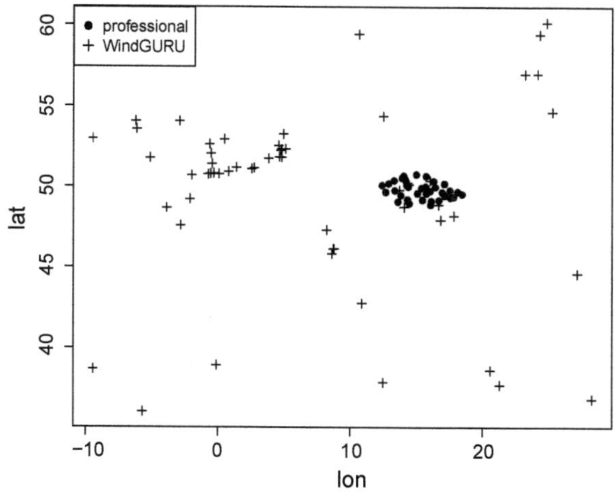

Fig. 1 Locations of the measurement points

points in longitude, latitude coordinates. Measurement consists of hourly averaged windspeeds in km per hour for the year 2015.

Since we have measurements from many stations in relatively fine time granularity, the data have a structure of many relatively long series. Since we are not going to model them spatially (for the reasons mentioned previously), it is natural to view their structure as an example of longitudinal data, [2] (or as a panel) where individual locations are approximately independent but within each site, we have to cope with a rich correlation structure.

We computed our own predictions from the WRF model (The Weather Research and Forecasting Model, [15]). They are evaluated at 0:00, 6:00, 12:00, 18:00 for different horizon lengths grouped into intervals $(0, 6), [6, 12), \dots, [78, 84)$ and denoted respectively by $0, 6, \dots, 78$. Spatial resolution was 9×9 kilometers and the gridded output was interpolated to the measurements' locations. From the NWP output, we use hourly wind speed forecasts. For computations, data manipulations beyond the NWP evaluation we use R package [11].

3 Behavior of the Raw NWP Predictor

Longitudinal data are commonly modeled by mixed effect models, [2, 8]. When using a statistical model for calibrating the NWP output, the main predictor will be obviously the NWP output. An "operational" approach to the calibration problem is, in practice, typically based on linear regression or on other relatively simple techniques which are often called model output statistics (MOS), [5].

As we can see from Fig. 2, the NWP output is far from being an ideal predictor even for the shortest prediction horizons. Due to various NWP imperfections, the NWP effect can show some departures from linearity (especially when it is used as

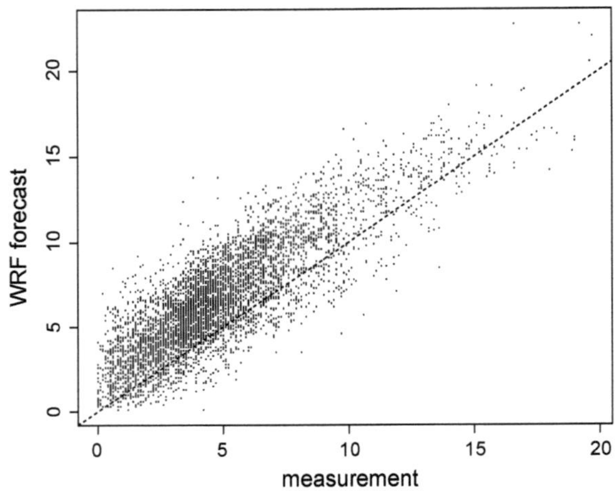

Fig. 2 Example of a nonlinear relationship between the NWP predictor and measurements (horizons shorter than 6 h, Kingsdown location)

a single predictor) and using the standard linear regression in the NWP output (N_t) for the windspeed (W_t), $W_t = \beta_0 + \beta_1.n_t + \epsilon_t$ might not be the best idea. The form of the nonlinearity might be complex and its exact functional form certainly cannot be assumed e.g. from physical principles.

Another way to view the imperfection of the NWP predictor is through basic time series properties. Figure 3 shows autocorrelation function (ACF) and spectrum of the prediction residuals, estimated from a typical location. We can see that both ACF and spectrum are far from the ideal of white noise. In fact, they both suggest that a lot of information is left in the prediction errors after the NWP. Besides non-trivial autocorrelation structure, there is a substantial circadial periodicity (also a shorter period of about 4 h is present). In the right panel, we can compare spectrum of the wind measurements (measurements) and that of residuals of after NWP predictions to different horizon lengths (0–72). Clearly, the NWP extracts a substantial part of the information contained in the data—the residual spectra are markedly flatter than the spectrum of the measurements. Nevertheless, the NWP also lefts a non-negligible unexplained part in the residuals. Quite surprisingly, this is true for all prediction horizons (although, there is clear decline in prediction quality for long horizons, say beyond two and half days). It is also interesting to note that shape of the residual spectrum is quite similar for all horizons and that it is dominated by low frequencies, daily and 4-hourly periodicity. It is precisely various systematic and time series properties of the prediction errors that will be of our interest in the rest of the paper—where we will try to improve the forecasting properties by calibration through structured statistical modeling.

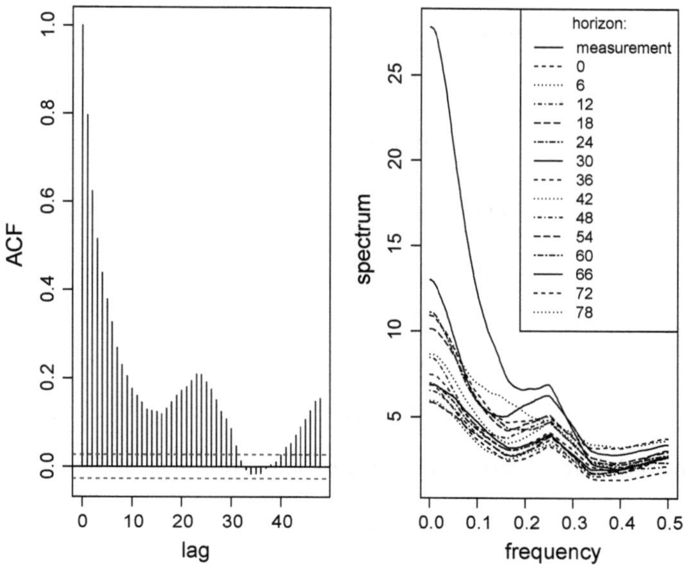

Fig. 3 Example of ACF (*left panel*) and spectrum (*right panel*) computed from residuals of the NWP predictor (Kingsdown location)

4 Statistical Modeling for NWP Calibration

4.1 Linear Models

Given the longitudinal structure of the data (many long time series corresponding to different locations), a natural starting point is a the linear mixed effects model (LME), [8] with random location effects. It captures both location heterogeneity with respect to being more or less windy, and self-similarity (or correlation) of the measurements from the same location. We start with a set of four simple LME models of different complexity:

$$\begin{aligned} W_{it} &= \beta_0 + b_{0i} + \beta_1 . n_{0,it} + \epsilon_{it} \\ b_{0i} &\sim N(0, \sigma_{b0}^2) \\ \epsilon_{it} &\sim N(0, \sigma_\epsilon^2) \ , \end{aligned} \tag{1}$$

where W_{it} and $n_{0,it}$ are, respectively, the measurement and NWP prediction (for horizons up to 6 h ahead) of the windspeed at location i and time t. b_{0i}'s are location-specific (additive) effects and $\beta_0 + b_{0i} + \beta_1 . n_{0,it}$ is the systematic part of the model constituting linear calibration of the NWP output.

$$\begin{aligned} W_{it} &= \beta_0 + b_{0i} + \beta_1.n_{0,it} + b_{1i}.n_{0,it} + \epsilon_{it} \\ \begin{pmatrix} b_{0i} \\ b_{1i} \end{pmatrix} &\sim N\left(\begin{pmatrix} 0 \\ 0 \end{pmatrix}, \begin{pmatrix} \sigma_{b0}^2 & 0 \\ 0 & \sigma_{b1}^2 \end{pmatrix}\right) \\ \epsilon_{it} &\sim N(0, \sigma_\epsilon^2) \ , \end{aligned} \tag{2}$$

where we have both intercepts (b_{0i}) and calibration slopes (b_{1i}) location-specific.

$$\begin{aligned} W_{it} &= \beta_0 + b_{0i} + \beta_1.n_{0,it} + \epsilon_{it} \\ b_{0i} &\sim N(0, \sigma_{b0}^2) \\ \epsilon_{it} &= \phi.\epsilon_{i,t-1} + \omega_{it} \\ \omega_{it} &\sim N(0, \sigma_\omega^2) \ , \end{aligned} \tag{3}$$

i.e. we have a (heteroscedastic) AR1 process in the residuals of the LME model.

$$\begin{aligned} W_{it} &= \beta_0 + b_{0i} + \beta_1.n_{0,it} + \epsilon_{it} \\ b_{0i} &\sim N(0, \sigma_{b0}^2) \\ \epsilon_{it} &= \phi.\epsilon_{i,t-1} + \omega_{it} \\ \omega_{it} &\sim N(0, \sigma^2.\exp(2.\alpha.n_{0,it})) \ , \end{aligned} \tag{4}$$

i.e. we have a heteroscedastic AR1 process in the residuals of the LME model.

AIC for (1), (2), (3), (4) are 774063, 761677, 648147, 633980, respectively. Calibration differs among locations mostly in the intercept and not much in slope. We have an indication of autocorrelation in the residuals and some hint of their heteroscedasticity. Residual autocorrelation in the calibration model is non-negligible—about 0.7. Comparing models (2) and (4), it turns out that it is better to model residual heteroscedasticity explicitly rather than to rely on the heteroscedasticity induced by random calibration slopes. We stratify on the station type when fitting the models (1), (2), (3), (4). Not surprisingly, calibration slope of the winning model (4) is smaller for professional than for amateur stations (proportional error tends to be smaller for professional stations).

Next question is whether we can improve the linear calibration based on LME. Being inspired by the circadial periodicity noted in Fig. 3, it is natural to include a seasonal term with 24 h period. But more information is readily available from the NWP for calibration purposes. In fact, for any given time, predictions from different past times (i.e. predictions with different horizons) are readily available. One can also try to employ them in the model simultaneously. The systematic part of the previous model then becomes:

$$\beta_0 + b_{0i} + \sum_{j=1}^{23} \gamma_j.I(\text{hour of } t \text{ is } j) + \beta_{1,0}.n_{0,it} + \beta_{1,6}.n_{6,it} + \cdots + \beta_{1,24}.n_{24,it} \ , \tag{5}$$

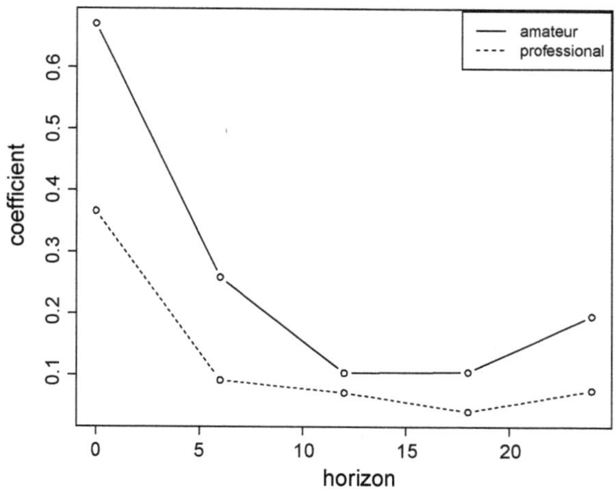

Fig. 4 $\hat{\beta}_{k,0}$ from model (5) plotted against k (prediction horizon)

where typical "baseline" parametrization in the ANOVA style is used for the periodic terms—with $I(.)$ denoting indicator function which assumes 1 if its argument is true and 0 otherwise. The remaining terms might superficially resemble distributed lags, [6]. Nevertheless, they are fundamentally different in that not lagged values of a predictor are used here. Instead, NWP predictors for the same time but computed at different timepoints (i.e. different horizon predictions) are used. Assuming that the information in more than day-old forecasts is essentially negligible, we use only $n_{0,it}$ to $n_{4,it}$, or 0–24 h horizons. The model (5) indeed performs better than the previous single-term calibration models. As expected, the periodic term is highly significant and has a flatter profile for professional stations. It is interesting to inspect the shape of the $\beta_{1,k}$ coefficient estimates when plotted against k (prediction horizon) in Fig. 4. As before, we have flatter regression for professional stations. What is new and quite surprising is that the weight is really distributed over several horizons. This means that some useful *additional* information is contained even in nominally outdated forecasts. Ideally, with a high quality NWP forecast, this should not happen and only the most recent should score in the model. While all horizons smaller than 24 h have nonzero (positive) coefficients, most of the weight is concentrated to the first two—corresponding to (0,6) and [6,12) hours horizons.

4.2 Semiparametric Models

So far, in the Sect. 4.1, we quietly assumed linearity in the NWP predictor. But is this reasonable? To this end, we will employ the generalized additive model (GAM), [14] class. To accomodate for location heterogeneity, we will still keep location-specific random effects (effectively invoking GAMM or generalized additive mixed models, but the random intercept can be easily taken as a special case of the quadratically penalized terms of GAMM). The NWP effect will be modeled nonparametrically—via penalized splines. Our model will be essentially a nonlinear expansion of (5), with the following three provisos: (i) the periodic term will be modeled not as a terms saturated in hours as before but as a cyclic cubic regression spline, (ii) only the first two horizons (i.e. those being the most important) are used, and (iii) model is reparametrized in terms of $n_{0,it}$ (newest forecast) and the difference $n_{0,it} - n_{6,it}$ (difference between the newest and previous forecasts), instead of using $n_{0,it}$ and $n_{6,it}$ directly. Systematic part then becomes semi-parametric:

$$W_{it} = \beta_0 + b_{0i} + f_{new}(n_{0,it}) + f_{dif}(n_{0,it}) , \quad (6)$$

where f_{new}, f_{dif} functions are estimated as cubic regression splines with penalty coefficients determined via crossvalidation (in the mgcv library, [14]). While the periodic component is still highly significant and has a shape very similar to what we have seen in previous (saturated w.r.t. the daily periodicity) models, it is of interest to inspect the shape of f_{new}, f_{dif} functions, shown in Fig. 5. The left panel shows estimate of f_{new} (that is of calibration with respect to the newest forecast—with horizon lower than 6 h) together with the 95% confidence intervals (constructed pointwise). From there, we can see that while in most of the NWP output range, the trend is pretty linear, but there is some evidence of nonlinearity close to the extremes (especially in the upper part). While at the lower end, the calibration function is convex for both station types, the shape differs qualitatively between amateur and professional stations at the upper end—convex for amateur and concave for professional stations. This is consistent with the generally flatter calibration of the professional stations. The right panel shows the estimate of f_{dif} function, that is of the smooth effect of difference between the most current and slightly outdated prediction (between forecasts with horizon (0,6) and [6,12) hours). Nonlinearity is somewhat more pronounced for amateur stations. It is interesting to note the asymmetry around zero, i.e. the flat part for negative differences and essentially linear decrease for positive differences. This can be interpreted as an interesting example of "selective smoothing" with substantial practical consequences. In fact, the model does differentiates strongly between situations in which (i) newest forecasts is higher than the previous and (ii) newest forecast is lower than the previous. In case (ii), the model essentially relies on the information from the newest forecast (leaves it as it is). Nevertheless, in case (i), it progressively disbelieves the newest forecast and inclines more and more to the older forecast as the difference between them becomes larger. This analysis offers a unique view into the nature of NWP biases (there is substantially more bias when the most recent forecast exceeds substantially the previous prediction).

Once we have the systematic part of the calibration model, we can explore what is left in the residuals. Specifically, we can look at their time-series properties in more detail than before when we just explored their simple Markovian property via first order autoregression. To this end, we formulate a comprehensive model which includes ARFIMA(5,d,0) with GARCH(1,1) in heavy tailed (t-distributed) residuals:

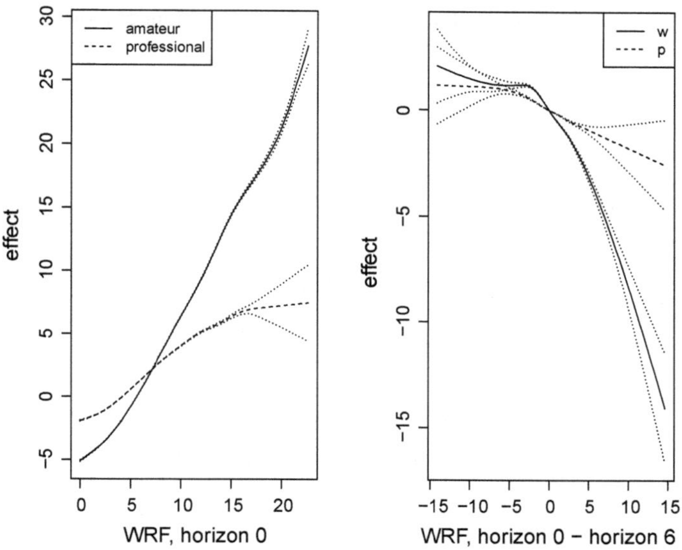

Fig. 5 Estimates of f_{new} (*left panel*) and f_{dif} (*right panel*) functions together with their 95% (pointwise) confidence intervals

$$\begin{aligned} W_{it} &= \mu_{it} + \epsilon_{it} \\ \mu_{it} &= \beta_0 + b_{0i} + f_{new}(n_{0,it}) + f_{dif}(n_{0,it}) \\ b_{0i} &\sim N(0, \sigma^2_{b0}) \\ \Phi_i(L)(1-L)^{d_i}\epsilon_{it} &= \omega_{it} \\ \omega_{it} &\sim t(0, \sigma^2_{\omega,it}, \nu_i) \\ \sigma^2_{\omega,it} &= \kappa_i + \alpha_i.\omega^2_{i,t-1} + \delta_i.\sigma^2_{\omega,i,t-1} \ , \end{aligned} \tag{7}$$

where AR, fractional differencing and GARCH parameters are estimated in the R package rugarch, [3]. The model allows us to investigate various subtle features of the wind process and compare its predictability via NWP between amateur and professional stations. Figure 6 shows the distribution of the estimated degrees of freedom for the t innovations over measurement locations. While there is variability in the estimates among the locations, the amateur stations show a clear tendency to heavier tails (lower degrees of freedom) compared to the much well-behaved professional stations. Figure 7 shows histograms of estimated GARCH δ parameters and compares them between station types. α estimates tend to be lower in magnitude. Once again, the amateur stations are somewhat more variable and less well-behaved. This is also observed with the fractional difference parameter d_i. All of these findings—together with the findings related to the systematic part of the calibration, are quite vital in exposing weak points of the idea of using amateur network results in the

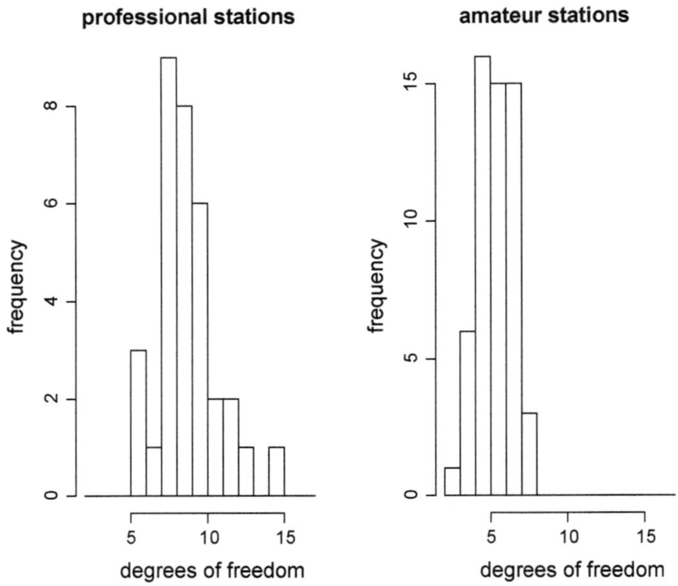

Fig. 6 Histograms of the estimates of degree of freedom parameter (ν_i's) for the t innovations of individual amateur measurement locations

style of "citizen science"—their use as a source of information for supplementing the professional network is quite complicated and has substantial risks.

After we inspected the structure of the models and interpreted some of their features with respect to important characteristics of practical interest, we are now interested in comparing prediction performance of different models. Table 1 compares different models in the prediction mode (to the shortest horizons below 6 h). We use the raw NWP model as a baseline and see that it is definitely worthwhile to calibrate rather than to use NWP directly. It is interesting to note that the random variability is dominating RMSE for raw NWP predictions of professional stations—but for amateur stations, the role of bias is much more substantial. Traditional simple linear calibration does indeed improves prediction ability. Curiously, ordinary least squares and quantile regression based results are quite similar not only in RMSE but also in MAE. The main morale is that a substantial further improvement in prediction performance can be brought by careful statistical modeling, considering both systematic part of the calibration and time-series properties of the residuals. Model (7) improves both RMSE and MAE. Not surprisingly, it does not incur any substantial bias. The RMSE enhancement is much better for the professional stations than for amateur measurements, quite in line with amateur stations' heterogeneity.

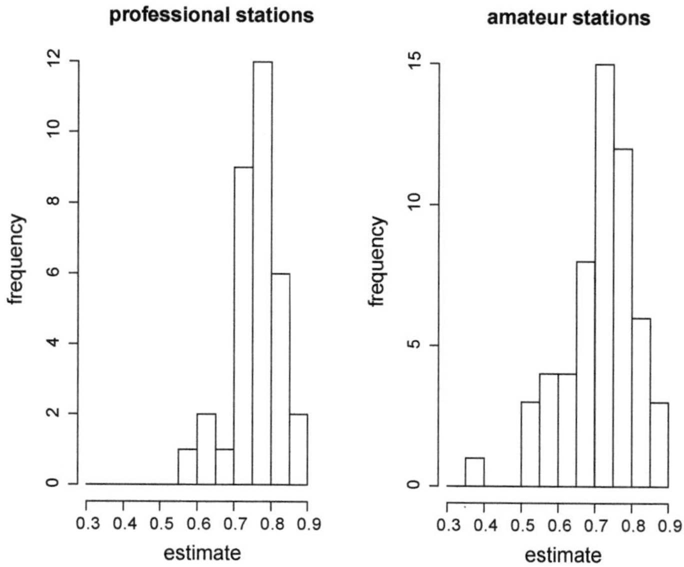

Fig. 7 Histograms of the estimated GARCH parameters δ

Table 1 Prediction results for different models: raw NWP, linear calibration via ordinary least squares, linear calibration via quantile regression, model (7)

Model	Station type	Bias	RMSE	MAE
Raw WRF	Amateur	2.38	5.70	4.12
Raw WRF	Professional	−0.39	2.19	1.59
Linear calibration of WRF (OLS)	Amateur	0.11	4.87	3.74
Linear calibration of WRF (OLS)	Professional	0.03	2.16	1.58
Linear calibration of WRF (QR)	Amateur	0.44	4.91	3.74
Linear calibration of WRF (QR)	Professional	0.27	2.18	1.55
Model (7)	Amateur	0.13	4.12	3.09
Model (7)	Professional	−0.01	1.48	1.13

5 Conclusions

We analyzed performance of the numerical weather prediction (NWP) model for wind speed forecasts and explored systematic ways to improve them through structured statistical modeling. Specifically, we investigated how useful is the hourly wind speed output of the WRF NWP model for predictions at a large set of spatially disparate locations coming from amateur and professional measurement networks. While simple calibration approaches do improve the prediction abilities aver the raw NWP output, the forecasts can be substantially enhanced when a formalized statistical semiparametric GAM model is employed to capture various systematic features of the NWP prediction errors. Furthermore, it is important to capture time-series properties in the residuals in order to cope with their autocorrelation, heteroscedasticity and heavy tails.

In fact, semiparametric statistical modeling offers far more than just a way to construct forecasting model. We show how important information about the nature of the NWP biases can be read off the structure and actual estimates of both systematic and random error parts of the model. This information is important not only as a guidance when constructing, calibrating and improving prediction models, but far more generally: as a feedback for numerical modelers and when considering risks related to the NWP predictions.

Our results show that there is substantial heterogeneity among different stations. There are large and quite structured differences between the prediction performance of NWP (and their calibrations) on professional and amateur stations. Professional stations tend to be more spatially representative, hence they show much better predictability by NWP output. Despite that, the behavior of amateur stations is much more heterogeneous, the NWP output tends to keeps to be quite undersmooth kind of predictor. Semiparametric modeling offers a systematic way to smooth the NWP output in a careful way tuned to the prediction performance. Calibration of the NWP output departs from linearity—especially in the extremes. Therefore is is not surprising that the shrinkage-like smoothing of NWP is beneficial. Structure of the calibration model shows that useful information might be not only in the most current prediction (there is some information left in slightly outdated forecasts) and that smoothing should be done asymmetrically with respect to the change between most current and previous forecasts.

Acknowledgements The work on this article was partly supported by the CVUT (Czech Technical University in Prague, Czech Republic) institutional resources for research and by the long-term strategic development financing of the Institute of Computer Science (RVO:67985807) and also by the Czech Science Foundation grant GA13-34856S, Advanced random field methods in data assimilation for short-term weather prediction.

References

1. Chen, N., Qian, Z., Nabney, I.T., Meng, X.: Wind power forecasts using gaussian processes and numerical weather prediction. IEEE Trans. Power Syst. **29**(2), 656–664 (2014)
2. Diggle, P.J., Heagerty, P., Liang, K.Y., Zeger, S.L.: Analysis of Longitudinal Data. OUP, Oxford (2002)
3. Ghalanos, G.: rugarch: univariate GARCH models. R package version 1.3-6 (2015)
4. Giebel, G., Brownsword, R., Kariniotakis, G.: The state-of-the-art in shortterm prediction of wind power: a literature overview. Project ANEMOS, Deliverable Report D1.1 (2003). http://anemos.cma.fr/download/ANEMOS_D1.1_StateOfTheArt_v1.1.pdf

5. Glahn, B., Gilbert, K., Cosgrove, R., Ruth, D., Sheets, K.: The gridding of MOS. Weather Forecast. **24**, 520–529 (2009)
6. Johnston, J.: Econometric Methods. McGraw-Hill, New York (1984)
7. Koenker, R.: Quantile Regression. Cambridge University Press, Cambridge (2005)
8. Laird, N.M., Ware, J.H.: Random-effects models for longitudinal data. Biometrics **38**(4), 963–974 (1982)
9. Lei, M., Shiyan, L., Chuanwen, J., Hongling, L., Yan, Z.: A review on the forecasting of wind speed and generated power. Renew. Sustain. Energy Rev. **13**, 915–920 (2009)
10. Mohandes, M., et al.: A neural networks approach for wind speed prediction. Renew. Energy **13**(3), 345–354 (1998)
11. R: A language and environment for statistical computing. R Foundation for Statistical Computing, Vienna, Austria (2016). http://www.R-project.org/
12. Riahy, G.H., Abedi, M.: Short term wind speed forecasting for wind turbine applications using linear prediction method. Renew. Energy **33**(1), 35–41 (2008)
13. Windguru project. http://www.windguru.cz/int/sitemap.php
14. Wood, S.N.: Generalized Additive Models: An Introduction with R. Chapman and Hall/CRC, London (2006)
15. Weather Research and Forecasting Model. http://www.wrf-model.org/index.php

Analysis of Time-Series Eye-Tracking Data to Classify and Quantify Reading Ability

Goutam Chakraborty and Zong Han Wu

Abstract Time series eye-tracking data, consisting of a sequence of fixations and saccades is a rich source of information for research in the area of cognitive neuroscience. With advanced eye-tracking equipments, many aspects of human perception and cognition are now analyzed from fixations and saccades data. Reading is a complex cognitive process involving a coordination of eye movements on the text and its comprehension. Reading necessitates both a vocabulary sufficient to cover the words in the text, as well as the ability to comprehend the syntax and composition of complex sentences therein. For rapid reading additional factors are involved, like a better peripheral vision. The motivation of this work is to pinpoint lacunae in reading, from her/his eye-tracking data while reading—whether the person lacks in vocabulary, or can not comprehend complex sentences, or needs scanning the text letter by letter which makes the reading very slow. Once the problem for an individual is identified, suggestions to improve reading ability could be made. We also investigated whether there is any basic difference how a native language (L1) and a second language (L2) are read? Is there any difference while reading a text in phonetic script and in logosyllabic script? Eye tracking data was collected while subjects were asked to read texts in their native language (L1) as well as in their second language (L2). Time series data of horizontal axis position and vertical axis position of the location where the fovea is focused, were collected. Simple features were extracted for analysis. For experiments with second language (in this work it is English) subjects belonged to 3 groups: expert, medium proficiency and poor in English. We proposed a formula to evaluate the reading ability, and compared scores with what they obtained in standardized English language test like TOEFL or TOEIC. We also find the correlation of a person's ability of peripheral vision (by Schultz's test) and reading speed. The final goal of this work is to build a platform for e-learning of foreign language, while eye-tracking data is analyzed in real-time and appropriate suggestions extended.

Keywords Time-series eye-tracking data · Features of fovea movement · Reading ability · Schultz's test

G. Chakraborty (✉) · Z. Han Wu
Faculty of Software and Information Science, Iwate Prefectural University,
152-52 Sugo, Takizawa 020-0693, Japan
e-mail: goutam@iwate-pu.ac.jp

I. Rojas et al. (eds.), *Advances in Time Series Analysis and Forecasting*,
Contributions to Statistics, DOI 10.1007/978-3-319-55789-2_26

1 Introduction

Even as early as the second half of 19th Century, it was suggested by Von Helmholtz that visual attention is an essential mechanism of visual perception. The dynamics of Visual attention and its relation to visual perception is an ever growing field of interdisciplinary research area for cognitive neuroscience and computer science. How the visual stimuli is filtered and processed in the brain for task in hand, how our training or prior knowledge to accomplish the task affect such selective attention, are important to understand human intelligence [1]. Traditionally, eye-tracking data, the sequence of fixations and saccades, was used to understand the underlying principle of human visual perception [2], and subsequently visual attention [3]. Recently, eye-tracking data is used in many practical applications. A few examples are market research like super market article arrangement, packaging design, advertising research, communication tools etc.

In this work, we concentrate on reading a text, asking subjects to comprehend the meaning of the text while reading. We do not verify whether the understanding is correct or not? We analyze the eye-tracking time-series data and correlate it with the physical structure of the text, to estimate reading ability and suggest ways to improve.

While viewing an object, the visual field is divided into 3 regions:

1. the central vision area subtending an angle of about 2° at fovea [2, 4]. The visual acuity is best here.
2. Peripheral vision extends about 3° on either side of the central vision area,
3. which is surrounded by the periphery.

Considering the reading material being held at a distance of 70–80 cm, known as resting point of accommodation (RPA) [5], a width of about 2.5 cm. is sharply focused. Depending on the font and number of characters within, it would cover 1–2 words. The peripheral vision will extend to a width of about 10 cm. The process of reading involves knowing the words as well as understanding the composition of a complex sentence. Yet, to speed up reading one needs to scan a chunk of 5/6 words or a whole line at the same time, which needs an improved peripheral vision. With equal command on a language, it is known that some person can read only 200 words per minute while a few can read over 1000 words per minute. We are interested to investigate from eye-tracking data whether a subject (i) has a poor vocabulary because of long fixations on certain difficult words, (ii) going back and forth over the words of a sentence because it is difficult to comprehend the meaning of the whole sentence, or (iii) the reading is smooth but word by word and not in a holistic fashion, making the reading process slow. Once we could answer the above questions from the eye-tracking data, we could suggest ways to improve reading ability.

For our experiment, we need an instrument to collect eye-tracking data. With advancement in technology, many startup companies are developing sophisticated and convenient-to-use eye-tracking devices, e.g. Tobii [6], SMI [7], SR research [8], Eyemark [9]. In our experiment we used Tobii EyeX.

The rest of the paper is organized as follows. Section 2 we will briefly review a few relevant previous works. Section 3 is the description of experimental setup, followed by Sect. 4 where we give details of the data collected. Section 5 is the analysis of eye-tracking data while reading texts. The paper is concluded in Sect. 6.

2 Previous Works and Proposed Idea

Though the study of visual perception started as early as the beginning of 20th Century, one of the seminal and widely cited work by Noton and Stark was published in 1971 [2]. Through experimental data he showed what features of an image attract long fixations. More recent works on visual attention and scene analysis by saccadic eye-movement are reported in [3, 10–12].

In recent years many works are reported in the journal of Studies in second language acquisition, Second Language Research, Language learning, Language learning & Technology, Study of language acquisition, Consciousness in second language learning, to name a few, where works are reported with eye-tracking data as a tool to measure attention of the subjects. In addition, Journal of Eye Movement Research, European journal of cognitive psychology etc. too publish works on measuring attention in second language [13], or relation between eye movement and word familiarity [14]. In [15], eye-tracking is used during IELTS to analyze readers cognitive process. One prominent work using eye-tracking data for analyzing reading ability and comprehension is by Prof. Rayner [16–18]. Rayner's works are more involved with psychology and comprehension, how inhibiting regression lowers comprehension and other experiments where subjects interact and are involved throughout the experiment. In our work, we plan to build the system where users can use it remotely over the internet. Our conclusions need to be based solely on the analysis of eye-tracking data.

In contrast to the above works, mainly by researchers from the field of linguistics and psychology, our approach is to find features of the saccades data, and how does that differ as the proficiency of the language changes, be it second language with varied proficiency or between first language and second language. The motivation is not to model the cognitive process, but to quantitatively evaluate the reading ability and suggest improvements by detecting wasteful idiosyncrasies. It is also interesting to study how the fovea moves as one reads a phonetic language text and logosyllabic languages, like English and Chinese. We did that experiment, but in our preliminary experiments we could not find any significant difference.

In previous works, some researches recorded reading speed by changing their word frequency [19, 20]. Some researches recorded fixation duration and reading time [21]. The result shown in those works is that when the word frequency is higher, the reading speed will decrease. Fixation duration and time of reading will be different too. In our experiment, we used texts where words are very simple and the construction of sentences are such that there is no ambiguity in meaning, where the meaning is clear only from the context. But, the level of English for some subjects was very poor. The same text is used for all subjects.

During our experiment we recorded the time series eye movement data, location where the fovea is focused. We have two sets of data, for the x-axis and the y-axis location on the text during the entire reading time. Subjects were divided into 3 groups: (i) those who are fluent in English, (ii) subjects with medium level of English proficiency, and (iii) with poor English proficiency.

We also performed experiment to find the correlation between peripheral vision and reading speed. The detail of the experimental set-up is explained in the next section.

3 Experimental Setup

In our experiment, we used Tobii EyeX [6] to record eye movement. It is shown in Fig. 1, Tobii is set below the monitor.

The Tobii EyeX Controller is an eye tracking peripheral based on Tobiis latest dedicated hardware. It connects to the computer via a USB 3.0 cable. It is mounted with a slim size magnetic mount on both desktop and laptop setups.

The Tobii EyeX Controller together with application software for eye gaze interaction allows to collect eye-tracking data, when the subject sits or stands in front of the monitor. After initialization, the system provides consistent and accurate data of viewing location at small intervals of time. The data is claimed to be independent of head movements, and changing light conditions, and there is no need for regular re-calibrations. The locations are x, and y coordinates on the monitor.

Before actual collection of data, first we need to calibrate the instrument. One need to focus on bright spots appearing on the monitor at corners. Initialization is

Fig. 1 Experiment setup

Fig. 2 Example of Schultz Table

10	05	11	09
15	04	03	16
02	14	06	13
12	01	07	08

completed in a few trials. The text to be read is displayed on the monitor. Subjects keep a distance of about 80 cm., i.e., an arm's length, from the monitor.

In the first set of experiments, we provide an English text in font 18 consisting of 14 lines. As mentioned, subjects kept a distance of about 80 cm. Subject are instructed to scan from left to right, and top to bottom of the text, if possible by movement of eyes only. Reading speed was different for different subjects, from 2000/60 s for an expert reader to 18000/60 s for a subject with poor English proficiency. Sampling rate per second is 60.

3.1 *Peripheral Vision and Reading Speed*

Most of the "Speed Reading" web-sites on the internet are for those with enough knowledge of the language but reading speed is low. The two main suggestions are to stop the habit of subvocalization (silent speech while reading) and to practice widen the eye-span, i.e., improve peripheral vision (to read not word by word but a chunk of words together). For improving peripheral vision, Schultz table is used [22], which is shown in Fig. 2. Once the red button is clicked, the matrix will be filled out with integers from 1 to 16 at random locations. Keeping the eye (fovea) fixed on the red dot, one has to find integers in serial order. One with better peripheral vision can do it quickly compared to someone with poorer peripheral vision. For this experiment we used 10 subjects with same mother tongue (Chinese). Their average time for Schultz test, for 20 runs, were collected. They were told to read the same simple text in their mother tongue. Corresponding reading time was noted. The results show a high positive correlation between them.

4 Experimental Results

Preliminary Experiment: The result of a subject from Group1 who is poor in English, is shown in Fig. 3. While reading a line from left to right, the x-axis angle will change say from $-\theta°$ to $+\theta°$. Tobii EyeX reading is clipped at $\pm 20°$ when the eye orients at an angle > $\pm 20°$, because the text display is within that range. The upper part of Fig. 3, with blue plot is the movement of the eye in x-direction. We see that while reading a line horizontally its orientation changes roughly from $-15°$ to $+15°$. Yet,

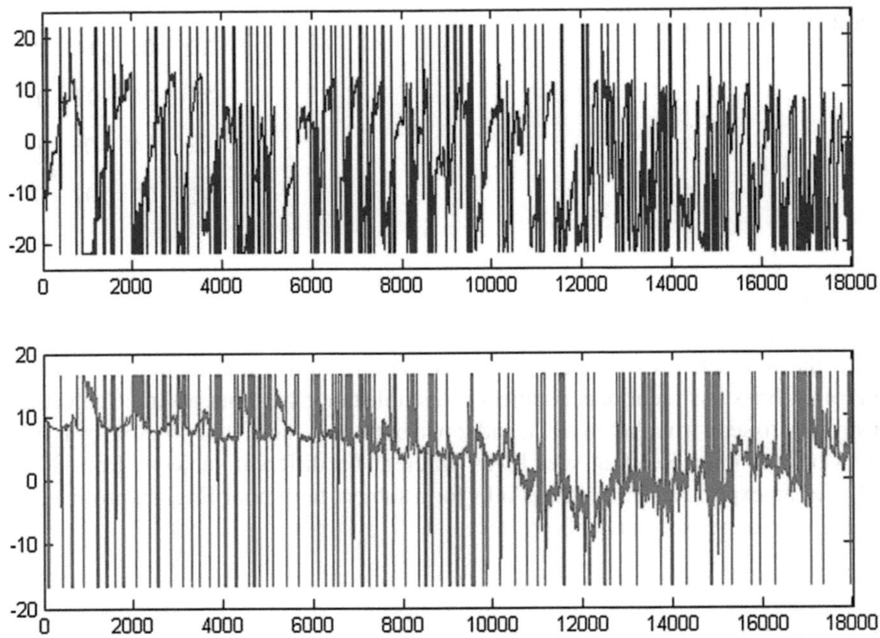

Fig. 3 Experiment result from subject whose English proficiency is poor

often it is confused with wasteful movements going unnecessarily to extreme left and right. This regression, or going back to refer to already read words, is a typical behavior for person whose comprehension of a composition is poor. The Subject took a lot of time (5 min) to read the text of just 184 words. Figure 4 shows the data for a subject from Group2. His native language is not English but he studied English as a second language for a few years. His eye movement has clear pattern, from beginning to end of a line. Still there are some confusion especially in y-axis movement. The total time taken, compared to the previous subject, is much less, about 1 min 20 s. Figure 5 shows data of a subject from group3 whose English proficiency is like native speaker. The eye movement pattern is regular, and start and finish of a line is clear. The total reading time was short, a little more than 30 s.

5 Analysis of Data

The most striking difference in patterns collected from group1, group2 and group3 subjects are the span and frequency of eye movement. For group1 subjects, the x-axis span is more than $\pm 22°$. For y-axis too it is $\pm 18°$. Moreover, the eye is moving from up to down, and from left to right frequently. For a subject from group2, the pattern is more regular but the x-axis and y-axis spans are large. Subject from group3 does

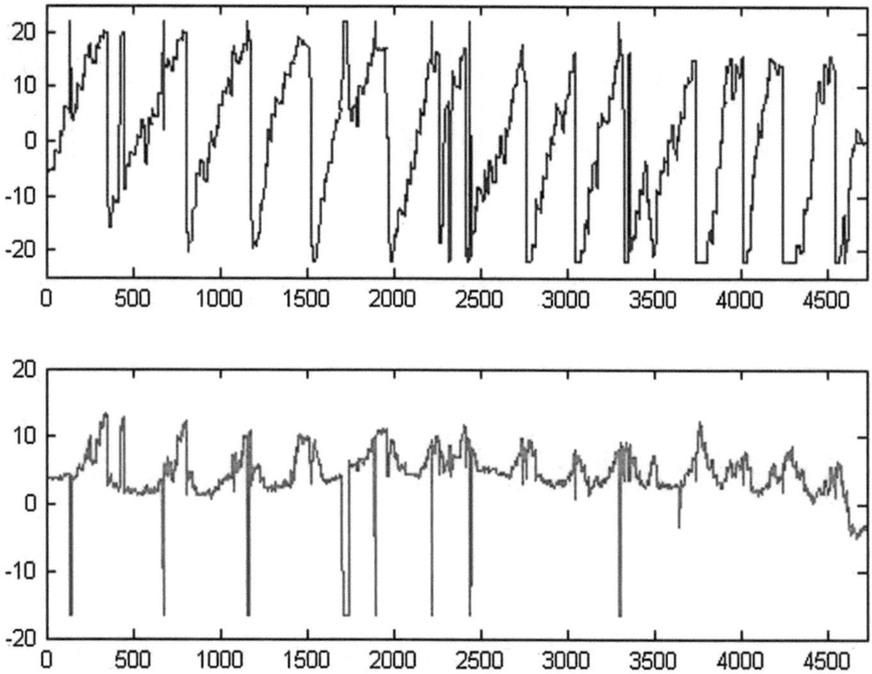

Fig. 4 Experiment result from subject whose second language is English

not move the eye so much, reading a line in a holistic fashion instead of character by character or even word by word. We focused on a single line of the text to zoom in our data and analyze eye fixations. The sentence was at the second line which read: *down upon him; this soon wakened the Lion, who placed his huge paw*.

Figure 6 shows the data for subject from Group1, whose English proficiency is poor. The range of angle of eye movement is 37°. In addition to going widely to the left and right, the eye is moving back and forth on a word as if trying to find a suitable meaning of a word. Though it was checked that all words in the text are within subjects' vocabulary, it is not clear whether the subject read the text character by character to form the meaning of a word. Further zoom-in will reveal that. The subject has to back-track several times to the previous word/s to comprehend the meaning of the whole sentence. The time to read this single line was 14.75 s. Figure 7 shows data of a subject from Group2 who learned English as a second language. The reading direction is more regular. The movement range is almost the same, 35°. The fixations are more than the number of words in the line, and a few times the eye strayed to the end of the line. Only a few times subject 2 looked back to previous words. The time to read this line was 7.36 s. Figure 8 shows the data for a subject from Group3. Here the movement range is only 18°. The fixations are less than the number of words in the line, i.e., more than one word is grasped at a time. There is no look back. This subject takes 2.46 s to read the line.

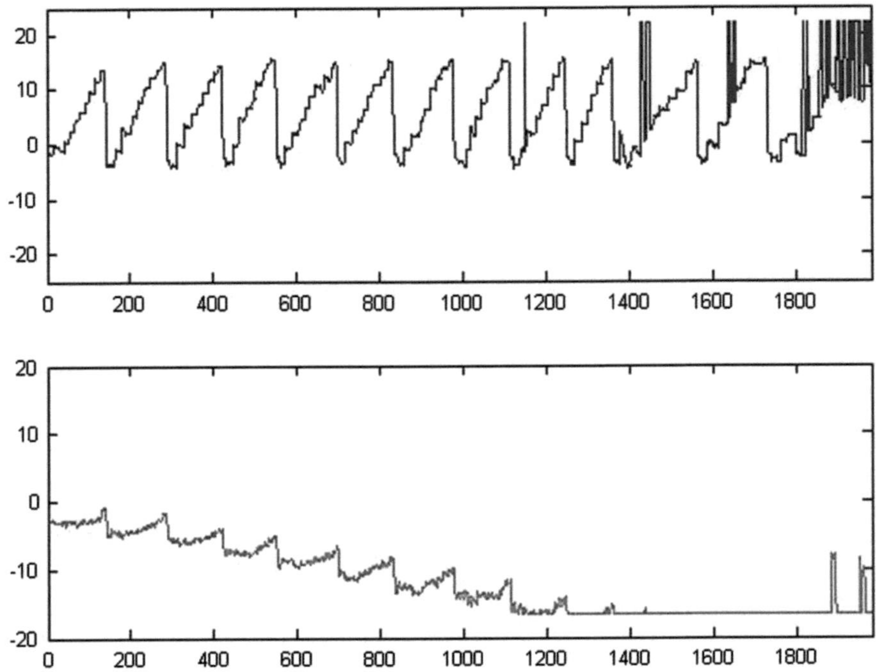

Fig. 5 Experiment result for subject whose native language is English

Fig. 6 Eye-movement data while reading a single line. Subject is poor in English

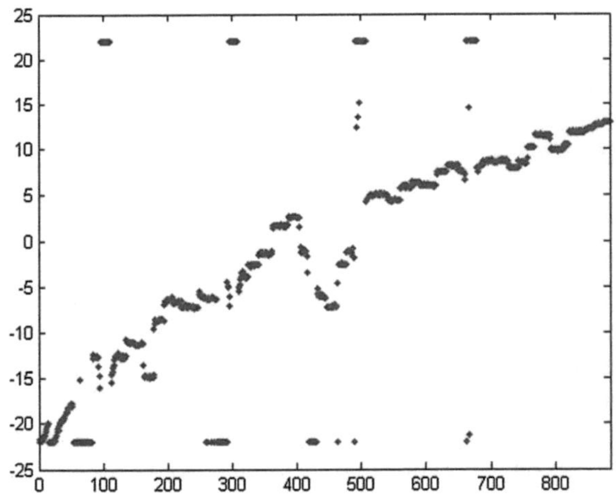

From the above results we know the features from the eye-tracking data, by which we can classify the level of proficiency of the language. The rate of reading is the slope of y-axis eye-tracking data. It is found by linear regression of the y-axis data points against time. In addition, the range of angle while reading a line is a measure of reading ability, shorter range means better ability. The other parameter is the times

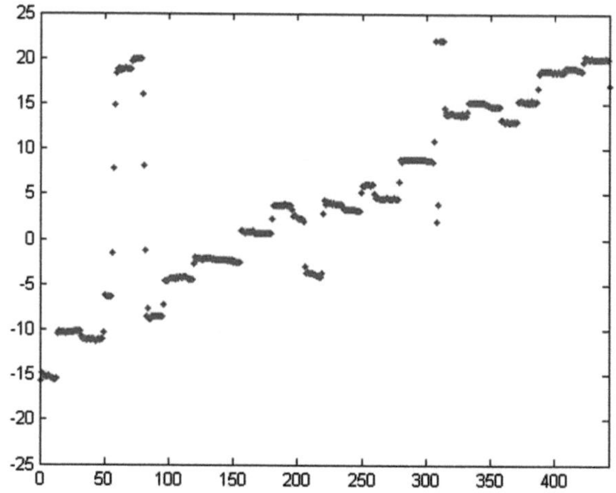

Fig. 7 Eye-movement data while reading a single line. Subjects second language is English

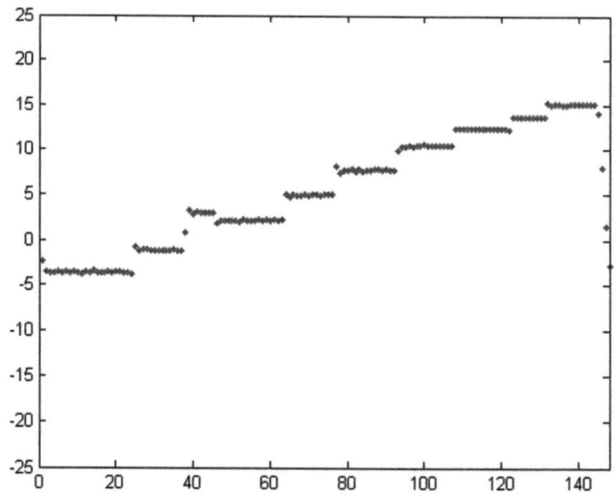

Fig. 8 Eye-movement data while reading a single line. Subject is native English speaker

the eye moves way beyond the word being read at a particular instant. The following parameters are calculated from the eye-tracking data. To clean the data of noise, we took the moving average. T is the total duration of time to read the text. m_a is the slope for y-axis data. We define threshold high T_h and threshold low T_l, to count the times eye moves way beyond the present reading point. The counter increases by one, every time the view point moves above T_h or below T_l. c represents this count. D_T is the excess time for reading (over average). Finally, the reading score is formulated as:

$$S = 100 - \left(\frac{c}{T} \times 100 + D_T\right)$$

In Table 1, we show the results for 3 subjects from 3 different groups. The final score is calculated using the above equation, and the scores they obtained in standardized English reading ability test (TOEIC) do have high correlation.

Table 1 Summarizing results of 3 subject with different language proficiencies

	Subject 1	Subject 2	Subject 3
Excess time (D_T)	84.7	64.6	0
Exceed count (c)	19.84	5.91	2.08
Slope (m_a)	−0.0029	−0.0034	−0.0062
Score (S)	51.9	72.6	97.9

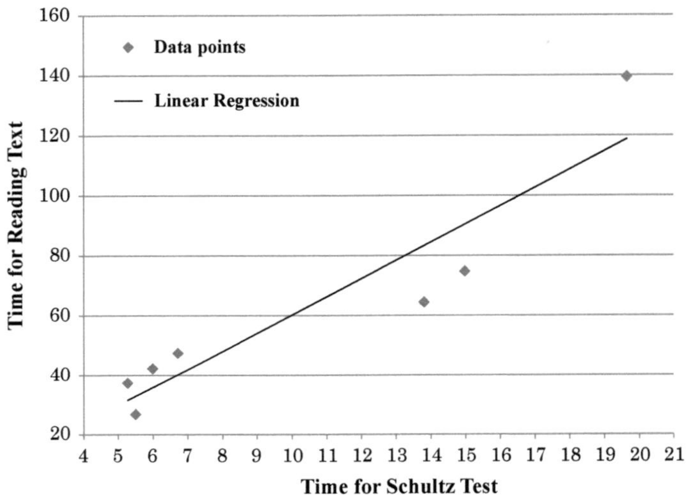

Fig. 9 Correlation between Schultz test and reading speed

We did experiment to find correlation between peripheral vision and time taken to read a text in mother tongue Data were taken from 10 subjects, all with the same mother tongue and the same simple text was read by them. They were asked to read Schultz table for 20 times, with different random initialization. The average time for Schultz tests was noted. The linear regression line is shown in Fig. 9. It is clear that the correlation between the two time durations is positive and high.

In Fig. 10, we show the eye-tracking time series data when a native Japanese speaker from Group2 (moderate English proficiency) reads an English text and a Japanese text. As shown in the key to symbols (too small to read), the blue line is the collected eye-tracking data, violet the moving average. The green line, called threshold is to count the number of times the eye moves away from the presently read location. Similarly, the slope of y-axis line (in red) represents the average reading speed. When we compared the lower part (for L1) of Fig. 10 (Japanese, a logosyllabic language) with Fig. 5 (English. a phonetic language), basically we could not notice any difference. Eye-tracking data shows similar smoothness when one reads a language of which the subject has strong command. For a conclusive result, we need to perform more experiments.

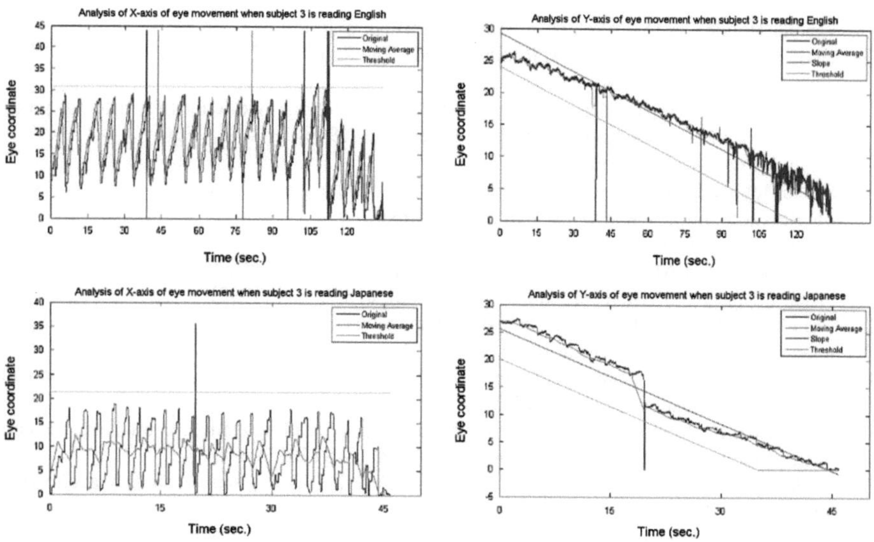

Fig. 10 Comparison of eye-movement while reading native logosyllabic language (Japanese) and second language (English)

6 Conclusion

We have shown that time-series eye-tracking data is a rich source of information about the reading idiosyncrasies and estimating proficiency of reading the language. Proposing a scoring function using different features of the data, we can quantify the reading ability and give a score. The proposed scoring function is heuristic, and designed to match with a few experimental results. To have a reliable score we need lots of data, especially from native speakers to define the bias of the scale. We identified the wasteful eye movements, pin-point the problem and suggest improvement. By further analysis, it is possible to identify whether the reader is poor in vocabulary, or find it difficult to associate different words to comprehend the semantics. Further experiments are necessary to verify our notions and results.

Acknowledgements This was partially supported by research grant from Iwate Prefectural University, iMOS research center.

References

1. Chakraborty, G., Kozma, R., Murata, T., Zhao, Q.: Awareness in Brain, Society and Beyond, A Bridge Connecting Raw Data to Perception and Cognition. IEEE Syst. Man Cybern. Mag. 9–16 (2015)
2. Noton, D., Stark, L.: Eye Movements and Visual Perception, Scientific American, pp. 35–43 (1971)

3. van der Heijden, A.H.C.: Selective Attention in Vision. Routledge (1992)
4. Godfroid, A.: Eye tracking. In: P. Robinson. (ed.) The Routledge Encyclopedia of Second Language Acquisition, pp. 234–236 (2012)
5. Ankrum, D.R.: Viewing Distance at Computer Workstations (guidelines for monitor placement). Work-Place Ergonomics, pp. 10–13, Sept/Oct (1996)
6. Tobii: http://www.tobii.com/
7. SensoMotoric Instruments(SMI): http://www.eyetracking-glasses.com/
8. SR Research: http://www.sr-research.com/eyelinkII.html
9. Eyemark: http://www.eyemark.jp/
10. Kosslyn, Stephen M., Thompson, William L., Alpert, Nathaniel M.: Neural systems shared by visual imagery and visual perception. Neuroimage **6**, 320–334 (1997)
11. Brandt, Stephan A., Stark, Lawrence W.: Spontaneous eye movements during visual imagery reflect the content of the visual scene. J. Cogn. Neurosci. **9**(1), 27–38 (1997)
12. Krieger, G., et al.: Object and scene analysis by saccadic eye-movements: an investigation with higher-order statistics. Spat. Vis. **13**(2, 3), 201–214 (2000)
13. Dolgunsoz, E.: Measuring attention in second language reading using eye-tracking: the case of the noticing hypothesis. J. Eye Mov. Res. **8**(5):4, 1–18 (2015)
14. Williams, R.S., Morris, R.K.: Eye movements, word familiarity, and vocabulary acquisition. Vis. Res. **46**, 426–437 (2004)
15. Juhasz, BJ.: The processing of compound words in english: effects of word length on eye movements during reading. Lang. Cogn. Process. **23**(7–8), 1057–1088 (2008)
16. Schotter, E.R., Tran, R., Rayner, K.: Dont Believe What You Read (Only Once): Comprehension is Supported by Regressions During Reading. UC San Diego Library Digital Collections (2015). doi:10.6075/J08G8HM2
17. Rayner, K., Slattery, T.J., Drieghe, D., Liversedge, S.P.: Data from: Eye movements and word skipping during reading: effects of word length and predictability. In: Rayner, K. (ed.) Eye Movements in Reading Data Collection. UC San Diego Library Digital. Collections (2013). doi:10.6075/J0F769G5
18. Rayner, K., Yang, Ji., Schuett, S., Slattery, T.J.: Eye movements of older and younger readers when reading unspaced text. In: Rayner, K. (ed.) Eye Movements in Reading Data Collection. UC San Diego Library Digital Collections. doi:10.6075/J0J10122
19. Rayner, K., Raney, G.E.: Eye movement control in reading and visual search: Effects of word frequency. Psychon. Bull. Rev. **3**(2), 245248 (1996)
20. Raney, G.E., Rayner, K.: Word frequency effects and eye movements during two readings of a text. Can. J. Exp. Psychol. **49**(2), 151172 (1995)
21. Raney, G.E., Campbell, S.J., Bovee, J.C.: Using eye movements to evaluate the cognitive processes involved in text comprehension. J. Vis. Exp. **83**, 1–7 (2014)
22. Schultz table to improve reading speed. http://www.ababasoft.com/wider_eye_span/shultc.html
23. EMR-9: http://eyemark.jp/product/emr_9/index.html

Forecasting the Start and End of Pollen Season in Madrid

Ricardo Navares and José Luis Aznarte

Abstract In this paper we approach the problem of predicting the start and the end dates for the pollen season of grasses (family Poaceae) and plantains (family Plantago) in the city of Madrid. A classification-based approach is introduced to forecast the main pollination season, and the proposed method is applied to a range of parameters such as the threshold level, which defines the pollen season, and several forecasting horizons. Different computational intelligence approaches are tested including Random Forests, Logistic Regression and Support Vector Machines. The model allows to predict risk exposures for patients and thus anticipate the activation of preventive measures for clinical institutions.

Keywords Forecasting · Time series · Pollen · Poaceae · Plantago · Support vector machines · Logistic regression · Random forests

1 Introduction

Airborne pollen levels have been associated to allergic rhinoconjunctivitis, asthma and the oral allergy-symptom in about 15 million people in Europe. Allergies have been continuously increasing in developed countries, not only in the number of affected patients but also in the severity of allergic reactions [20]. The establishment and the prediction of a pollen calendar is essential to reduce the exposure of allergic patients to pollen during the days of higher pollen concentration. It is also important to enable the development of other preventive measures.

There is no consensus on how to define the pollination season [9] which is the period where airborne concentrations of pollen are measured. Some authors define it based on the cumulative daily pollen counts [1, 7, 13] and other authors define it based on predefined threshold levels over which the season is considered to be started

R. Navares · J.L. Aznarte (✉)
Department of Artificial Intelligence, Universidad Nacional de Educación a Distancia (UNED), Madrid, Spain
e-mail: jlaznarte@dia.uned.es

I. Rojas et al. (eds.), *Advances in Time Series Analysis and Forecasting*, Contributions to Statistics, DOI 10.1007/978-3-319-55789-2_27

and ended [18]. This study visits both approaches in order to define the season which is going to be forecast.

Climate directly or indirectly defines the vegetation and acts on two levels: (1) during the stages prior to flowering [4, 14], and (2) during the pollen season [12, 17]. In this study, we characterize different features of the pollen season in order to determine the effect of meteorological parameters on the incidence of Poaceae and Plantago pollen in Madrid, Spain. Once the features are defined, several computational intelligence techniques are applied and compared according to their performance on this problem. We cast the season predicting problem into a binary classification one, in order to obtain the most accurate estimates for the start and end of the pollination season with special attention to the threshold at which allergy reactions might appear.

The rest of this paper is as follows. Section 2 deals with data preprocessing, including its cleansing, formatting and set up. Then in Sect. 3, we summarize the different approaches of what is considered a peak season and present its definition in order to identify the data points which belong to it. The computational intelligence models considered are described in Sect. 4, which walks through the system design and the definition of the features which will be tested according to its forecasting relevance. Section 5 contains the results and analysis of the different experiments. Finally, Sect. 6 draws the conclusions and the future lines of work in this line.

2 Data Description

The study uses observations of Poaceae and Plantago pollen from the Faculty of Pharmacy of Complutense University of Madrid, Spain (40°26′52.1″ N, 3°43′41.1″ W) from 1994 to 2013, provided by Red Palinológica de la Comunidad de Madrid. Meteorological data is provided by weather stations located in Barajas, Cuatro Vientos, Getafe and Colmenar and consists of hours of sunlight per day, the speed of wind in km/h, rainfall in mm/h and daily maximum, minimum and average temperature in degrees Celsius.

A first look at the pollen observations reveals the presence of missing data points. Bearing in mind the season start problem and the minimization of the loss of information, it is clear that those missing data points which appear around the months of February, March and April have a more severe impact, as they are the months in which usually the pollen season start is usually recorded. Thus, a long sequence of consecutive missing data points might multiply the forecasting errors as it may artificially delay the predicted season start. If we find a sequence of missing data around the season start date, the use of the traditional 'last observation carried forward' (LOCF) method may lead to an incorrect prediction of the season start. These reasons support the initial hypothesis that interpolation within each year is not enough.

Consequently, we propose to redistribute the data into a matrix of dimensions $N \times 365$, where N denotes the year. As there are leap years in the data sample, a first check has been done to verify whether there is any data point on the 29th of February which is missing. As it is not the case, each data point which lays on that date is not taken into account to interpolate. Later on, the data point will be plugged into the correspondent year. With this format, missing data points can be regressed using data within the year and between years.

From this matrix, two new matrices are generated, one with the missing data estimated using regression by rows (within the year) and another with the data regressed by columns (by years). Given the different years' conditions due to factors which directly influence pollen concentrations, it is important to avoid over-influence of data from previous or subsequent years when estimating a data point. High concentration of grains the same day in other years as the one to be estimated does not imply high concentration on day that day. In order to give more importance to most recent data, it is within the year, the final estimation is weighted.

Meteorological data, on the other hand, presented very few missing data points, so they were directly linearly interpolated.

3 Definition of Season Start and End

There is no consensus on the definition of the main pollination season, but the different proposals lie in two main categories: those based on cumulative daily pollen counts, which define the period with respect to a percentage of yearly total sum of daily concentrations, and those which rely upon a consistent pollen threshold breach [9].

Table 1 shows how different the effective computed dates are for our data depending on the season definition. It is noticeable that the definitions which use thresholds, such as [6, 18], instead of cumulative concentrations, such as [1, 7, 13], tend to limit the season to the period where the peak concentrations appear. They are also sensitive to out of period isolated peak concentrations.

Fig. 1 shows the pollen concentrations, both for Plantago and for Poaceae, for the same years considered in Table 1, as well as the limits of the season defined according to [13, 18]. In the case of Poaceae (bottom row), it is interesting to see how the latter approach (based on a threshold of 30 grains/m^3) restricts the pollination season to a few days around the main peak in 2006. The same applies to Plantago pollen (top row). It is noticeable how concentrations differ for each species, being Plantago less prolific compared to Poaceae, which motivates a threshold level adjustment based on the pollen class as proposed by [21].

In general, the proposal of [18] seems much more restrictive than [13]. This can be mitigated by relaxing the threshold condition by reducing the threshold to 15 grains/m^3, which produces a more realistic result in the cases studied (shown in the graph as a shaded rectangle).

Cumulative approaches imply forecasting, before the season start, the expected total yearly accumulation, which is an entirely different problem. Henceforth, we will limit this work to threshold-based definitions. In order to establish a systematic approach which allows for a more informed decision about the threshold, we will study a set of thresholds allowing the experts to choose the most influential defini-

Table 1 Considered definitions for the start and end of the Poaceae pollination season, with examples for some years

Approach	Definition	Year	Start	End
Nilsson et al. [13]	The day in which the sum of daily pollen concentration reaches a value over 5% (start) and 95% (end) of the total yearly sum	1999	26 Feb	06 Aug
		2001	17 Mar	09 Jul
		2006	16 Feb	05 Jul
		2010	12 Apr	27 Jul
Galán et al. [7]	The day in which the sum of daily pollen concentration reaches a value over 1% (start) and 99% (end) of the total yearly sum	1999	24 Jan	19 Oct
		2001	13 Feb	24 Sep
		2006	02 Feb	02 Sep
		2010	18 Feb	16 Sep
Andersen et al. [1]	The day in which the sum of daily pollen concentration reaches a value over 2.5% (start) and 97.5% (end) of the total yearly sum	1999	08 Feb	13 Sep
		2001	22 Feb	08 Aug
		2006	09 Feb	29 Jul
		2010	20 Mar	17 Aug
Sánchez-Mesa et al. [18]	The first day in which the daily pollen concentration reaches values over (start) and below (end) 30 grains/m^3	1999	16 May	20 Jun
		2001	17 May	03 Jun
		2006	01 May	02 Jun
		2010	23 May	24 Jun
Feher et al. [6]	The first day in which the daily pollen concentration reaches values over (start) and below (end) 3 grains/m^3 for 4 consecutive days	1999	09 Feb	11 Jul
		2001	02 Feb	05 Sep
		2006	03 Feb	17 Jul
		2010	29 Mar	08 Sep

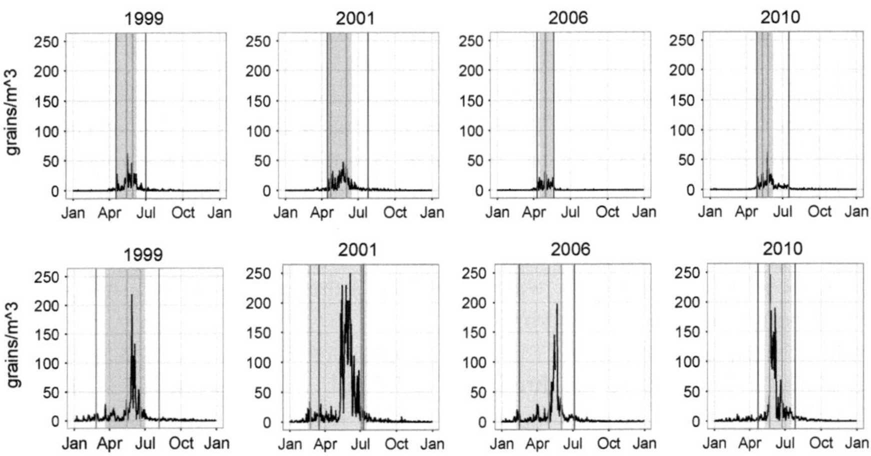

Fig. 1 Plantago (*top row*) and Poaceae (*bottom row*) pollen concentrations for years 1999, 2001, 2006 and 2010 and definition of the season according to [13] (*vertical red line*) and [18] (*vertical green line*). The shaded rectangle represents the latter approach relaxing the threshold to 15 grains/m^3

tion according to the relevance on their field. In what follows, let u be a fixed daily pollen concentration threshold, then the pollen season start as the first (last) day that surpasses u.

4 Methods

The final aim of this work is to help allergy patients in knowing in advance between which dates the pollen concentrations will be at risk levels. Given the above definitions of pollination season start and end, we aim at developing a model which forecasts these dates.

As seen in Sect. 3, there is no consensus as to which are the pollen concentrations considered as risk levels. Hence, several thresholds, ranging from 5 to 50 grains/m^3 for Poaceae and 5–15 grains/m^3 for Plantago [21], will be used in this work in order to provide a variety of options and to compare them.

Another important element that needs to be fixed is the forecasting horizon, which corresponds to the number of days in advance pollen concentrations will be forecast. There is always a trade off between precision and anticipation, and in the literature we can find predictions of the pollen season which range from 1 to 10 days in advance. In order to test its predictive capacities, the model will produce forecasts for several forecasting horizons ranging from 1 to 15 days.

Finally, for each combination of thresholds and horizons, different derived meteorological and pollen features are computed to set up the instances on which different machine learning algorithms will be trained.

Our approach is based on the idea that one can cast the forecasting problem into a binary classification problem where the featured instances represent influential factors for the predictions. Hence, daily pollen concentrations are mapped to $\{0, 1\}$ depending on whether they are above the threshold (1) or not (0). Given the definition of season start, the first data point classified as 1 will indicate the start of the season.

4.1 Feature Generation

In order to build such a classification system, the instances of each class should contain the most relevant data for that class. This relevant data can be meteorological conditions or pollen levels themselves, either for the day in which the prediction is to be made or for previous days, weeks or months, as it is generally assumed that those are the values that play a role in the development of the pollination process. At the same time, we need to avoid data which might not be related with the problem, for example we can assume that the average maximum temperatures of 5 years ago might not carry much information for the pollination period of the actual year.

According to previous works [16, 19], it is important to include the influence of most recent data, and hence cumulative pollen observations until the forecast day is defined as a synthetic variable. A 10 and 30 days cumulative sums of pollen daily concentrations prior to the forecast date are defined as features along with the prior 7 daily concentrations and the total sum of the pollen concentrations within the year.

Some authors assume that there is a linear relationship between the energy a plant receives and the growth state of buds [4]. This energy is represented in several ways, for example it is usual to consider that the sum of temperatures up to some point can be of help to forecast the state of the flowers [1, 4, 17].

On the other hand, some authors use the concepts of *chilling temperatures* and *forcing temperatures*, which are the weighted sum of the temperatures below and above certain thresholds given a period. Instead of using a predefined period, our approach is intended to capture all possible relevant periods which might influence the state of buds. In [14] the start of the chilling period is defined as the 1st of October and the start of the forcing period as of the 1st of February, which consequently it is also called the end of the chilling period. The forcing period ends when the pollination season start, and according to this approach is when the pollen concentration surpasses certain threshold. Instead, our approach calculates for each month the forcing and chilling temperatures and generate for each instance features that represent previous forcing and chilling temperatures, computed for previous months, quarters and so on.

Given the non-fixed definition of the chilling and forcing period, we decided not to apply any weight to the temperatures so the calculation of the forcing temperature sum is as follows:

$$F_{\text{sum}}(d) = \sum_{i=d-n}^{d} R_{\text{forc}}(i), \tag{1}$$

where

$$R_{\text{forc}}(i) = \begin{cases} 0 & \text{if } T(i) < T_{\text{forc}} \\ T(i) - T_{\text{forc}} & \text{if } T(i) \geq T_{\text{forc}} \end{cases} \tag{2}$$

being d the forecast date, n the number of days which define the calculation period for the sum of forcing temperatures, $T(i)$ the temperature for day i, and T_{forc} the base temperature for forcing (all temperatures are in degrees Celsius). The same applies for the chilling. In order to determine the base temperatures for T_{forc} and T_{chill} the levels proposed by [14] are used as a reference. The authors proposed a base temperature for the forcing period of 1 °C and 16 °C for the pollen thresholds of 10 grains/m^3 and 50 grains/m^3, respectively, and −6 °C and 8 °C for the chilling period. As this study uses different threshold, it is fair to approximate the values using simple geometrical relations setting the new values accordingly. Given a definition of the threshold of 30 grains/m^3 the corresponding base temperatures are 8 °C for the forcing and 6 °C for the chilling. Cumulative temperature parameterization is widely used to capture the energy induced to the plant during the early stage of bud states.

Table 2 Number of features generated by variable

	i	10	y	m	q	Q	std	MA_5	MA_{10}
Pollen	7	1	1	1	1	1	–	–	–
T[a]	21	–	–	–	–	–	–	3	3
T_{forc}	–	–	–	1	1	1	–	–	–
T_{chill}	–	–	–	1	1	1	–	–	–
Wind	7	1	1	1	1	1	1	1	1
Rain	7	1	1	1	1	1	1	1	1
Sun	7	1	1	1	1	1	1	1	1

i previous $i \in [1, 7]$ day observation; *10* previous 10 day cummulative sum; *y* year to date cummulative sum; *m* previous month cummulative sum; *q* previous 90 day cummulative sum; *Q* previous 180 day cummulative sum; *std* previous 15 days standard deviation; MA_5: previous 5 days moving average; MA_{10}: previous 10 days moving average
[a]accounts for 3 variables ($T_{min,max,avg}$)

Finally, it is known that pollen release is more prolific during dry weather rather than in rainy periods, even during cooler weather. Thus, it makes sense to take the cumulative approach introduced for temperatures in order to capture the prolonged rain periods. We need to capture heavy rains as well so the approach is based on the standard deviation of the last 15 days rainfall before the forecasting day. On the other hand, there are long term issues with heavy rains which need to be captured. For example, heavy spring rains are known to cause grass species to become more abundant as they grow more rapidly. Heavy rains during fall and winter cause pollen level increases in spring. Having said that, it is logical to include the accumulation of previous meteorological seasons. The same applies to the daily sun hours which is used as a proxy for dryness. For all climate data, similar as for the pollen counts, the prior 7 daily data observations are included. All variables are summarized in Table 2.

4.2 *Setting up the Data*

The aforementioned feature generation process leaves us with a total of 90 features. Depending on the desired threshold and forecast horizon, the data is set up according to the parameters in order to transform it into a classification problem. The first step consists on discretising the class and then assigning the class to the correspondent instance based on the forecast horizon defined.

$$\left[\begin{array}{cccc|c} x_{1,1} & x_{1,2} & \cdots & x_{1,90} & p_1 \\ \vdots & \vdots & \ddots & \vdots & \vdots \\ x_{n,1} & x_{n,2} & \cdots & x_{n,90} & p_n \end{array}\right] \rightarrow \left[\begin{array}{cccc|c} x_{1,1} & x_{1,2} & \cdots & x_{1,90} & c_{1+t} \\ \vdots & \vdots & \ddots & \vdots & \vdots \\ x_{n-t,1} & x_{n-t,2} & \cdots & x_{n-t,90} & c_n \end{array}\right] \quad (3)$$

$$c_i = \begin{cases} 0 & if \ p_i < u \\ 1 & if \ p_i \geq u \end{cases}, \tag{4}$$

where p_i is the daily pollen observation at time i, t the forecast horizon in number of days and u is the threshold as defined in Sect. 3. The observations are split into two subsets which will be used to train the correspondent algorithm and test its prediction accuracy. The test set consist on the observations which belong to the years 2011, 2012 and 2013 and the rest of the available years belong to the training set. As well, to avoid the over-fitting phenomenon, common to many machine learning models, a cross-validation procedure is performed on each year of the training set using as an error measure the absolute value of number of days between the estimated and the observed season start and end.

4.3 Feature Selection

Some models are highly sensitive to collinearity in the variables. In order to provide equal competitiveness to the algorithms, we need to reduce the number of features to those which are relevant for the class.

Hence, a filter algorithm based on [8] and on the definition of feature relevance by [10] is applied to rank subsets of features according to a correlation based evaluation function. This algorithm will select subsets that contain features highly correlated with the class and uncorrelated with each other. A feature is accepted when it predicts the class in areas of the instance space not already predicted by other features. The features are treated uniformly by discretisation in a pre-processing step, and then a correlation based heuristic is repeatedly applied to test the merit of a subset, defined as

$$M_s = \frac{k\overline{r_{cf}}}{\sqrt{k + k(k-1)\overline{r_{ff}}}}, \tag{5}$$

where M_s the merit of a subset S containing k features and $\overline{r_{cf}}$ is the mean feature-class correlation and $\overline{r_{ff}}$ the average feature-feature correlation.

4.4 Computational Intelligence Models

Different classification approaches are trained using the training set in order to forecast the start and end of the season for test set. Concretely we compare Random Forests (RF) [3], Logistic Regression (LR) [11] and Support Vector Machines (SVM) [15].

Proposed in 2001 by Leo Breiman [3], a Random Forest is the name for an ensemble approach which leverages the performance of many decision trees to produce predictive models. It is a supervised learning procedure which combines several randomized decision trees and aggregates their predictions by averaging. The procedure operates over sample fractions of the data, grows a randomized tree predictor on each one and aggregate these predictors together.

With respect to logistic regression, is a widely used regression model used in Statistics where the dependent variable is categorical. The model predicts the probability that a given example belongs to one class via the sigmoid function. A ridge estimator [11] was introduced to add penalty on weights learned to avoid over-fitting.

RF and LR make different assumptions about the data and has different rates of convergence. On the one hand, RF assumes that the decision boundaries are parallel to the axes based on whether a feature is $\geq, \leq, <$ or $>$ to certain value so the feature space is chopped into hyper-rectangles. On the other hand, LR finds a linear decision boundary in any direction by making assumptions on $P(C|X_n)$ applied to weighted features so non-parallel to the axes decision boundaries are picked out. This trade off motivates to take into account SVM as an alternative.

The current SVM standard algorithm, proposed by Cortes and Vapnik [5] in 1995, is a learning method used for binary classification which finds a hyper-plane which separates the d-dimensional data perfectly into its two classes. However, since sample data is often not linearly separable, SVM's introduces the notion of a *kernel induced feature space* which casts the data into a higher dimensional space where the data is separable.

In sum, the experiments are tailored to compare the models and compute their forecasts for each threshold and time horizon previously defined. Both parameters, threshold and horizon, define a set up of the data presented to the models according to Eqs. (3) and (4). Then a three step process applies consisting on feature selection and evaluation of the learning algorithm over the training set and prediction on the test set.

5 Results

One of the objectives of the experiments was to evaluate, in terms of their predictive ability in the framework of forecasting the pollen season in Madrid, different general purpose machine learning or statistic methods based on different paradigms. Hence, in order to select the best suited model of those described in Sect. 4.4, we tried them against the data described in Sect. 2.

For a set of thresholds and for a set of forecast horizons $h = \{1, 2, 5, 7, 10, 15\}$, we trained the three methods using the training set and checked their performance against the test data set.

Fig. 2a shows the start and end of the season for Poaceae pollen along with the predicted values for each combination of algorithm, threshold and forecast horizon.

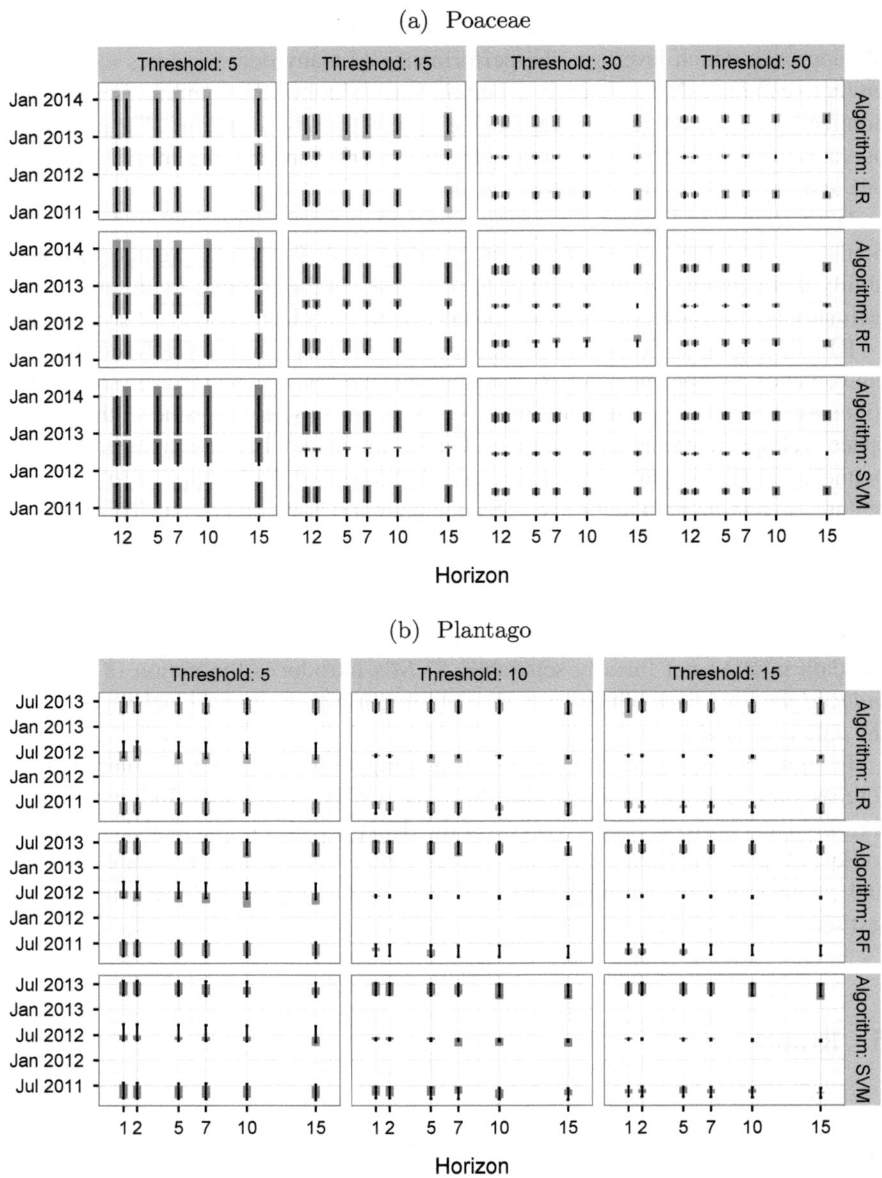

Fig. 2 Predicted (*coloured rectangles*) and observed (*black vertical lines*) season start and end dates for 2011, 2012 and 2013, by algorithm and threshold

Table 3 Test data set average errors for $u = 30$ and $u = 15$ for Poaceae and Plantago respectively, in number of days, of the predictions for the start of the season

Poaceae	Horizon					
Algorithm	1	2	5	7	10	15
LR	1.00	8.00	8.67	9.00	9.00	19.33
RF	0.33	1.00	10.67	12.33	15.67	23.50
SVM	1.33	1.67	1.33	1.67	1.33	3.33
Plantago	Horizon					
Algorithm	1	2	5	7	10	15
LR	1.00	1.67	0.67	0.78	5.18	12.47
RF	12.22	12.72	15.50	15.83	12.50	16.06
SVM	10.39	9.61	10.17	11.89	12.28	18.53

It can be clearly seen the highly dependence between the season duration and the definition of the threshold.

The results with the test data set might derive from the fact that the models do not have enough data to properly generalize, as we only have 20 years, which means only 20 season starts and ends. However, it is clear that high threshold levels lead to more satisfactory results, enabling the classifier to identify the patterns which influences the season start and end even for long forecasting periods.

On the other hand, Fig. 2b shows comparatively very short pollination seasons for Plantago pollen. This is due to the fact that this species is not as common in metropolitan areas as in rural regions [21]. For this reason, the threshold levels were relaxed according to the findings in [21]. However, the proposed models are in this case tested with a small set of data that are effectively classified as main pollination season, and this plays against the computational intelligence models as they need a high number of training observations. For instance, RF strives to identify the main pollination season in 2012 for thresholds over 5 grains/m^3.

From a clinical point of view, predicting the moment in which most of the patients will start having symptoms is of a greater interest than predicting the moment when they will experience relief. Hence, Table 3 shows the error obtained by each model for all the horizons considered at predicting the start of the season. Only the threshold $u = 30$ is considered for Poaceae, following [2] (all patients experience moderate or severe symptoms) and $u = 15$ for Plantago [21]. It is clear that SVM outperforms the other algorithms for horizons over 5 days, while RF is the best for 1 or 2 days ahead forecasts of Poaceae pollen season. Conversely, LR is shown as the best performer for Plantago given the limited amount of training samples in this case. This situation leads to an increase in robustness for LR compared to the other proposals, which need a higher number of observations over the threshold in order to obtain the inner information from the data.

6 Conclusions and Future Works

This study introduces a new approach to foresee the start and end of the pollination season, which might help allergic patients as well as public health institutions. It is shown that tackling the problem from a purely data-driven point of view produces good results and gets accurate forecasts of the pollination season even in years with particularly odd characteristics as it is 2012, which shows a specially short main pollination period with a sudden start.

We have seen SVM as the most general model for prediction on this problem having accurate results for horizons within a week. The definition of the threshold, which dictates the start and end of the pollination season, takes an important role on the performance of the models. This study shows that levels above 20 grains/m^3 allow an accurate prediction in the case of Poaceae. It is to note that previous works set the threshold at 30 grains/m^3 or above [9].

Regarding Plantago pollen, the season definition produced a limited number of observations over the threshold above 15 grains/m^3, and LR was the most robust approach.

The proposed approach provide forecasts based on the data and making no assumptions on the phenology of the plant. Thus, it can be applied to any kind of pollen regardless its origin. The results are presented in a way to be easily interpreted either by experts from other fields or patients.

The results are promising but some ideas are worth deeper exploration. For example, the generation and selection of features could be improved by using bio-inspired algorithms. As well, the introduction of numerical weather predictions should enhance the prediction results. As well, predictions which account for uncertainty for the start date, like probabilistic predictions, could also be of interest.

Acknowledgements This work has been partially funded by Ministerio de Economía y Competitividad, Gobierno de España, through a *Ramón y Cajal* grant awarded to Dr Aznarte (reference: RYC-2012-11984).

References

1. Andersen, T.B.: A model to predict the beginning of the pollen season. Grana **30**, 269–275 (1991)
2. Antépara, I., Fernández, J.C., Gamboa, P., Jauregui, I., Miguel, F.: Pollen allergy in the Bilbao area (European Atlantic seaboard climate): pollination forecasting methods. Clin. Exp. Allergy **25**(2), 133–140 (1995)
3. Breiman, L.: Random forest. Mach. Learn. **45**, 5–32 (2001)
4. Cannell, M.G.R., Smith, R.I.: Thermal time, chill days and prediction of budburst in Picea sitchensis. J. Appl. Ecol. **20**, 269–275 (1983)
5. Cortes, C., Vapnik, V.N.: Support-vector networks. Mach. Learn. **20**, 273–276 (1995)
6. Feher, Z., Jarai-Komlodi, M.: An examination of the main characteristics of the pollen seasons in Budapest, Hungary (1991–1996). Grana **36**, 169–174 (1997)

7. Galan, C., Emberlin, J., Dominguez, E., Bryant, R.H., Villamandos, F.: A comparative analysis of daily variations in the gramineae pollen counts at Cordoba, Spain and London, UK. Grana **34**, 189–198 (1995)
8. Hall, M.A.: Correlation-based feature selection for machine learning. Ph.D. thesis. University of Waikato (1999)
9. Jato, V., Rodriguez-Rajo, F.J., Alcazar, P., De Nuntiis, P., Galan, C., Mandrioli, P.: May the definition of pollen season influence aerobiological results? Aerobiologia **22**, 13–25 (2006)
10. Kohavi, R., John, G.H.: Wrappers for feature subset selection. Artif. Intell. **97**, 273–324 (1997)
11. le Cessie, S., van Howelingen, J.C.: Ridge estimators in logistic regression. Appl. Stat. **41**, 191–201 (1992)
12. Myszkowska, D.: Predicting tree pollen season start dates using thermal conditions. Aerobiologia **30**, 307–321 (2014)
13. Nilsson, S., Persson, S.: Tree pollen spectra in the stockholm region (Sweden), 1973–1980. Grana **20**, 179–182 (1981)
14. Pauling, A., Gehrig, R., Clot, B.: Toward optimized temperature sum parametrizations for forecasting the start of the pollen season. Aerobiologia **30**, 45–57 (2014)
15. Rakotomamonjy, A.: Variable selection using SVM-based criteria. J. Mach. Learn. **3**, 1357–1370 (2003)
16. Ribeiro, H., Cunha, M., Abreu, I.: Definition of main pollen season using logistic model. Ann. Agric. Environ. Med. **14**, 259–264 (2007)
17. Rodriguez-Rajo, F.J., Frenguelli, G., Jato, M.V.: Effect of air temperature on forecasting the start of the Betula pollen season at two contrasting sites in the south of Europe (1995–2001). Int. J. Biometeorol. **47**, 117–125 (1983)
18. Sanchez-Mesa, J.A., Smith, M., Emberlin, J., Allitt, U., Caulton, E., Galan, C.: Characteristics of grass pollen seasons in areas of Southern Spain and the United Kingdom. Aerobiologia **19**, 243–250 (2003)
19. Smith, M., Emberlin, J.: A 30-day-ahead forecast model for grass pollen in North London, UK. Int. J. Biometeorol. **50**, 233–242 (2006)
20. Sofiev, M., Bergmann, K.C.: Allergenic Pollen: A Review of the Production, Release, Distribution and Health Impacts. Springer Science and Business Media (2012)
21. Tobías, A., Sáez, M., Galán, I., Benegas, R.: Point-wise estimation of non-linear effects of airborne pollen levels on asthma emergency room admissions. Allegy **64**, 961–962 (2009)

Statistical Models and Granular Soft RBF Neural Network for Malaysia KLCI Price Index Prediction

Dusan Marcek

Abstract Two novel forecasting models are introduced to predict the data of Malaysia KLCI price index. One of them is based on Box-Jenkins methodology where the asymmetric models, i.e. EGARCH and PGARCH models were used to form the random component for ARIMA model. The other forecasting model is a soft RBF neural network with cloud Gaussian activation function in hidden layer neurons. The forecast accuracy of both models is compared by using statistical summary measures of model's accuracy. The accuracy levels of the proposed soft neural network are better than the ARIMA/PGARCH model developed by most available statistical techniques. We found that asymmetric model with GED errors provide better predictions than with Student's t or normal errors one. We also discuss a certain management aspect of proposed forecasting models by their use in management information systems.

Keywords ARIMA and ARCH/GARCH models • Neural networks • Forecast accuracy • Daily KLCI-price index prediction

1 Introduction

Statistical models and artificial neural networks as machine learning techniques are a modern part of quantitative prediction models which are one of the most used approaches for finance in making right decisions. This paper discusses and compares the forecasts from ARIMA/ARCH-class models (AutoRegressive Integrated Moving Average/AutoRegressive Conditional Heteroscedastic) and RBF NN (Radial Basic Function Neural Network). The aim is to examine, whether potentially non-linear neural networks outperform latest statistical methods or generate prognoses which are at least comparable with those of statistical models. First one is

D. Marcek (✉)
Research Institute of the IT4Innovations Centre of Excellence,
The Silesian University Opava, Bezruc. Square 33, Opava, Czech Republic
e-mail: dusan.marcek@fpf.slu.cz

I. Rojas et al. (eds.), *Advances in Time Series Analysis and Forecasting*,
Contributions to Statistics, DOI 10.1007/978-3-319-55789-2_28

based on latest statistical methods makes use of the ARIMA/ARCH-class models. The second one is the neural network based on radial basic activation function that makes uses both supervised and unsupervised learning. After that, we discuss certain management aspects of proposed forecasting models such as capabilities and interests of the people who will make and use the forecast in their decision processes.

We organize our paper as following: in Sect. 2 we shall introduce some necessary theoretical background of ARIMA/ARCH family models considered for this paper. In Sect. 3 we describe the models we used in RBF neural networks and in Sect. 4 we present the data, conduct some preliminary analysis of the time series and demonstrate the forecasting abilities of both ARIMA/ARCH and RBF class models applied to the data taken from the Malaysia KLCI (Kuala Lampur Composite Index) price index time series. Section 5 includes an empirical comparison and proposes future work.

2 Statistical Models

Time series models have been initially introduced either for descriptive purposes like prediction or for dynamic control. For more than 20 years Box-Jenkins ARIMA models have been widely used for time series modelling. The econometric approach adopted from early days of econometrics is referred to as "AER" or Average Economic Regression [1] is concerned with the functional form of the multiple regression models in the form

$$y_t = \beta_0 + \beta_1 x_{1t} + \cdots + \beta_p x_{pt} + u_t \qquad (1)$$

where represent a series of independent variables, β_i regression intercept, partial regression coefficients, for $i = 1, \ldots, p$, u_t is the independent random error term, for $t = 1, \ldots, N$.

2.1 Box-Jenkins Time Series-Class Models

Box and Jenkins [2] introduced new time series models derived from linear filter which usually are called ARIMA models. The fundamental aim of time series analysis is to understand the underlying mechanism that generates the observed data and, in turn, to forecast future values of the series. Given the unknowns that affect the observed values in time series, it is natural to suppose that the generating mechanism is probabilistic and to model time series as stochastic processes.

ARIMA models combine autoregressive (AR) and moving average (MA) part. AR part is a linear combination of previous values, 'I' is an operator for differencing a time series and MA part is a linear combination of previous errors. An ARMA(p, q) model of orders p and q is defined as

$$y_t = \phi_1 y_{t-1} + \phi_2 y_{t-2} + \cdots + \phi_p y_{t-p} + \varepsilon_t + \theta_1 \varepsilon_{t-1} + \theta_2 \varepsilon_{t-2} + \cdots + \theta_q \varepsilon_{t-q} \tag{2}$$

where $\{\phi_i\}$ and $\{\theta_j\}$ are the parameters of the autoregressive and moving average parts, respectively, and ε_t is the white noise with mean zero and variance σ^2. We assume that ε_t is normally distributed, that is $\varepsilon_t \sim N(0, \sigma)$. Then, the ARIMA($p$, d, q) represents the d-th difference of the original series as a process containing p autoregressive and q moving average parameters.

2.2 Asymmetric ARCH-GARCH Class Models for Financial Data

Among the field of applications where the standard ARIMA fit is poor are financial and monetary problems. Another weakness of ARMA models is the inability to model non-constant variance. This type of variance is very common in stock indexes, currency pairs and so on. In this context, ARCH models introduced by Engle [3] arose as an appropriate framework for studying these problems. Bollerslev [4] proposed an extension of Engle's ARCH model known as the generalized ARCH (GARCH) model of order (p, q) for time sequence $\{y_t\}$ or $\{h_t\}$, respectively in the following form

$$h_t = \alpha_0 + \sum_{i=1}^{m} \alpha_i y_{t-i}^2 + \sum_{j=1}^{s} \beta_j h_{t-j} \tag{3}$$

where $\{\alpha_i\}$ and $\{\beta_j\}$, are the ARCH and GARCH parameters, *ht* represent the conditional variance of time series. The primary restriction of GARCH-type models is that they force a symmetric response of volatility to positive and negative news. The asymmetric response of good and bad news to future volatility is known as the leverage effect. Nelson [5] proposed the following exponential asymmetric GARCH (Generalized ARCH) model abbreviated as EGARCH to allow for leverage effects in the form

$$\log h_t = \alpha_0 + \sum_{i=1}^{p} \alpha_i \frac{|\varepsilon_{t-i}| + \gamma_i \varepsilon_{t-i}}{\sigma_{t-i}} + \sum_{j=1}^{q} \beta_j \log h_{t-j} \tag{4}$$

Note if ε_{t-i} is positive or there is "good news", the total effect of ε_{t-i} is $(1 + \gamma_i)\varepsilon_{t-i}$. However, contrary to the "good news", i.e. if ε_{t-i} is negative or there is “bad news”, the total effect of is $(1-\gamma_i)|\varepsilon_{t-i}|$. Bad news can have a larger impact on the volatility. Then the value of γ_i would be expected to be negative, see [6].

The basic GARCH model can be extended to allow for leverage effects. This is performed by treating the basic GARCH model as a special case of the power GARCH (PGARCH) model proposed by Ding, Granger and Engle [7].

$$h_t^d = \alpha_0 + \sum_{i=1}^{p} \alpha_i(|\hat{\varepsilon}_{t-i}| + \gamma_i \hat{\varepsilon}_{t-i})^d + \sum_{j=1}^{q} \beta_j h_{t-j}^d \tag{5}$$

3 Neural Networks

3.1 Mathematical Model of Neural Network

Neural networks can be understood as a system which produces output based on inputs the user has defined. In this system the user has no knowledge about internal working of the system of ANN. This principle is named as Black Box Principle, see Fig. 1.

If we look at the ANN in Fig. 2 on the left, from mathematical point of view we can see, that each neuron in hidden layer computes nonlinear function F of its inputs. Each neural network is then a composition of large number of such partial function with different parameters. We can then be looking on the whole network with k inputs and one output as a vector mapping function F: $x_t \in R^k \to y_t \in R^1$, i.e. as a projection assigning k-dimensional vector of inputs $x_t^T = (x_{1t}, x_{2t}, \ldots, x_{kt})$ one dimensional output y_t in specific time t. The process of learning then seems as an attempt to approximate this function by setting its parameters with an error function $E(\mathbf{w}_t) = \sum_{x_t, y_t \in R^k_{train}} G(\mathbf{x}_t, \mathbf{w}_t - y_t)^2$, which represents desired behavior of a learned network. The task is then to find the values of $\mathbf{w}_t$ so that the error function will reach a minimum. Then, of course we will be interested in what accuracy can the networks be approximated. A key achievement in this area was Kolmogorov theorem, which says, that any continuous function of multiple variables can be expressed by the sum and composition of function with one variable. On this basis Hornik [8, 9] proved results that neural networks can learn with arbitrary precision to imitate any behavior that can be described with continuous function. If we apply the nonlinear mapping function for neurons in the hidden and output layer, we obtain an explicit expression for the complete function represented by the network

Fig. 1 *Black Box* principle of ANNs architecture (*Source* Author’s own processing)

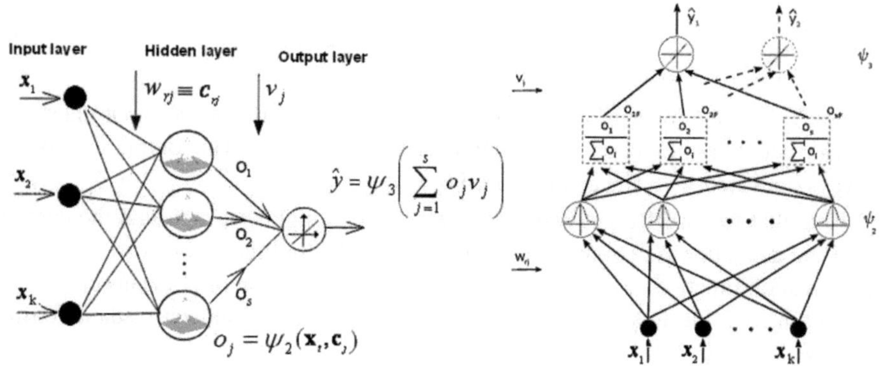

Fig. 2 On the *left* classic RBF NN, on the *right* soft RBF NN architecture (*Source* Author's own processing)

diagram in Fig. 2 on the left in the form $y_t = \psi_3[\mathbf{v}_t^T \psi_2(\mathbf{w}_t \mathbf{x}_t)]$ where ψ_2, ψ_3 denote transfer function for neurons in the hidden and output layer; $\mathbf{w}_t, \mathbf{v}_t$ denote the weights in the hidden and output layer, respectively.

3.2 *RBF Neural Networks*

The most known representatives of feed-forward networks are perceptrons and their upgraded version called RBF network.

RBF neural networks are described by their architectures, see Fig. 2. In Fig. 2 on the left each circle or node represents the neuron. This neural networks consists an input layer with input vector $\mathbf{x}$ and an output layer with the output value $\hat{y}_t$. The layer between the input and output layers is normally referred to as the hidden layer and its neurons as RBF neurons. Here, the input layer is not treated as a layer of neural processing units. RBF network defines potential of *j*-th hidden neuron u_j as a difference of Euclidean distance between the input vectors and the weight vectors w_j given by formula

$$u_j = \|x - w_j\|, \quad \text{for } j = 1, 2, \ldots, s \tag{6}$$

where s denotes the number of neurons in the hidden layer.

The output signals of the hidden layer are

$$o_j = \psi_2(\|\mathbf{x}_t - \mathbf{w}_j\|) \tag{7}$$

here $\mathbf{x}_t$ is a *k*-dimensional neural input vector, $\mathbf{w}_j$ represents the hidden layer weights, ψ_2, are radial basis (Gaussian) activation functions defined as

$$\psi_2(u_j) = e^{\frac{-(x-w_j)^T \sum^{-1}(x-w_j)}{}} \quad (8)$$

where $\sum^{-1}$ is the inverse of the variance-covariance matrix of the input data.

The activation function of output neuron is different. The output neuron is always activated by a linear function, i.e. $\hat{y} = \sum_{j=1}^{s} o_j v_j$.

Note that for an RBF network, the hidden layer weights $\mathbf{w}_j$ represent the centres $\mathbf{c}_j$ of the activation functions in the hidden layer. To find the weights $\mathbf{w}_j$ or centres of activation functions we use the following adaptive version of *K*-means clustering algorithm for *s* clusters. In step 2 this algorithm uses the Kohonen's adaptive rule [10].

Step 1 Randomly initialize the centres of RBF neurons $c_j^{(t)}$, $j = 1, 2, \ldots, s$ where *s* represents the number of chosen RFB neurons (clusters).

Step 2 For each vector $x^{(t)} = (x_1, x_2, \ldots, x_k)$ of the training data set find the nearest centre to $x^{(t)}$ and replace its position as follows

$$c_j^{(t+1)} = c_j^{(t)} + \lambda(t)(x^{(t)} - c_j^{(t)})$$

where $\lambda(t)$ is the learning coefficient selected as the linearly decreasing function of *t* by $\lambda(t) = \lambda_0(t)(1 - \frac{t}{N})$ where $\lambda_0(t)$ is the initial value, *t* is the presented learning cycle and *N* is the number of learning cycle.

Step 3 After reaching a selected number of epochs, terminate learning, otherwise go to **Step 2**.

The competitive clustering algorithm is regarded as one of the granular methods presenting bottom-up granulation, i.e. the input data are combined into larger granules. The standard deviations of clusters are calculated by the formula

$$\sigma_j = \left(1/M \sum_{i=1}^{M} \left\|c_j - x_m\right\|^2\right)^{1/2}, \; j = 1, 2, \ldots, s \quad (9)$$

where x_m is the *m*-th input vector belonging to cluster $\mathbf{c}_j$.

The granules extracted from available data, see Fig. 3 on the left, are then described by the three digital characteristics of the normal cloud model, namely expected value, entropy and hyper-entropy. The mean of a granule is regarded as the expected value of normal cloud model, and the standard deviation of a granule provides the entropy of normal cloud model. Both characteristics are calculated in the process of competitive learning. The hyper-entropy of a granule is a measure of dispersion of the cloud drops and it can be calculated using backward algorithm [11] or set up manually. Figure 4 on the right illustrates the cloud activation function of granules in the hidden layer. Then, in case of soft RBF network, the Gaussian activation function ψ_2 (%) has the form [12]

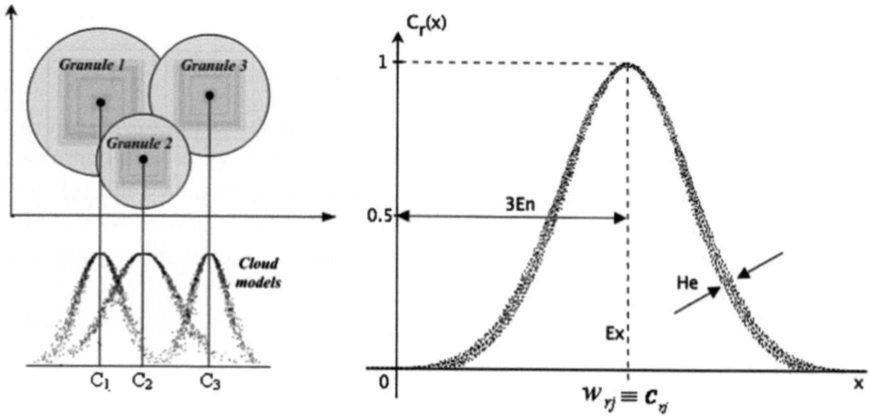

Fig. 3 On the *left* a demonstration of clusters (granules) extracted from data and described by cloud concept. On the *right* a possible model of cloud concept with the Gaussian function (*Source* Author's own processing)

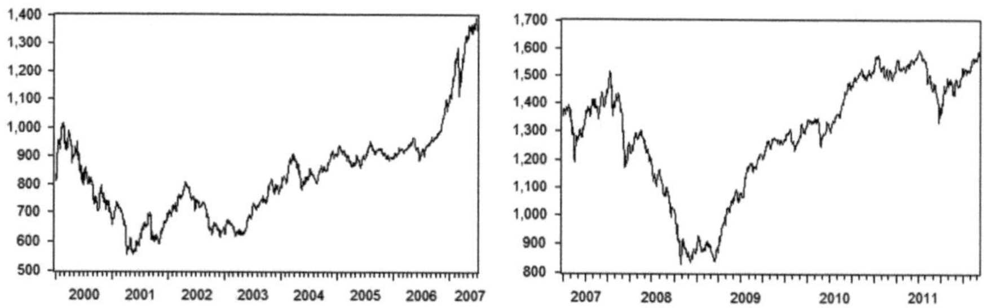

Fig. 4 On the *left* KLCI price index values, period 1/2000–6/2007; on the *right* KLCI price index values, period 7/2007–3/2012 (*Source* Author's own processing)

$$\psi_2(x_t, c_j) = \exp\left[-(x_t - \mathrm{E}(x))^2 / 2(En')^2\right] = \exp\left[-(x_t - c_j)^2 / 2(En')^2\right] \quad (10)$$

where En' is a normally distributed random number with mean En and standard deviation He, E is the expectation operator, $\mathbf{x}_t$ denotes the input data vector, $t = 1, \ldots, N$.

The output layer consists of one output neuron with linear activation function given by formula $\hat{y}_t = \sum_{j=1}^{s} o_j v_j$ where o_j are the outputs from hidden neurons, v_j are the weights between hidden and output layer, s is the number of hidden neurons and $\hat{y}_t$ is the output from the RBF network.

If in the structure of RBF network according to Fig. 2 on the left the scalar output values $o_{j,t}$ from the hidden layer will be normalized where the normalization means that the sum of the outputs from the hidden layer is equal to 1, then the RBF network will compute the output data set as follows

$$y_t = \sum_{j=1}^{s} v_{j,t} \frac{\psi_2(x_t, c_j)}{\sum_{j=1}^{s} \psi_2(x_t, c_j)}, \ t = 1, 2, \ldots, N \tag{11}$$

The network with one hidden layer and normalized output values $o_{j,t}$, see Fig. 3 on the right, is the soft or fuzzy logic network [13]. The network with one hidden layer and normalized output values $o_{j,t}$ is the fuzzy logic model or the soft RBF network. The architectures of both networks in Fig. 2 are the same in the sense that each has just one hidden layer. Bat, in fuzzy logic, the parameters to be learned, are the output layer weight changes v_j, $j = 1, 2, \ldots, s$.

4 Building the Statistical and G RBF NN Prediction Models

We will develop two prediction models for the time series of daily Malaysia KLCI price index. Empirical analysis is performed on daily data of KLCI price index in period from January 3, 2000 to March 8, 2012,[1] which includes total of 2918 observations. The first period (as the training data set) was defined from January 2000 to the end of June 2007, i.e. the time before the global financial crisis or pre crisis period, and the second one so called crisis and post crisis period (as the validation data set or ex post period) started at the beginning of July 2007 and finished by the March 8, 2012. We use post-crisis period in order to imply dynamics and volatilities into our model. Visual inspection of the time plot of the daily values of KLCI price index is shown in Fig. 4. The daily time series exhibits non-stationary behaviour. However, after its first differencing form become stationary, which was confirmed by ADF test, see Table 1.

ARIMA model was chosen based on traditional statistical analysis: these included the raw KLCI price index values and lags thereof. The relevant lag structure of potential inputs was analyzed using traditional statistical tools, i.e. by using auto-correlation function (ACF), partial auto-correlation function (PACF) and Akaike information criterion (AIC): we looked to determine the maximum lag for which the PACF coefficient was statistically significant and the lag given the minimum AIC. According to these criterions the ARIMA(1, 1, 0) model was specified in the form

$$\Delta y_t = \xi + \phi_1 \Delta y_{t-1t} \tag{12}$$

Estimated mean Eq. (12) is shown in Table 2 on the left.

However, there are some aspects of the model which can be improved so that it can better capture the characteristics and dynamics of a particular time series. For

[1]This time series can be obtained from http://www.bloomberg.com/quote/FBMKLCI:IND.

Table 1 Results of ADF test applied to KLCI time series (*Source* Author's own processing)

ADF test	*t*-statistics	p-value
Original KLCI time series	−0.1094	0.9467
KLCI time series after differencing	−49.301	0.0001

Table 2 On the *left* estimated mean Eq. (12); on the *right* re-estimated mean Eq. (12) for KLCI price index values assuming that the random component follow PGARCH(1, 1) GED process (*Source* Author's own processing using E-views software)

Co-eff.	Value	St. dev.	p-value	D-W	Co-eff.	Value	St. dev.	p-value	D-W
ξ	3.938	5.899	0.505	1.994	ξ	0.304	0.160	0.058	1.996
$\phi 1$	0.047	0.019	0.0134		$\phi 1$	0.175	0.022	0.000	

example, the R system[2] assists in performing residual analysis (computes the Gaussian, Student's and generalized residuals with generalized error distribution—GED). For this purpose the Akaike information criterion (AIC) and Maximum-Likelihood (ML) functions as model selection criteria was applied to select the fittest model to the data. The model with lowest value of AIC fits the data best. Thus, we re-estimated the models after having eliminated restrictive assumption that the error terms follow a normal distribution. In order to accomplish this goal, we assumed that the residuals of model (12) follow successively Normal, Student's distribution and also a Generalized Errors distribution (GED). Table 3 presents AIC and Log likelihood functions in all cases. From Table 3 it is seen that the model PGARCH with GED error distribution fits the data best.

After these findings we re-estimated the mean Eq. (12) assuming that the random component follow PGARCH(1, 1) with GED. Re-estimated parameters are given in Table 2, on the right.

Finally to test for nonlinear patterns in KLCI price index time series the fitted standardized residuals $\hat{\varepsilon} = e/\sqrt{h_t}$ were subjected to the BDS test [14]. The BDS test (at dimensions N = 2, 3, and tolerance distances $\varepsilon = 0.5, 1.0, 1.5, 2.0$) finds no evidence of nonlinearity in standardized residuals of KLCI price indexes.

Now, we can make predictions for the validation data set (ex post period). These predictions are calculated by the dynamic forecast means of the E-views software[3] which means that future values of lagged residuals are generated using the forecasted values of the dependent variable and are shoved in Fig. 5 on the left.

For the investigation of the neural networks a granular RBF NN with architecture given in Fig. 2 on the right was employed.

[2]http://cran.r-project.org.

[3]http://www.eviews.com.

Table 3 Information criteria and maximum-likelihood function for error distribution models (*Source* Author's own processing using R system software)

Criterion model	AIC	ML	AIC	ML	AIC	ML
Distribution	Normal		Student's		GED	
GARCH(1, 1)	6.5025	−6348.0	6.4178	−6264.2	6.4171	−6263.7
PGARCH(1, 1)	6.5018	−6345.3	6.4182	−6262.6	**6.4170**	**−6262.8**
EGARCH(1, 1)	6.5047	−6349.1	6.4194	−6264.8	6.4184	−6264.0

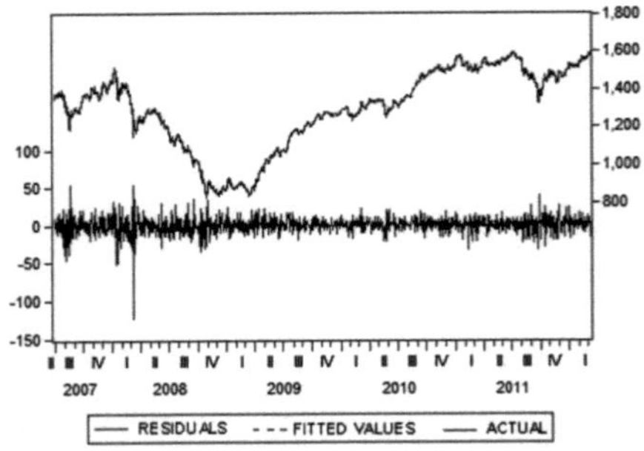

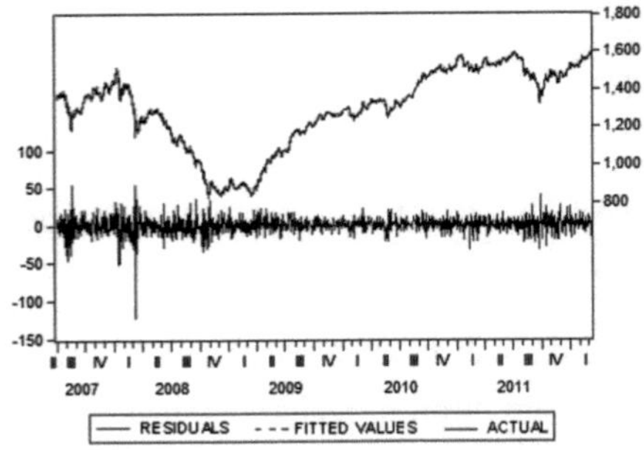

Fig. 5 On the *left* actual and fitted values of KLCI price index (validation data set) for re-estimated statistical model (12), residuals are at the *bottom*. On the *right* actual and fitted values of KLCI price index (validation data set) for G RBF NN model (*Source* Author's own processing using E-views software)

To test whether neural network with implemented cloud concept is able to produce more accurate outputs and hence to lower risk finance in better way than latest statistical models we used the some high-frequency time series data—daily close prices of the KLCI price index. We use own application of the feed forward neural network of soft RBF type with one hidden layer. The weights of network o_j were initiated randomly generated from the uniform distribution (0, 1). The weights v_j were trained by using *back-propagation* algorithm. The learning rate of *back-propagation* was set to 0.001 to avoid the easy imprisonment in local minimum. The final results were taken from the best of 5000 epochs and not from the last epoch, in order to avoid overfitting of the neural network. The transfer function in the hidden layer were a Gaussians RBF with cloud concept (10), whereas for the output unit a linear transfer function was applied.

The output values o_j from the hidden layer were normalized. To find the centres c_j of radial basis functions we used the on-line version of K-means clustering algorithm described in Sect. 3. The initial value $\lambda_0(t)$ in the adaptive K-means clustering algorithm was set to 0.98.

Table 4 Comparison of forecast summary statistics for KLCI price index time series statistical and neural approach: ex post period (*Source* Author's own processing)

Measures of the model's forecast accuracy: model	RMSE	MAPE
ARIMA(1,1,0)	10.99619	0.605779
Soft G RBF NN	8.68245	0.492541

5 Empirical Comparison and Conclusion

In this paper we suggested a soft artificial neural network with implemented cloud concept forecasting model as an alternative technique to Box-Jenkins methodology where the asymmetric models EGARCH and PGARCH were used to form the random component for ARIMA model.

Table 4 presents the summary statistics of each model based on RMSE and MAPE calculated over the validation data set (ex post period). We see from Table 4 that the best performing method is soft G RBF NN. From Table 4 it is also shown that both forecasting models used are very accurate. The development of the error rates on the validation data set showed a high inherent deterministic relationship of the underlying variables. Though promising results have been achieved with both approaches, for the chaotic financial markets a purely linear (statistical) approach for modelling relation-ships does not reflect the reality. For example if investors do not react to a small change in price index at the first instance, but after crossing a certain interval or threshold react all the more, then a non-linear relationship between y_t and y_{t-1} exist in model (12). As could we seen, neural networks are usually used in the complicated problems of prediction because they minimize the analysis and modelling stages and the resolution time. In our case, they omit diagnostic checking, significantly simplify estimation and forecasting. Thus, we can expect more interests of the managers who will make and at the some time use the forecast.

Finally, on base of our experiments we can say that the artificial neural network has a big potential to perform even better forecasts and therefore, can be helpful in financial risk management by providing more accurate predictions.

In future more advanced and enhanced techniques like RBF neural network based on incorporation of an error-correction mechanism will be used which is sometime included in dynamic regression, or exploring other ways of combining the prediction by self-adapting multi-agent learning systems with evolution development features, or generally concepts, representing the causal relationships among the information granules, taking in the account dynamic nature of the cause with uncertainties which is very important for prognoses construction.

Acknowledgements This work was supported by the Ministry of Education, Youth and Sports from the National Programme of Sustainability (NPU II) project IT4Innovations excellence in science—LQ1602.

References

1. Kennedy, P.A.: Guide to Econometrics. Basil Blackwell, Oxford (1992)
2. Box, G.E.P., Jenkins, G.M.: Time Series Analysis: Forecasting and Control. Holden-Day, San Francisco (1976)
3. Engle, R.F.: AutoRegressive conditional heteroscedasticity with estimates of the variance of United Kindom inflation. Econometrica **50**, 987–1007 (1982)
4. Bollerslev, D.: Generalized autoregressive conditional heteroscedasticity. J. Econ. **31**, 307–327 (1986)
5. Nelson, D.B.: Conditional heteroscedasticity in asset returns: a new approach. Econometrica **59**(2), 347–370 (1991)
6. Zivot, E., Wang, J.: Modeling Financial Time Series with S-PLUS®. Springer, NY (2005)
7. Ding, Z., Granger, C.W., Engle, R.F.: A long memory property of stock market returns and a new model. J. Empir. Financ. **1**, 83–106 (1993)
8. Hornik, K., Stinchcombe, M., White, H.: Multilayer feedforward networks are approximators. Neural Netw. **2**(5), 359–366 (1989)
9. Hornik, K.: Some new results on neural network approximators. Neural Netw. **6**(8), 1069–1072 (1993)
10. Kohonen, T.: Self-Organization and Associative Memory. Springer, Berlin (2012)
11. Li, D., Du, Y.: Artificial Intelligence with Uncertainty. Chapman & Hall/CRC, Taylor & Francis Group, Boca Raton (2008)
12. Marcek, M., Marcek, D.: Granular RBF neural network implementation of fuzzy systems: application to time series modeling. J. Multi-Valued Log. Soft Comput. **4**(3–5), 401–414 (2008)
13. Kecman, V.: Learning and Soft Computing: Support Vector Machines, Neural Networks and Fuzzy Logic. The MIT Press, Cambridge, MA (2001)
14. Brock, W.A., Dechert, W.D., Scheinkman, J.A., LeBaron, B.: A test for independence based on the correlation dimension. Econom. Rev. (1996)

Author Index

I. Rojas et al. (eds.), *Advances in Time Series Analysis and Forecasting*,
Contributions to Statistics, DOI 10.1007/978-3-319-55789-2

MIX
Papier aus verantwortungsvollen Quellen
Paper from responsible sources
FSC® C105338

If you have any concerns about our products,
you can contact us on
ProductSafety@springernature.com

In case Publisher is established outside the EU,
the EU authorized representative is:
Springer Nature Customer Service Center GmbH
Europaplatz 3, 69115 Heidelberg, Germany

Printed by Libri Plureos GmbH
in Hamburg, Germany